Digital Electronics

Robert K. Dueck
Red River College
Winnipeg, Manitoba

Kenneth J. Reid
Ohio Northern University
Ada, Ohio

DELMAR
CENGAGE Learning™

Australia • Brazil • Japan • Korea • Mexico • Singapore • Spain • United Kingdom • United States

Digital Electronics
Kenneth J. Reid and Robert K. Dueck

Vice President, Editorial: **Dave Garza**

Director of Learning Solutions: **Sandy Clark**

Senior Acquisitions Editor: **James DeVoe**

Managing Editor: **Larry Main**

Product Manager: **Mary Clyne**

Editorial Assistant: **Cris Savino**

Vice President, Marketing: **Jennifer Baker**

Marketing Director: **Deborah Yarnell**

Senior Marketing Manager: **Erin Brennan**

Senior Production Director: **Wendy Troeger**

Production Manager: **Mark Bernard**

Senior Content Project Manager: **Glenn Castle**

Senior Art Director: **Casey Kirchmayer**

Cover design: **Joe Devine/Red Hangar Design**

Cover image: **©Gettyimages/Photographer's Choice RF/Suk-Heui Park**

Library of Congress Control Number: 2011931769

ISBN-13: 978-1-4390-6000-1

ISBN-10: 1-4390-6000-2

Delmar

5 Maxwell Drive
Clifton Park, NY 12065-2919
USA

Cengage Learning is a leading provider of customized learning solutions with office locations around the globe, including Singapore, the United Kingdom, Australia, Mexico, Brazil, and Japan. Locate your local office at: **international.cengage.com/region**

Cengage Learning products are represented in Canada by Nelson Education, Ltd.

For your lifelong learning solutions, visit **delmar.cengage.com**

Visit our corporate website at **cengage.com.**

Notice to the Reader

Publisher does not warrant or guarantee any of the products described herein or perform any independent analysis in connection with any of the product information contained herein. Publisher does not assume, and expressly disclaims, any obligation to obtain and include information other than that provided to it by the manufacturer. The reader is expressly warned to consider and adopt all safety precautions that might be indicated by the activities described herein and to avoid all potential hazards. By following the instructions contained herein, the reader willingly assumes all risks in connection with such instructions. The publisher makes no representations or warranties of any kind, including but not limited to, the warranties of fitness for particular purpose or merchantability, nor are any such representations implied with respect to the material set forth herein, and the publisher takes no responsibility with respect to such material. The publisher shall not be liable for any special, consequential, or exemplary damages resulting, in whole or part, from the readers' use of, or reliance upon, this material.

Printed in the United States of America
2 3 4 5 15 14 13 12

Brief Contents

Contents

Chapter 5
Boolean Algebra and Combinational Logic 102

Chapter 6
Combinational Logic Functions 152

Chapter 7
Digital Arithmetic and Arithmetic Circuits 220

PREFACE

*D*igital Electronics is a new textbook for teachers who want to inspire their students to explore career pathways in this challenging and exciting discipline. By presenting the principles and concepts that engineers and design professionals use to shape today's digital environment, *Digital Electronics* will help students develop the problem-solving skills and technological literacy they need to succeed in post secondary engineering and engineering technology programs.

DIGITAL ELECTRONICS WITH PROJECT LEAD THE WAY, INC.

This text resulted from a partnership forged between Delmar Cengage Learning and Project Lead The Way® Inc. in February 2006. As a nonprofit foundation that develops curriculum for engineering, Project Lead The Way® provides students with the rigorous, relevant, reality-based knowledge they need to pursue engineering or engineering technology programs in college.

The Project Lead The Way® curriculum developers strive to make math and science relevant for students by building hands-on, real-world projects into each course. To support Project Lead The Way's® curriculum goals, and to support all teachers who want to develop project/problem-based programs in engineering and engineering technology, Delmar Cengage Learning is developing a complete series of texts to complement all of Project Lead The Way's® nine courses:

Gateway to Technology
Introduction to Engineering Design
Principles of Engineering
Digital Electronics
Aerospace Engineering
Biotechnical Engineering
Civil Engineering and Architecture
Computer Integrated Manufacturing
Engineering Design and Development

To learn more about Project Lead The Way's® ongoing initiatives in middle school and high school, please visit: http://www.pltw.org.

HOW THIS TEXT WAS DEVELOPED

This book's development began with a focus group that brought together experienced teachers and curriculum developers from a broad range of engineering disciplines. Two important themes emerged from that discussion: (1) teachers need a single resource to help them teach a rigorous yet high school–appropriate course in digital electronics, and (2) teachers want an engaging, interactive resource to support project/problem-based learning.

For years, teachers have struggled to fit conventional textbooks to STEM-based curricula. *Digital Electronics* addresses that need with an interactive text organized around the principles and applications essential to success in this discipline. For the first time, teachers will be able to choose a single text that addresses the challenges and individuality of project/problem-based learning while presenting sound coverage of the essential concepts and techniques used in digital electronics design.

This book is unique in that it was written by a team of authors with pedagogy and engineering expertise to ensure a textbook that students can understand and one that contains valid engineering content. *Digital Electronics* supports project/problem-based learning by:

▶ Creating an unconventional, show-don't-tell pedagogy that is driven by *concepts*, not traditional textbook content. Objectives are outlined at the beginning of each chapter and clearly addressed as students navigate the chapter.

▶ Reinforcing major concepts with Applications, Projects, and Problems based on real-world examples.

▶ Providing a text rich in features designed to bring digital electronics to life in the real world. Resources for extended learning include a set of Multisim® files students can use to simulate and test example circuits. Multisim files are available on the Instructor's Resource CD.

Features

Teachers want an interactive text that keeps students interested in the story behind digital electronics. Here are some examples of how this text is designed to keep students engaged in a journey through the design process.

▶ Plentiful **Notes** in text and in the margins highlight facts and points of interest on the road to new and better circuits and applications.

Note . . .
Deriving a POS expression from a truth table:
1. Every line on the truth table that has a LOW output corresponds to a maxterm in the truth table's Boolean expression.
2. Write all truth table variables for every maxterm in true or complement form. If a variable is 1, write it in complement form (with a bar over it); if it is 0, write it in true form (no bar).
3. Combine all maxterms in an AND function.
Note that these steps are all opposite to those used to find the SOP form of the Boolean expression.

Note . . .
If two product terms in an SOP expression differ by exactly one variable, the differing variable can be canceled:

$$xyz + xy\bar{z}$$
$$= xy(z + \bar{z}) = xy$$

► **Multisim Examples with data files** provide students with an opportunity to simulate and test circuits used in text examples.

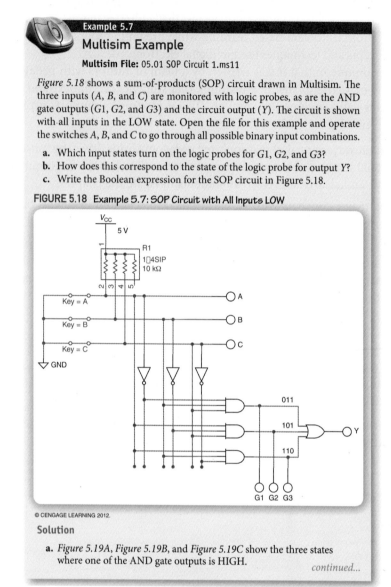

Example 5.7

Multisim Example

Multisim File: 05.01 SOP Circuit 1.ms11

Figure 5.18 shows a sum-of-products (SOP) circuit drawn in Multisim. The three inputs (A, B, and C) are monitored with logic probes, as are the AND gate outputs (G1, G2, and G3) and the circuit output (Y). The circuit is shown with all inputs in the LOW state. Open the file for this example and operate the switches A, B, and C to go through all possible binary input combinations.

   a. Which input states turn on the logic probes for G1, G2, and G3?
   b. How does this correspond to the state of the logic probe for output Y?
   c. Write the Boolean expression for the SOP circuit in Figure 5.18.

FIGURE 5.18 Example 5.7: SOP Circuit with All Inputs LOW

© CENGAGE LEARNING 2012.

Solution

   a. *Figure 5.19A*, *Figure 5.19B*, and *Figure 5.19C* show the three states where one of the AND gate outputs is HIGH.

continued...

► **Key Terms** are defined throughout the text to help students develop a reliable lexicon for the study of engineering.

KEY TERMS

Logic gate network Two or more logic gates connected together.

Logic diagram A diagram, similar to a schematic, showing the connection of logic gates.

Combinational logic Digital circuitry in which an output is derived from the combination of inputs, independent of the order in which they are applied.

Combinatorial logic Another name for combinational logic.

Sequential logic Digital circuitry in which the output state of the circuit depends not only on the states of the inputs, but also on the sequence in which they reached their present states.

▶ **Bring It Home:** Activities and problems are provided at the end of each chapter. The activities progress in rigor from simple, directed exercises and problems to more open-ended projects.

BRING IT HOME

5.1 Boolean Expressions, Logic Diagrams, and Truth Tables

5.1 Write the unsimplified Boolean expression for each of the logic gate networks shown in *Figure 5.58*.

5.2 The circuit in *Figure 5.59* is called a majority vote circuit. It will turn on an active-HIGH indicator lamp only if a majority of inputs (at least two out of three) are HIGH. Write the Boolean expression for the circuit.

FIGURE 5.58 Problem 5.1: Logic Circuits

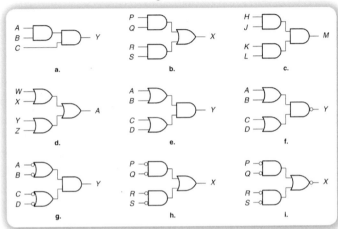

continues...

▶ **Extra Mile:** These problems and activities at the end of each chapter provide extended learning opportunities for students who want an additional challenge.

EXTRA MILE

5.1 Boolean Expressions, Logic Diagrams, and Truth Tables

5.31 Write the unsimplified Boolean expression for each of the logic gate networks shown in *Figure 5.67*.

5.32 Redraw the logic diagrams of the gate networks shown in Figure 5.67a, e, f, h, i, and j so that they conform to the bubble-to-bubble convention. Rewrite the Boolean expression of each of the redrawn circuits.

5.33 a. Write the unsimplified Boolean equation for each of the logic diagrams in *Figure 5.68*.

    b. Redraw each of the logic diagrams in Figure 5.68 so that they conform to the bubble-to-bubble convention. Rewrite the Boolean expression for each of the redrawn circuits.

5.34 a. Write the unsimplified Boolean equation for the logic diagram in *Figure 5.69*.

    b. Redraw the logic diagram in Figure 5.69 so that it conforms to the bubble-to-bubble convention. Rewrite the Boolean expression of the redrawn circuit.

5.35 Draw the logic circuit for each of the following Boolean expressions:

    a. $Y = \overline{AC} + \overline{B + C}$

    b. $Y = \overline{\overline{AC} + B} + C$

    c. $Y = \overline{\overline{\overline{ABD} + \overline{BC}} + \overline{A} + C}$

    d. $Y = \overline{\overline{AB} + \overline{\overline{AC}} + \overline{BC}}$

    e. $Y = \overline{AB} + \overline{AC} + \overline{\overline{BC}}$

Supplements

A complete supplements package accompanies this text to help instructors implement twenty-first–century strategies for teaching engineering design:

▶ An **Instructor's e-resource** includes answers and solutions to text problems, instructional outlines and helpful teaching hints, a STEM mapping guide, PowerPoint presentations, computerized testing options, and all Multisim® files.

PUBLISHER'S ACKNOWLEDGMENTS

The publisher wishes to acknowledge the invaluable wisdom and experience brought to this project by our focus group and review panel.

Focus Group:

Connie Bertucci, Victor High School, Victor, NY
Omar Garcia, Kearny High School, San Diego, CA
Brett Handley, Wheatland-Chili Middle/High School, Scottsville, NY
Donna Matteson, State University of New York at Oswego
Curt Reichwein, North Penn High School, Lansdale, PA
George Reluzco, Mohonasen High School, Rotterdam, NY
Mark Schroll, Program Coordinator, The Kern Family Foundation
Lynne Williams, Coronado High School, Colorado Springs, CO

Reviewers:

Chad Weaver
Canton City Schools
Canton, Ohio

Barry Witte
South Colonie Central School District
Albany, New York

The publisher also wishes to thank our special consultants for this text:
George Zion, Rochester Institute of Technology, PLTW® affiliate professor and lead curriculum writer for *Digital Electronics*.
Aaron Clark, North Carolina State University, Raleigh, NC.

Finally, the publisher extends special thanks to Project Lead The Way's® curriculum directors Sam Cox and Brian Kind for reviewing chapters at manuscript stage.

AUTHORS' ACKNOWLEDGMENTS

Thanks to everyone at the Cengage editorial and production team, especially Jim DeVoe for initiating this project and Mary Clyne for her kindness, patience, and encouragement. My students and colleagues continue to be an inspiration—I always learn something new from you whenever I teach. More than I can say, I want to thank my wife, Joan Duerksen. Joan, thanks for your love, your support, your wisdom, and all that you've given me over the years.

—*Bob Dueck*

A text like *Digital Electronics* could not be produced without the patient support of family and friends and the valuable contributions of a dedicated educational community. Many friends and colleagues have supported us during the development of this book. First, thank you to coworkers and friends who have been supportive of this effort from the beginning, especially the Dean of the College of Engineering at Ohio Northern University, Eric Baumgartner for his ongoing support. Thanks also to some good friends for their input: Chris Floyd, Barbara Christe, and Sami Khorbotly to name a few.

Most of all, I want to thank my family and my wife Jenny. She is the most supportive and friendliest person I've ever known. There is no way a book or any other major project would be complete without her love, prayers, help, and support. She is my wife and best friend, and she has the most beautiful smile I've ever seen.

—*Ken Reid*

ABOUT THE AUTHORS

Robert Dueck received his B.Sc. degree in electrical engineering from the University of Manitoba in Winnipeg, Canada, and worked for several years as a design engineer at Motorola Canada in Toronto. He began teaching in 1986, specializing in digital and microcomputer subjects in the Electronics and Computer Engineering Technology programs at Seneca College in Toronto. His first book, *Fundamentals of Digital Electronics,* was published in 1994, and he has written several additional textbooks. He now teaches digital electronics and related courses at Red River College in Winnipeg. Mr. Dueck is a member of the Association of Professional Engineers and Geoscientists of Manitoba (APEGM) and the Institute of Electrical and Electronics Engineers (IEEE). He served as chair of the Winnipeg Section of IEEE in 2002 and was branch counselor of the Red River College Student Branch from 1997–2006.

Ken Reid received his B.S. degree in computer and electrical engineering from Purdue University in 1988 and his M.S.E.E. degree in 1994 from Rose-Hulman Institute of Technology. He was also one of the first in the nation to receive his Ph.D. in engineering education from Purdue University in 2009. Prior to beginning his teaching career in 1996, he worked for the U.S. Navy in electronics manufacturing research. In addition to publishing and presenting in the field, Dr. Reid has extensive experience in teaching digital electronics and computer programming, and he is currently Director of Freshman Engineering and Director of Engineering Education at Ohio Northern University. He has received awards including the ASEE IL/IN Region Outstanding Teaching Award, the IEEE Third Millennium Medal for service and was codeveloper of the Best Middle School Curriculum for the Tsunami Model Eliciting Activity in 2009. He is a Senior Member of the Institute of Electrical and Electronics Engineers (IEEE) and is active in the American Society of Engineering Education (ASEE).

Digital Electronics

CHAPTER 1
Electrical Safety and Components

| GPS DELUXE | | START LOCATION | DISTANCE | END LOCATION |

Menu

Chapter Objectives

Upon successful completion of this chapter, you will be able to:

1 Describe the basic logic characteristics of static electricity.

2 Describe why some sources of voltage can be dangerous (power supplies) while some are not (AA batteries, for example).

3 Describe the difference between polarized and non polarized components.

4 List potential safety issues in working with electronic assemblies.

5 Describe the function of a resistor in a circuit.

6 Find the value of a resistor based on colored bands on its body.

7 List some of the functions of a capacitor in a circuit.

8 Find the value of a capacitor based on numbers stamped on its body.

9 Describe solder: list its components and describe its function.

10 List a step-by-step method for properly hand soldering an electronic connection.

11 Describe the appearance of a good soldered connection.

PHOTO: © ISTOCKPHOTO.COM/TEBNED.

Electricity surrounds us in every aspect of our lives, from game systems to power for our homes to lightning. Electricity is generated in power plants and transmitted through wires almost everywhere throughout the country and the world, to the outlets in your home and school. It is used when our cell phones send a text message, it is used when we turn on a lamp at night, it is used to power spacecraft, and it is used to power amusement park rides. If your alarm clock woke you this morning, the radio station playing the music was powered by electricity as was your alarm clock. While we can barely imagine life without electricity, we also need to know to handle electricity safely and with respect.

Static electricity is not necessarily dangerous to humans, but can be dangerous to electronics. Sources of electricity that generate a voltage and produce a current can be very dangerous. Even very small amounts of current can be dangerous to humans. Digital logic circuits, like those discussed in this textbook, are very rarely dangerous to humans if built properly. Other safety rules apply as well: circuit boards are made from a mixture of epoxy and fiberglass and can cause splinters or cuts. Polarized components that are inserted backward in a circuit can overheat quickly and can burn the skin if touched.

Resistors are very common components and are found on nearly every circuit board. Axial-leaded resistors are usually marked with colored stripes to indicate the value of resistance. Capacitors are also very common on nearly every circuit board. Some capacitors are polarized and must be properly inserted in the circuit. Resistors resist the flow of current through a circuit, while capacitors smooth voltage levels within a circuit and store charge. These components will be further investigated later in the text.

Soldering is the act of making an attachment between the pads on the circuit board and the lead of the component. When properly formed, the soldered connection (or solder joint) provides a mechanical and electrical connection between the circuit board and the component. Solder is a metal alloy that melts at a significantly lower temperature than the metals it is to connect.

1.1 ELECTRICAL SAFETY

Electricity is used in our lives from the moment our alarm clock wakes us up, we switch on a lamp, and we turn on some music. On the way to school, electricity is used to control traffic lights and help power cars and buses. Lights, computers, vending machines, projectors, calculators, cameras, microphones, scoreboards, and even many pencil sharpeners at school are powered by electricity. At home, electricity is used when we heat and cool our house, use an oven to cook, do homework with a computer, and watch our favorite shows on TV.

Common sense tells us that electricity can be dangerous and must be respected. Everyone has experienced a quick "shock" from static electricity; hopefully, you have not experienced an electrical shock from a household outlet. If you have, you know it can be painful and very dangerous—and possibly deadly. Because the heart also relies on electricity to beat regularly, an electrical shock can do tremendous damage. Electrocution is death caused by an electrical shock, usually by stopping the heart.

If electricity is so dangerous, how can we safely work with it and build electrical circuits? Respecting electricity and following safe procedures are important. Fortunately, most of these rules are "common sense." Let's review some basic safety rules.

Working with Electricity

KEY TERMS

Static electricity A stored charge in an item that dissipates quickly when the item comes into contact with something at a different electrical potential.

Volts Unit of voltage.

Voltage Electrical potential.

Lightning Discharge of static electricity between the atmosphere and ground.

Current The flow of electricity.

FIGURE 1.1 Static Electricity

Static Electricity

Nearly everyone has scuffed his or her shoes over a carpet and touched a handrail or another person, and experienced a quick shock. This shock might be painful (and surprising), but it's usually not dangerous. This is from static electricity (*Figure 1.1*). Your body can build a charge as you move across carpet or get up from a plastic chair. As this charge builds, it is attracted to the opposite charge (positive charge is attracted to negative charge and vice versa). When your positively charged body comes in contact with a neutral or negatively charged surface, this charge jumps to the other surface, causing the shock. Technically, electricity has flowed, but once the charge is neutralized, the flow of electricity stops.

In cases where you have built up enough charge to feel the shock, your body may be charged up to 5,000 V (**volts**). We will examine what **voltage** is in the next chapter, but this sounds like a very high number. We might be familiar with 9-V batteries; other voltages are AA and AAA batteries (1.5 V), car batteries (12 V), and household voltage for anything we plug into the wall (120 V).

The ultimate demonstration of static electricity is **lightning**, shown in *Figure 1.2*. The Earth's atmosphere can become very highly, positively charged. When the difference in charge becomes great enough, the positive charge will jump to a neutrally charged site; sometimes, this is a tall building, a tree, or the ground. The voltage difference between the atmosphere and the ground can be in the millions of volts.

While a lightning strike is often fatal, a typical static shock from a chair, sweater or carpet isn't. There is a huge difference in the voltage, which means a huge difference in the **current**. Current is the flow of electricity; we will discuss current in the next chapter in more detail. Basically, a voltage causes a current to flow if there is a path for the charge to get back to neutral (ground). The more voltage applied to a person's body, the more current will flow, and the more dangerous the situation becomes.

There is even more to the story. If a static shock is caused by 400 V or more, shouldn't that be dangerous? The nature of static electricity, and the cause of the shock, is the rapid discharge. In other words, the charge jumps from your body to the metal object quickly and current stops flowing—almost instantly.

Therefore, static electricity can be annoying, and sometimes painful, but it is rarely dangerous.

IMAGE COPYRIGHT MARTIN FISCHER, 2011. USED UNDER LICENSE FROM SHUTTERSTOCK.COM.

FIGURE 1.2 *Lightning*

FIGURE 1.3A *Static Shield Bag, Used to Protect Electronics*

PHOTO: © ISTOCKPHOTO.COM/TOBIAS OTT.

FIGURE 1.3B *Electrostatic Discharge (ESD) Protective Symbol*

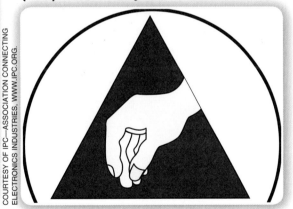

COURTESY OF IPC—ASSOCIATION CONNECTING ELECTRONICS INDUSTRIES, WWW.IPC.ORG.

Note . . .

Static electricity isn't really dangerous to humans, but it is definitely dangerous to electronics. Even small static discharges can damage delicate electronic components. Electronic circuit boards and components are usually packaged in static dissipative bags or tubes (*Figure 1.3*). When handling electronics, circuit boards, or components, it is often a good idea to wear a static wrist strap: this allows built-up charges to dissipate through the attached strap instead of through the electronics.

Voltage and Current in Electronics

Electronics need a source of power to function; unlike static electricity, there must be a source of continuous voltage to generate a current. Smaller electronic devices require batteries to operate; even electronics that you plug into the wall to recharge have a battery inside. Larger electronic devices plug into the wall, and get power from outlets. As we have seen earlier, static electricity results from a charge rapidly dissipating. If we have a steady source of power, we have a steady source of current, and this can create a more dangerous situation.

You are probably familiar with 9-V batteries, and may know that the voltage generated by a AA or AAA battery is 1.5 V. You also know that these are safe to handle (although you may have touched a 9-V battery to your tongue to feel the shock). These batteries are not capable of generating large currents, especially through the human body. Typically, small electronics such as cell phones or handheld devices that run off of AA or AAA batteries have circuits requiring very little current.

How much current is dangerous? Even small amounts of current can be dangerous. According to the National Institute for Occupational Safety and Health (NIOSH), currents as low as 5 mA (5 milli-amperes, or 50/1000 amperes) can be painful. Currents in the range of 9 mA to 30 mA for men, or 6 mA to 25 mA for women result in a painful shock, and may "freeze" muscles; in other words, it may not be possible to let go. At currents as low as 50 mA, death is possible.

Current is produced by applying a voltage to a circuit—or to a body. If you apply a voltage to your body, it will tend to produce a current through you to the ground. The voltage applied is a factor in determining how much current passes through you and how much damage is done. The path and duration of the current and the amount of power the source can deliver also play a role.

Generally, voltages you encounter when using household batteries (AA, AAA, 9 V) do not cause dangerous conditions. Likewise, voltages within digital circuits, like those in this textbook, are usually not dangerous to handle, as they generate very small currents (sometimes in the μA or micro-ampere range: one-millionth of an ampere). However, sources of power plugged into an outlet can easily be dangerous. For example, boards and other electronic equipment may be powered by an AC adapter; these plug into the wall and usually have a large cube at one end. This transforms the high voltage and current from the wall to the voltage for use on the board. They may be capable of generating a current large enough to cause injury. Any internal power supply can also be dangerous. Power supplies inside of a computer, for example, store a high amount of charge and can output large currents. Such power supplies can continue to be dangerous for a period of time after they have been unplugged because their internal components may take time to discharge their electrical charge and "settle down" to a safe level. You should always be very careful not to touch sources of high voltage, even if they are not powered.

FIGURE 1.4 Epoxy-Glass Circuit Board

HARRIS SHIFFMAN, 2011. USED UNDER LICENSE FROM SHUTTERSTOCK.COM.

Always be careful around electricity. Some rules to follow include:

- Always have a knowledgeable adult assisting you.
- In general, do not take equipment apart if it is powered or has been powered recently. If you are working inside of any equipment, turn off the power and unplug it for a few minutes.
- Always pay attention to warning labels: many electrical devices such as computer monitors and televisions have components capable of dealing fatal amounts of current and should not be disassembled.
- Do not eat or drink while working with electricity. Any liquid on an electrical circuit is dangerous.

Note . . .

Wiring mistakes can cause components to become very hot—be careful!

Other Safety Issues

KEY TERMS

Polarized part A component which must be inserted in a circuit with correct polarity: a component with a positive and negative side.

You may notice that laptop computers and other electronic devices can get very warm during operation. Electrical devices all generate heat, and small devices with large voltages or currents can become very hot. While you are working with electronics, there is always a possibility of making a wiring mistake, using the wrong part, or inserting an integrated circuit or polarized part backward. A **polarized part** is one which must be inserted with the correct polarity: that is, inserted so the "+" side is at a higher voltage than the "−" side. In most cases, these small mistakes can result in an extremely hot component—easily hot enough to cause a burn. When wiring circuits, be extremely careful of very hot components! Some components such as resistors heat gradually; some components, like some types of capacitors, heat rapidly and can explode. If you notice smoke or a burning smell, turn off the circuit as quickly as possible. Do not check if a component is too hot by touching it with your finger! Hold a hand over the component to see if it is hot.

Other safety issues include preventing injury from sharp wires, components, or leads. Integrated circuits, resistors, capacitors, and other components can have very sharp legs (or leads). Circuit board edges can be rough and have sharp edges, and can splinter easily. Be careful to avoid cuts or puncture wounds.

Your Turn

1.1 Which of the following employees is violating a safety rule?
- **a.** An employee smells a burning smell, so gently touches each component on a board until he finds the one that is overheating.
- **b.** In order to make "a good contact," an employee licks the end of a power supply that is plugged into the wall.
- **c.** A hurrying employee decides to take a power supply apart without unplugging it first.
- **d.** An employee plans to replace the batteries in the TV remote in the lounge, and grabs AAA batteries from the package without wearing protective gloves.

1.2 COMPONENT IDENTIFICATION

There are many different electronic components: we'll look at two of the more common parts here. Some of the parts we will use will be polarized and must be inserted into the circuit in the right orientation. Some are not polarized, and can go in the circuit either way. Components also come with different tolerance levels, or an amount that the specified value may not be accurate. In general, components with a tighter tolerance (more accurate) are more expensive.

Resistors

KEY TERMS

Resistor A circuit component which resists the flow of current.

Ohm (Ω) Unit of resistance.

Axial-leaded component A component package with a cylindrical body and leads from both ends.

Resistors (*Figure 1.5*) are by far the most common component in electronics. They can be found on any circuit board. We will investigate the functions of resistors and how resistors behave in a circuit in the next chapter. The simplest description of resistors is that they resist current flow. Resistors with a higher value of resistance (in ohms) resist current flow more.

Imagine using a straw, but pinching it in the middle. It is much more difficult to drink through the straw, because you are adding resistance. If you have been stuck in traffic because a lane was closed for construction, picture the nice, smooth flow of traffic before the construction zone. Adding a resistance to the flow of traffic (construction zone) impedes (slows) the flow of traffic.

Resistors are typically found as **axial-leaded packages**; that is, cylindrical in shape with leads (wires) coming from each end. Electrical current passes through the resistor from one end to the other. Resistors are not polarized; in other words, they can go into a circuit in either direction.

Resistor values are marked on the package by a series of colored stripes or bands. Each color corresponds to a digit as shown in *Table 1.1*. Memorizing the table can be easy if you use a mnemonic: the first letter of each word is also the first letter of the color in the resistor color code; for example, **Better Be Right Or Your Great Big Venture Goes Wrong.** Values for resistors with four stripes (the most common package) are read as follows:

▶ The first band: the first digit of the value
▶ The second band: the second digit of the value
▶ The third band: the multiplier, or the number of zeros on the end of the value
▶ The fourth band: the tolerance

The first two bands give the value, then add the number of zeros shown by the third band. For example, a

---

Note . . .

The color of the resistor itself varies with the material from which it is made, and isn't usually important in finding the value of the resistor . . . look at the stripes.

Note . . .

The resistance of a resistor is measured in ohms, with the symbol Ω (the Greek letter omega).

FIGURE 1.5 *Assorted Resistors*

resistor as shown in *Figure 1.6* has a yellow, then a violet, then a brown band (then a band that tells us the tolerance). Yellow = 4, violet = 7 ("47" so far); brown = 1, so add 1 zero for a value of 470 Ω (470 ohms). The gold band at the end tells us that this resistor has a 5% tolerance; in other words, the value can be 5% too high or 5% too low (and still be acceptable). For example, this 470-Ω resistor can actually be between 446.5 Ω to 493.5 Ω.

Figure 1.7 shows another resistor with three red stripes (then a gold band for the tolerance). The value is: red (2), red (2) or 22, then add two zeros for a total of 2,200 Ω. With a 5% tolerance, the actual value can be anything between 2,090 Ω to 2,310 Ω. One more example is shown in *Figure 1.8*, where we see a resistor with brown (1), black (0), and another black (0) stripe (before the tolerance stripe). Brown−black−black gives us 1, then 0, (10) with 0 more zeros at the end: therefore, 10 Ω.

TABLE 1.1 Resistor Color Code

Color	Significant Figures	Multiplier	Tolerance
Black	0	$\times 10^0$	
Brown	1	$\times 10^1$	
Red	2	$\times 10^2$	
Orange	3	$\times 10^3$	
Yellow	4	$\times 10^4$	
Green	5	$\times 10^5$	
Blue	6	$\times 10^6$	
Violet	7	$\times 10^7$	
Gray	8	$\times 10^8$	
White	9	$\times 10^9$	
Gold			±5%
Silver			±10%
None			±20%

© CENGAGE LEARNING 2012.

Your Turn

1.2 What is the value of a resistor with the following bands?
   a. Brown Black Red Gold
   b. Brown Black Green Gold

Capacitors

Capacitors (*Figure 1.9*) are also found on nearly every electronic circuit board and inside each device. While resistors resist the flow of current, capacitors have two main functions: they are used to store charge, and they are used to ensure that other electronic components receive a smooth, steady supply voltage. Integrated circuits (or chips) on a circuit board usually have a "bypass capacitor" mounted very close to the chip; this protects the chip from some voltage spikes or irregularities. Capacitance is measured in **Farads**, although values are typically very small: in the micro-Farad (μF; 10^{-6} or one-millionth of a Farad), nano-Farad (nF; 10^{-9} or one-billionth of a Farad), or pico-Farad (pF; 10^{-12} Farads).

FIGURE 1.6 470-Ohm Resistor © CENGAGE LEARNING 2012.

FIGURE 1.7 2,200-Ohm Resistor © CENGAGE LEARNING 2012.

FIGURE 1.8 10-Ohm Resistor

DELMAR CENGAGE LEARNING 2012.

FIGURE 1.9 Collection of Capacitors

PHOTO: iSTOCKPHOTO.COM/GETHIN LANE.

FIGURE 1.10 *Electrolytic Capacitor*

FIGURE 1.11 *Tantalum Capacitors*

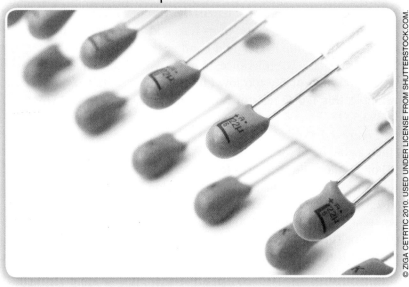

FIGURE 1.12 *Ceramic Capacitors*

Most resistors are similar in appearance, but there are a few different types of capacitors. Some capacitors are polarized. In these cases, it is extremely important to put them in the circuit properly; with the "+" indicator at the higher voltage, or the "−" toward the lower voltage in a circuit. If they are inserted backward, they can literally explode.

Electrolytic capacitors (*Figure 1.10*) and tantalum capacitors (*Figure 1.11*) are manufactured in such a way that they can have relatively large values of capacitance in a small package. Electrolytic capacitor packages usually appear like a can with legs (leads) from the bottom. Tantalum capacitors might remind you of a teardrop shape. These types of capacitors are polarized, and can "pop" or explode if inserted into the circuit incorrectly.

Ceramic capacitors are shaped more like a disk (*Figure 1.12*). These are not polarized so, like resistors, they can go into a circuit in either direction.

The value of these capacitors is sometimes written right on the capacitor itself, if the body of the capacitor is large enough. Some smaller ceramic capacitors have a code very similar to the resistor code printed on the body, indicating the value of the capacitance of the capacitor. For example, a capacitor with the number 103 printed on the body is 10, then 3 zeros, or 10,000 pico-Farads.

1.3 ELECTRONIC ASSEMBLIES

KEY TERMS

Soldering The act of attaching electronic components to a circuit board using solder.

Solder A metallic alloy comprised of a mixture of tin and lead or other metals with a melting point lower than the metals to be joined.

Solder joint A soldered connection between an electronic component and a circuit board.

When building a circuit board, we have the board itself and the components that belong on the board. By far, the most common means to attach components to the board is **soldering**. Soldering is the act of attaching electrical components to a circuit board using solder. *Figure 1.13* shows a circuit board with components attached: each place a connection is made, the visible silver connection is a solder joint, or a place where solder was used to make the attachment.

FIGURE 1.13 **Circuit Board with Components**

Solder

> **KEY TERMS**
>
> **Alloy** A combination of metals which gives it a lower melting point than other metals.
>
> **Flux** A rosin-based material contained in or used with solder to clean the surfaces as the solder joint forms.
>
> **Rosin** A key component in flux; made from pine tree sap.

Solder is a metallic **alloy**—a combination of metals which gives it a lower melting point than other metals. One common type of solder known as tin-lead solder is composed of 63% tin and 37% lead (also called 63/37 solder). This specific mixture of components is a eutectic mixture: that is, as the solder melts, it goes directly from a solid to a liquid at a specific temperature; it doesn't go through a softening process. Tin-lead solder melts at 183° C. Compare this to the melting point of copper which is 1083° C, and it's easy to see that solder melts at a lower temperature. Because lead is toxic, there has been a global effort to stop using lead in soldering, so lead-free varieties are typically used in manufacturing; these contain silver, copper, antimony, bismuth, or indium instead of lead. The disadvantage of lead-free solder is that it melts at a higher temperature, so it either requires a longer process or the temperature must be turned up to form a good solder joint.

FIGURE 1.14 **Wire Solder**

Wire solder (*Figure 1.14*), usually used when building boards by hand, contains solder and **flux**. Flux is **rosin**-based, which is made from the sap of pine trees. It is also very sticky, so you may have used similar rosin if you play baseball. The rosin is a cleaning agent, and helps the solder stick to the metal surfaces much better.

Soldering

> **KEY TERMS**
>
> **Lead** (pronounced "leed") The leg of the device: the connection piece of the component to be attached to the circuit board.
>
> **Pad** Copper on a circuit board meant to be a point where a component is attached. On a through-hole board, this copper will surround the hole where the lead passes through.
>
> **Soldering iron** A device with a heated tip used to manually solder components to circuit boards.
>
> **Tip** The heated end of the soldering iron—the tip should make contact with the lead and pad when soldering.
>
> **Oxidation** The process of oxygen reacting with the heated tip causing discoloration and affecting the ability of the tip to transfer heat.
>
> **Tinning the tip** The process of applying a light coat of solder to the tip of an iron for storage to prevent oxidation.

FIGURE 1.15 *Soldering Iron in a Complete Soldering Station*

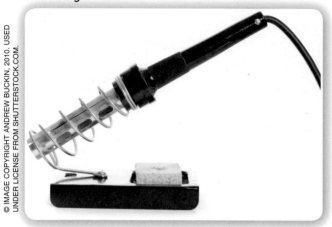

FIGURE 1.16 *Bottom of an Electronic Circuit Board with Solder Joints*

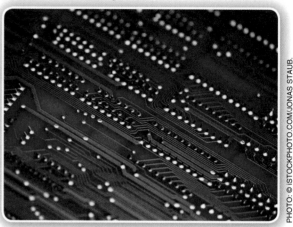

Soldering is done to attach components such as resistors and capacitors to circuit boards. The basic procedure is to heat the **lead** of the component and the **pad** on the circuit board hot enough to melt the solder; the solder then sticks to both surfaces and creates a connection. This keeps the part on the board, and establishes an electrical connection between the board and the component.

A **soldering iron** (*Figure 1.15*) is the tool used to apply the heat to these surfaces. While it is safe to touch the handle, the **tip** of the iron becomes very hot—from 300° C to 800° C.

Assembling a circuit board begins by inserting the component into the circuit board. The circuit board will have holes surrounded by copper pads; the pads look like edges of the hole. The component almost always belongs on the top of the board, with the legs sticking out of the holes where the pad is, on the bottom of the board (*Figure 1.16*). The solder will attach the lead to the pad on the bottom of the board; not on the side with the component.

Use a clamp to hold the board whenever possible. Do not hold the component with your hands, as the metal you solder will become very hot.

The secret to soldering successfully is to heat the lead and the pad enough to melt the solder. If the solder touches the tip of the soldering iron, it melts easily, and therefore, this might seem to be the easiest way to solder, but the surfaces must be hot enough to cause the solder to attach and form the desired shape. If the solder is melted quickly by the soldering iron, the solder will not adhere to the lead or pad, and the circuit may not work—or may work only for a short period of time. If you have ever owned an electronic device that simply quit working, odds are pretty good that there was a bad solder joint inside the device.

The steps to creating a good solder joint follow. Like many things in life, soldering correctly takes some practice, but becomes easy to do.

The steps to proper soldering include:

1. Prepare your soldering iron: every few solder joints, wipe the tip on the sponge and tin the tip. (See below for instructions on tinning the tip.)
2. Touch the tip of the soldering iron to the lead and the pad at the same time; hold for about 3 seconds (*Figure 1.17*).
3. Touch the end of the wire solder to the opposite side of the lead, touching the lead and the pad (but try to avoid touching the soldering iron tip).
4. When the solder begins to melt (which should be in 1–3 seconds), "push" the solder into the connection. Roughly ½ inch of solder should be used (depending on the thickness of the solder you use).

Safety Note

Because the tip of a soldering iron is very hot,
- Never touch it to any person, to your finger, to the top of your desk, or anywhere else except the circuit you are working with or its holder.
- Never point with the soldering iron: small drops of solder can fly off the tip, even if you can't see solder on the tip.
- Never hand the iron to anyone: put it back in the holder and let the other person pick it up on his or her own.
- The lead will also become very hot, very quickly when soldering. Do not touch the lead until you are done soldering and it has cooled.

FIGURE 1.17 *Soldering Iron Tip, Component Lead, and Solder*

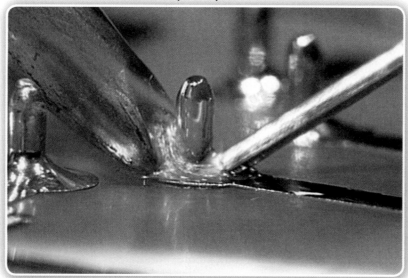

COURTESY OF IPC.

5. When the correct size solder joint appears, pull the solder away.
6. Wait a second or so; then pull the iron away.

Let's examine each of these steps in more detail:

1. *Prepare your iron*: The soldering iron tip becomes dull and gray or black if it sits in the holder unused for any amount of time. The heat causes **oxidation**, which darkens the tip and makes it less effective. Removing the oxidation and restoring the tip to its normally clean, shiny condition is important. There should be a sponge with your soldering iron. The sponge should be moistened before use. A quick swipe of the iron on the sponge cleans the tip briefly.

   If the tip is extremely discolored or dirty, or if you are about to put away the iron for a few minutes, you should **tin the tip**. To tin the tip, wipe it with a sponge, then carefully touch some solder on the end of the tip to "paint" a light coating of solder over the tip. The next time the tip is used, quickly wipe it on the sponge again. This will care for your tip and you should see excellent results.

2. *Touch the tip to the lead and pad*: The tip should be in contact with the pad and the lead. Imagine the lead was not in the hole, and try to put the tip of the iron in the hole instead. If the tip touches both surfaces for about 3–5 seconds, both surfaces should become hot enough to melt the solder. Keep the iron here through the rest of the steps.

3. *Touch the solder to the lead and pad*: Touch the solder to the opposite side of the lead (see Figure 1.17).

4. *Push the solder into the solder joint*: The lead and pad should be hot enough to melt the solder, and the solder should flow into the joint. Roughly ½ inch of solder should flow into the joint.

5. *Pull the solder away*: Once the size of the solder joint is right, remove the solder. It's important to pull the solder away first, not the soldering iron! Note that, if you do pull away the iron first, the solder will be stuck in the joint. This isn't a problem: just touch the iron back on the joint, and the solder will separate.

6. *Pull the iron away*: Wait a second or so after removing the solder, then remove the iron. Return the iron to its proper holder.

That's all there is to it! Once you have the timing down, soldering correctly is easy to do.

FIGURE 1.18 Quality
Soldered Connection

COURTESY OF IPC.

Quality Solder Connections

Figure 1.18 shows an example of a quality solder joint. Note that it's generally the same shape as a chocolate chip. If you look closely, you can see that the solder has adhered to the lead and to the pad.

An example of a solder joint where the tip of the iron melted the solder is shown in *Figure 1.19A*. This is shaped more like a beach ball than a chocolate chip. This is a cold solder joint, so named because the pad was too cold to melt the solder and it didn't form a connection. A case where the lead itself wasn't hot enough is shown in *Figure 1.19B*. The solder formed a shape more like a doughnut than a chocolate chip.

The problem with these cold solder joints isn't just that they make a bad physical connection, but they also don't have a good electrical connection. When we finish a circuit board with cold solder joints, the circuit may work intermittently, but eventually, the lead through the cold joint may come loose, and the electrical path through the device may stop. While the repair for a problem like this can be very easy, *finding* the bad solder joint among hundreds of thousands of joints can be tricky.

FIGURE 1.19A Soldered
Connection: Not Enough Heat
Applied to the Pad

COURTESY OF IPC.

FIGURE 1.19B Soldered Connection: Not
Enough Heat Applied to the Lead

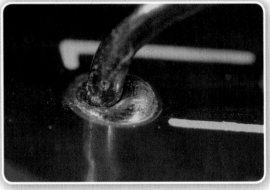

COURTESY OF IPC.

SUMMARY

1. Electricity surrounds us—it is an essential component of our lives. However, it can be dangerous and must be respected.
2. Electrocution is death caused by electrical shock.
3. Static electricity results when a stored charge rapidly discharges.
4. Static electricity usually poses no danger to humans (with the exception of lightning), but can be very damaging to electronic circuits.
5. Voltage is electrical potential energy.
6. Typical sources of voltage include batteries and wall outlets.
7. Voltage can be dangerous to humans and to electronic devices.
8. Current is the flow of electricity.
9. Even small amounts of current can be extremely dangerous to humans and electronics: currents as low as 5 mA can be very painful, and currents as low as 50 mA can be fatal.
10. Always be careful around electricity. Some rules to follow include:
    - Always have a knowledgeable adult assisting you.
    - In general, do not take equipment apart if it is powered or has been powered recently. If you are working inside of any equipment, turn off the power and unplug it for a few minutes.
    - Always pay attention to warning labels. Many electrical devices such as computer monitors and televisions have components capable of dealing fatal amounts of current and should not be disassembled.
    - Do not eat or drink while working with electricity. Any liquid on an electrical circuit is dangerous.
11. Polarized components are those components that must be inserted into a circuit in the proper direction to function.
12. Components that are inserted into circuits improperly, especially polarized components, can overheat enough to cause a burn. Do not check a component by touching it; instead, feel near the component to see if it is hot.
13. Circuit boards are usually made from a mixture of fiberglass and epoxy, and can have sharp edges. Be careful handling boards and components.
14. Resistors are the most common electronics component. They resist the flow of current in an electrical circuit.
15. The value of resistance for a resistor (in ohms, or Ω) is typically shown with a series of stripes (or nands) around the body of the resistor. If the resistor has four bands, the first two show a value (write those down), the third is a multiplier (add that many zeros), and the fourth is the tolerance. The values are shown in Table 1.1.
16. Capacitors are also found in almost every electronics circuit. They are used to store charge and/or "smooth" voltages. Their value (in pico-Farads, or pF) is typically printed on the component, sometimes using a code with three digits: two are values (write those down) and the last one is a multiplier (add that number of zeros).
17. Solder is used to connect electronic components to circuit boards. Soldering is the act of heating the lead and pad on the circuit board (with a soldering iron) to melt solder, forming a solder joint (or solder connection).
18. Tin-lead solder is a eutectic alloy: a mixture of metals that melts at a lower point than the metals it will connect and goes directly from a solid to a liquid without becoming soft. Because lead is toxic, other metals (indium, bismuth, etc.) are used in place of lead in lead-free solder.

continues...

continued...

19. The tip of a soldering iron becomes very hot; do not touch the tip, do not point with the tip, and do not touch the tip to anything other than the connection to be soldered.

20. Successful soldering involves following these steps:

    – Prepare your soldering iron: every few solder joints, wipe the tip on the sponge and tin the tip.
    – Touch the tip of the soldering iron to the lead and the pad at the same time: hold for about 3 seconds.
    – Touch the end of the wire solder to the opposite side of the lead, touching the lead and the pad (but try to avoid touching the soldering iron tip).
    – When the solder begins to melt (which should be in 1–3 seconds), "push" the solder into the connection. Roughly ½ inch of solder should be used (depending on the thickness of the solder you use).
    – When the correct size solder joint appears, pull away the solder.
    – Wait a second or so; then pull away the iron.

21. Quality soldered connections take the shape of a chocolate chip. The solder should be clearly attached to the pad and to the lead.

BRING IT HOME

1.1 Electrical Safety

1.1 List 10 items that use electricity in your home.

1.2 List 10 items that use electricity in your school.

1.3 How much voltage can a AAA battery supply?

1.4 Explain why you sometimes feel a static shock when touching a doorknob.

1.5 What is the reason that a 400-V static shock is less dangerous than improperly touching a 12-V car battery?

1.6 Write a sentence explaining the difference between voltage and current.

1.7 How much current can be dangerous to a human?

1.8 List two potential sources of a burn when working with electronics.

1.2 Component Identification

1.9 Which would allow less current to flow: a higher value of resistance or a lower value of resistance? Why?

1.10 Write a sentence describing an axial-leaded component.

1.11 What is the resistance of a resistor with the following stripes or bands around the body?
   a. Red Red Red Gold
   b. Brown Brown Orange Gold
   c. Yellow Violet Black Gold
   d. Orange Orange Green Gold
   e. Green Black Black Gold

1.12 What is the resistance of a resistor with the following stripes or bands around the body?
   a. Brown Black Blue Gold (*Figure 1.20*)
   b. Brown Black Red Gold
   c. Blue Red Brown Gold
   d. Orange Orange Orange Gold
   e. Brown Black Brown Gold

FIGURE 1.20 *Problem 1.12: Resistor: Brown Black Blue Gold Bands*

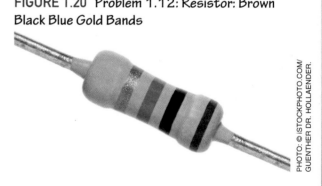

PHOTO: © ISTOCKPHOTO.COM/ GUENTHER DR. HOLLAENDER.

1.13 What is the maximum value of resistance that a resistor with bands of Red Red Red Gold can measure considering the tolerance?

1.14 What is the minimum value of resistance that a resistor with bands of Orange Orange Orange Gold can measure considering the tolerance?

1.15 How many ohms is a 10 M Ω resistor?

1.16 How many Farads is a 4.7-μF capacitor?

1.17 How many Farads (in pF) is a capacitor marked with:

  a. 220

  b. 103

  c. 206

  d. 475

1.18 How many Farads is a 4.7-μF capacitor?

1.3 Electronic Assemblies

1.19 Write a sentence describing solder.

1.20 What metals form eutectic solder?

1.21 What features of eutectic solder make it useful for soldering?

1.22 What is the melting point for tin-lead solder?

1.23 Write a sentence describing an excellent solder connection.

1.24 List the steps in making a good soldered connection.

1.25 What does it mean to "tin the tip" of a soldering iron? Why is it important?

1.26 Why is a cold solder joint a bad thing in an electronics circuit?

EXTRA MILE

1.2 Component Identification

1.27 What other tolerance bands (other than gold) can be found on a resistor?

1.28 What is the maximum value of a resistor with the following color bands:

  a. Brown Black Red Silver

  b. Red Red Red

  c. Red Red Black Orange Brown (*Figure 1.21*)

1.3 Electronic Assemblies

1.29 List metals that may be used in solder instead of lead.

FIGURE 1.21 Problem 1.28: Resistor: Red Red Black Orange Brown Bands

PHOTO: © ISTOCKPHOTO.COM/GETHIN LANE.

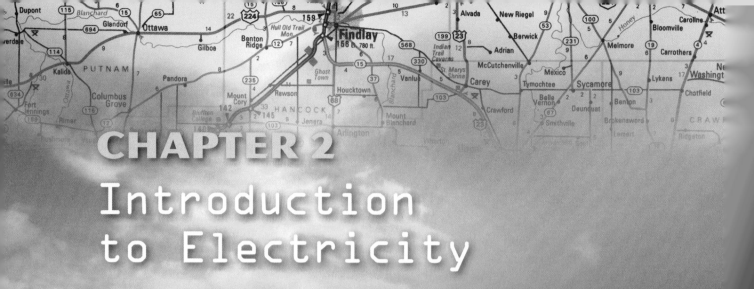

CHAPTER 2
Introduction
to Electricity

Menu

CHAPTER OBJECTIVES

Upon successful completion of this chapter, you will be able to:

1   Have a basic understanding of molecules, atoms, protons, neutrons, and electrons.

2   Describe voltage, current, and resistance.

3   Describe the difference between conductors and insulators.

4   Use Ohm's Law to find either voltage, current, or resistance if you know two of the three.

5   Describe Kirchhoff's Current Law (KCL) and Kirchhoff's Voltage Law (KVL).

6   Identify resistors that are in series or are in parallel.

7   Describe the process of creating a schematic, simulating, and downloading to a programmable device.

8   Identify how to build a small circuit using a breadboard.

PHOTO: © ISTOCKPHOTO.COM/ROOB.

You may have seen this band in concert, on TV, or on the Internet. The instruments, microphones, amps, and speakers at this concert are all powered by electricity. If you watched them on the Web, your computer, monitor, speakers, and the server were all probably plugged in, using electricity. But what is electricity?

Everything is made up of particles too tiny to see without extreme magnification. These particles are called molecules. Each molecule is made up of atoms of individual elements. In turn, each atom is made up of a nucleus surrounded by orbiting electrons. The nucleus is made up of protons and neutrons. Protons have a positive charge that attracts orbiting electrons, which have a negative charge. When electrons flow from one atom to the next, we have electric current (the flow of electrical charge).

Voltage is electrical potential energy; that is, it is capable of generating a current if it has a circuit through which to send the current. This is similar to a large water tank that has the potential to send water gushing through attached pipes. We can think of the water flowing through the pipes in a way similar to

current. Resistance is a characteristic that limits the amount of current possible through a circuit, similar to squeezing a straw as you drink through it.

Ohm's Law is one of the fundamental equations in electronics: $V = R \times I$. This says the voltage across a component is equal to the current through it multiplied by its resistance. If we know two of the three quantities, we can calculate the third. Kirchhoff's Current Law (KCL) explains why current entering a node must equal the current leaving the node. No current is "lost" because it all has to go somewhere; it cannot be created or destroyed. Kirchhoff's Voltage Law (KVL) allows us to see that, in an electrical circuit, voltage supplied from one or more "sources" (such as a battery) must be the same amount as the voltage "dropped" across one or more circuit elements.

Finally, we can design a circuit using a computer program that lets us draw a schematic, or a circuit diagram. This schematic can be simulated, then downloaded to a programmable device to test its function. We can also wire up individual components using a breadboard to test a circuit.

2.1 SI NOTATION

KEY TERMS

Units A term representing a precise quantity or measure.
SI system The International System of Units; the modern form of the metric system.
Metric system The system of units based on multiples of 10.

We use units to describe many things in our lives. You might have to walk 2 miles to school, you may weigh 165 pounds, and you may be 6 feet tall. Even if these numbers don't describe you, you probably have an idea of what this person might look like or how far the person has to walk. Miles, pounds, and feet are **units**. If your school was 321,869 centimeters from home, would it be better to walk or take a bus? Just how tall is someone who is 4 cubits tall? Which is heavier: 74.8 kilograms or 2,406 troy ounces? In reality, they are both the same—165 pounds. Which units should we use?

The most widely used system of units is the **SI system**, or the International System of Units. This is the modern, standard form of the **metric system**, with quantities based on powers of 10. The SI system has seven base units, five of which are shown in *Table 2.1*. These units can be used to derive other units, many of which are shown in *Table 2.2*. Each of these units can be used with metric prefixes as shown in *Table 2.3*. These metric prefixes often simplify any math we need to do: for example, to convert a resistance of 3.3 kΩ to ohms, multiply by 1,000 (k = 1,000) for 3,300 Ω.

We briefly saw units for current (ampere, or amp), voltage (volt), and resistance (ohm) in Chapter 1; soon, we will look at how they are related.

TABLE 2.1 *Selected SI Base Units*

Quantity	Symbol	Unit	Abbreviation
Length	L	Meter	M
Mass	M	Kilogram	Kg
Time	T, t	Second	S
Electric current	I, i	Ampere (Amp)	A
Temperature	T	Kelvin	K

© CENGAGE LEARNING 2012.

TABLE 2.2 *Selected SI Derived Units*

Quantity	Symbol	Unit	Abbreviation
Force	F	Newton	N
Energy	W	Joule	J
Power	P, p	Watt	W
Voltage	V, v (sometimes E, e)	Volt	V
Charge	Q, q	Coulomb	C
Resistance	R	Ohm	Ω
Capacitance	C	Farad	F
Inductance	L	Henry	H
Frequency	f	Hertz	Hz

© CENGAGE LEARNING 2012.

TABLE 2.3   Metric Prefixes and Powers of 10

Prefix	Symbol	Power of 10	Multiplier
tera	T	10^{12}	1,000,000,000,000
giga	G	10^{9}	1,000,000,000
mega	M	10^{6}	1,000,000
kilo	k	10^{3}	1,000
milli	m	10^{-3}	0.001
micro	μ	10^{-6}	0.000001
nano	n	10^{-9}	0.000000001
pico	p	10^{-12}	0.000000000001

© CENGAGE LEARNING 2012.

Your Turn

2.1 Convert the following quantities to the unit given:
  a. 1,000,000 μA to amps (A)
  b. 400 kΩ to ohms (Ω)
  c. 14.4 V to mV
  d. 8.68 km to cm

2.2   ATOMIC STRUCTURE

KEY TERMS

Molecule Combination of atoms; the smallest particle of a substance other than a pure element.

Atom Fundamental building block of molecules; smallest particle of an individual element.

Nucleus Center of an atom, containing neutrons and protons. Electrons orbit around the nucleus.

Proton Positively charged particle found in the nucleus of the atom.

Neutron Particle in the nucleus of an atom having no charge (or a neutral charge).

Electron Negatively charged particle orbiting the nucleus of an atom.

Orbital path Paths of electrons orbiting the atom's nucleus. Each orbital path holds a certain number of electrons.

Atomic number The number of protons in the nucleus of an atom.

Noble elements Elements with full orbital paths. It is very difficult for these elements to combine to form molecules.

Electric current Flow of charge.

All material is made from **molecules**, which are comprised of **atoms**. Atoms have a **nucleus**, consisting of **protons**, or positively charged particles, and **neutrons**, or neutrally charged particles, as shown in *Figure 2.1*. **Electrons**, which have a negative charge, orbit around this nucleus. Because opposite charges attract, the negatively charged electrons are attracted to the positively charged nucleus.

FIGURE 2.1 Atom with a Nucleus and Orbiting Electrons

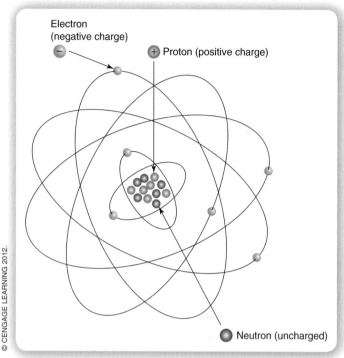

Electron (negative charge)

Proton (positive charge)

Neutron (uncharged)

There are **orbital paths** where these electrons orbit around the nucleus, much like planets revolve around the sun. Each orbital path can only hold so many electrons. For example, up to two electrons can be in the first orbital path, and up to eight can be in the second orbital. Once an orbital is full, it is very difficult to remove an electron from that orbital.

The **atomic number** of an element is the number of protons in the nucleus of an atom of that element. It is the property that determines what type of element an atom is. Elements with atomic numbers of 1 (Hydrogen, or H) or 2 (Helium, or He) have electrons only in the first orbital, and elements with numbers less than 10 have electrons only in the first or second orbital paths. Because these are the closest, the attraction is great. In fact, Helium (He) has an atomic number of 2, meaning the innermost orbital is full, and the element Neon (Ne) has an atomic number 10, which means it has 10 electrons, filling the first two orbitals. These are known as **noble elements**. Because the orbitals are full, the elements are very stable and don't easily combine with other elements.

Element 3, Lithium (Li), and element 11, Sodium (Na), both have one extra electron, which is forced to orbit in the next level, where the attraction isn't as strong. The outer electron in each element makes the element easy to combine with, since it's easy to break the bond of that extra electron. Element 9, Fluorine (F), has two electrons in its innermost orbit and only 7 in the next orbital, which can hold eight. If Fluorine comes in contact with Sodium, they bond to form Sodium Fluoride (NaF): you can think of the Fluoride atom "accepting" the extra electron of the Sodium atom into its empty spot in the second orbital, as the two lone atoms join together to form one molecule. The molecule that is formed, NaF, is very stable.

This is a very simplified view of how these atoms and molecules behave; you will be able to learn much more about the formation of molecules in your chemistry class.

FIGURE 2.2 Battery Sending Charge (Electrons) through a Wire

Atoms and Electricity

How does this relate to electricity? When a series of copper atoms is grouped together, like we would find in a piece of wire, some of the outer electrons jump back and forth between atoms of copper. If we apply a *positive* charge to one end, the negatively charged electrons find the attraction to the positive charge irresistible, and we can attract the electrons toward that charge, as we see in *Figure 2.2*. This movement of the charge is **electric current**. For current to flow, we need to replenish the electrons at the negative end so they can flow to the positive end; as electrons flow, the charge flows. Therefore, if we touch one end of a wire to the negative side of a 9-V battery and touch the other end of the wire to the positive side of the battery, the electrons move toward the positive side and flow back into the wire from the negative side. There is a chemical reaction inside the battery, which allows this charge to flow through the battery; the copper atoms in the wire simply allow their outer electrons to jump to the next copper atom heading toward the positive charge, as long as the copper on the other side sends an electron over to replace it.

How much charge can flow using a battery and a wire? In theory, an infinite amount of current can flow (which we'll see shortly). However, there is a limit in the amount of charge the battery can generate, therefore, a limit on the amount of current the battery can provide.

2.3 VOLTAGE, CURRENT, AND RESISTANCE

KEY TERMS

Voltage Electrical potential.

Potential energy Stored energy that has potential to do work.

Current The flow of electricity.

Resistance Characteristic of electrical components that resists current flow.

We have briefly looked at current, voltage, and resistance. When dealing with electricity or electrical circuits, we begin with a source of voltage. Voltage is electrical potential energy: that is, if we have something with a voltage (such as a battery), we have the potential to generate current. However, if we simply hold a battery in our hands, or look at batteries in a package in a store, no charge is moving, so current is flowing. It may be helpful to think of voltage like other sources of potential energy.

A large water tank on a tall hill has the potential to supply water to houses and businesses, but only if pipes are attached and the valve is open, allowing water to flow. Without a path for the water to follow, without pipes, or without open valves, we simply have a large water container on a tall hill with the potential to send water out. A boulder on a hill has potential energy. As it sits, waiting to roll down a hill, no work is being done and no energy is generated. If the boulder begins to roll, work is being done as the boulder rolls and crushes things in its path. Until it rolled, it only had *potential* to do work. The voltage in a 9-V battery has the potential to generate current, but remember, if it sends electrons out of one terminal, it must receive electrons in the other terminal, or it will not be able to continue to send charge (electrons) into the circuit. In order for current to flow, we need a complete circuit.

If we complete the circuit, or create a path from one battery terminal back to the other, we let charge flow out of one terminal and back into the other. This flow of charge is current. As we've discussed, we need to be sure we don't ask the battery to send an infinite amount of current through our circuit. When we add resistance to the circuit, we limit the amount of current that can flow in the circuit. In reality, anything that allows electricity to flow through it has some resistance, but for some things, resistance can be very, very low. Resistance (as the name implies) describes a device that *resists* the flow of current. The higher the resistance, the harder it is for current to flow.

We can think of current like water flowing through a hose. If you attach a garden hose to a faucet and turn on the faucet, water flows through the hose. For now, let's ignore that water goes out the end of the hose, since electricity certainly doesn't just go out the end of a wire. This is similar to attaching a wire to a battery: current flows through the wire.

Now, instead of a garden hose connected to a faucet, imagine attaching a fire hose to a fire hydrant. Both the faucet with the garden hose and the hydrant with the fire hose get water from the same source—the water tower, the original source of the water potential energy. However, the fire hose certainly lets more water through at a much higher pressure than the garden hose. Why? The fire hose is much larger, so it's easier for water to flow through the fire hose than the garden hose; therefore, more water flows through the fire hose. Less resistance means more flow, whether it's water or electricity.

You could try an experiment at home, at a restaurant, or at lunch: if you drink through a straw, give the straw a squeeze and continue to drink. What happens? You made it harder for the liquid to flow through the straw; in other words, you increased the resistance to the liquid.

In electrical circuits, we typically have a source of voltage (usually a battery or a wall outlet), and resistors in the circuit to regulate the amount of current that can flow.

Conductors and Insulators

> **KEY TERMS**
>
> **Conductor** A material that allows current to easily flow through it, with a very low resistance.
>
> **Insulator** A material that does not allow current to pass through, with a very high resistance.

If you have seen a cut or frayed electrical cord, you probably noticed that on the inside is a copper wire. On the outside, the cord is covered by a protective layer of rubber. Why cover the cord with rubber? If we didn't cover the copper wire with rubber, and we plugged a copper wire into an electrical outlet, we would get shocked and could be killed. Yet, because this same cord is coated in rubber, we don't get shocked at all.

Copper (and other metals) are **conductors**. Conductors allow electricity to pass through them—they *conduct* electricity—with very little resistance. In fact, the resistance of metal wires is so low, we don't even consider their resistance when calculating how much current is flowing or how much voltage is needed in a circuit. On the other hand, **insulators** (such as rubber) don't allow electricity to flow through—they don't allow charge to flow. Their resistance is very high—so high that we usually simply say they block electricity. You may have noticed technicians working on transformers on a power line: they usually wear thick rubber gloves and boots to protect them from being shocked. You may be surprised to hear that pure water itself is not a good conductor of electricity; however, impurities in water can make the water very conductive. Humans can be shocked fairly easily because, for one reason, the human body is approximately 60% water by weight.

2.4 OHM'S LAW

> **KEY TERMS**
>
> **Ohm's Law** $V = R \times I$: the voltage across a component equals its resistance multiplied by the current through it.
>
> **Series** Resistors in series require all current flowing through one resistor to also flow through the next resistor, and so on.

We have claimed that, by hooking a copper wire across the terminals of a battery, we are asking the battery to supply an infinite amount of current. Also, we've said

Note . . .

Ohm's Law: $V = R \times I$

Note . . .

We can write Ohm's Law in different ways, but it is always the same equation:

$$V = I \times R$$
$$I = V/R$$
$$R = V/I$$

are all forms of Ohm's Law.

that increasing the resistance in a circuit means less current is able to flow. What is the relationship?

Ohm's Law gives us the relationship between voltage, current, and resistance. Ohm's Law states that the voltage across a component is equal to the resistance of the component multiplied by the current through the component. Mathematically, Ohm's Law is

$$V = R \times I$$

This is one of the most common equations in electronics. If we know any two of the values, we can find the third.

For example, if we have a 9-V battery and use a piece of copper wire to touch both terminals, how much current are we asking the battery to produce? If the voltage is 9 V and the resistance of 1 inch of thin copper wire is 0.002 ohm, the current through the wire is:

$$V = R \times I$$
$$9\,V = 0.002\,\Omega \times I$$
$$I = 9\,V/0.002\,\Omega = 4{,}500 \text{ amps}$$

You might recall that less than 1 amp of current can easily be fatal. Fortunately, a battery cannot generate this much current. In fact, household batteries are designed to supply current in the mA (milliamp, or 1/1,000 amp) range.

If we examine more reasonable values of voltage, current, and resistance, we may want to find the current supplied by a 9-V battery if we use a 1,000-Ω resistor.

$$V = R \times I, \text{ or } I = V/R$$
$$I = 9\,V/1{,}000\,\Omega = 0.009\,A, \text{ or } 9\,mA$$

We may also use Ohm's Law to find the voltage across a resistor if we know the current through the resistor.

In *Figure 2.3* we see a 1.5-V battery sending current through two resistors. The resistors are in **series**: that is, the current flowing through one of the resistors must flow through the other as well. We can find the total resistance of a series circuit by adding together all the resistor values.

FIGURE 2.3 **1.5-V Battery with Two Resistors in Series**

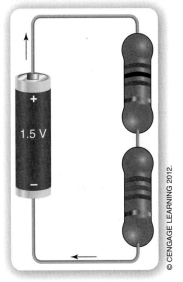

1.5 V

© CENGAGE LEARNING 2012.

Example 2.2

Use Ohm's Law to find the voltage across one resistor if we put two resistors in the circuit shown in *Figure 2.4*.

FIGURE 2.4 **Example 2.2: Two Resistors in Series**

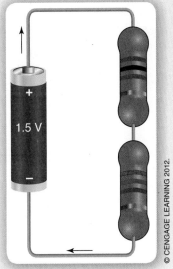

1.5 V

© CENGAGE LEARNING 2012.

Solution

We can use Ohm's Law to find the total current through the resistors:

$$I = V/R = 1.5\,V/(270\,\Omega + 100\,\Omega)$$
$$= 1.5\,V/370\,\Omega = 0.00405\,A, \text{ or } 4.05\,mA$$

Knowing the current through the 270-Ω resistor, we can find the voltage across this resistor:

$$V = R \times I = 270\,\Omega \times 0.00405\,A = 1.095\,V$$

We can also find the voltage across the 100-Ω resistor:

$$V = R \times I = 100\,\Omega \times 0.00405\,A = 0.405\,V$$

Note . . .

Notice that in Figure 2.3 the current flows from the positive terminal toward the negative terminal, while in Figure 2.2, the electrons flow from the negative to the positive terminals. The current direction shown in Figure 2.3 (and the rest of the book) is called conventional current, which is actually opposite to the direction of electron flow, but used most often. Blame Benjamin Franklin for getting the polarity wrong!

An interesting observation: the voltage across the 270-Ω resistor (1.095 V) added to the voltage across the 100-Ω resistor (0.405 V) is 1.5 V, the original voltage of the battery!

Example 2.3

Multisim Example

Multisim Files: 02.03 resistors in series.ms11, 02.03 resistors in series without multimeters.ms11

Figure 2.5 shows a Multisim circuit with a battery (shown as a 1.5-V voltage source) and two resistors in series; in fact, this is the same circuit as Example 2.2. The circuit as shown is a complete, fully functional circuit. We will often want to measure current through a device and voltage across a device; we would use a multimeter for these measurements. A multimeter is capable of measuring current through a circuit and voltage across an element.

Use Multisim software to open the file 02.03 resistors in series.ms11. This circuit is complete and ready to run: to begin the simulation, click the power switch on the top right side of the program to "I" instead of "O." You should see values for the current through the circuit and voltage across each resistor (which add up to 1.5 V, the voltage of the battery).

If you want to try adding the meters yourself, open the file 02.03 resistors in series without multimeters.ms11, and add multimeters as shown in *Figure 2.6* and *Figure 2.7*. *Figure 2.8* shows that the voltages across each resistor do add up to the original voltage of 1.5 V.

The voltage reading on meter 1 + meter 2 = the reading across meter 3.

FIGURE 2.5 Example 2.3: Two Resistors in Series

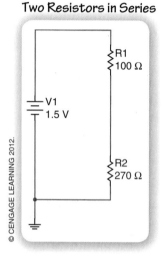

© CENGAGE LEARNING 2012.

FIGURE 2.6 Example 2.3: Circuit with Added Multimeters

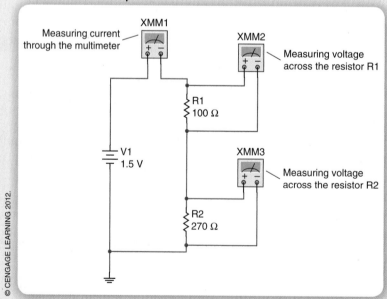

© CENGAGE LEARNING 2012.

Note . . .

A multimeter can be used to measure current through a component or voltage across the component. To measure current, the current must flow through the meter, so the meter must be part of the current path. Measuring voltage is done by placing each probe on one side of the component. Since voltage is much more easily measured, we will measure voltage most of the time.

continued...

FIGURE 2.7 Example 2.3: Circuit after Double-Clicking on Multimeters

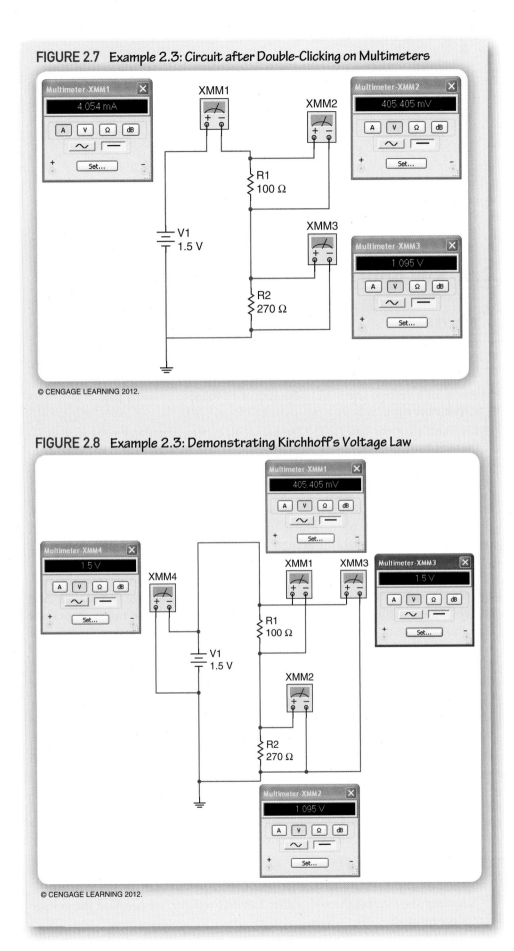

FIGURE 2.8 Example 2.3: Demonstrating Kirchhoff's Voltage Law

Your Turn

2.2 What is the voltage across a 1,000-Ω resistor with 0.05 A of current flowing through it?

2.3 What is the current through a 10-kΩ resistor with 4.6 V across it?

2.4 What is the resistance of a device with 6.7 V across it and 0.1 A through it?

2.5 What is the voltage across a 1.3-MΩ resistor with 4.7 μA through it?

2.5 KIRCHHOFF'S VOLTAGE AND CURRENT LAWS

FIGURE 2.9 *Two Resistors in Parallel*

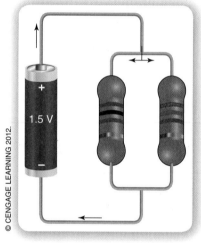

KEY TERMS

Parallel Resistors in parallel are connected to each other at both ends. They have the same voltage across them, but any current flowing into the parallel combination of these resistors splits among them and joins back together after passing through.

Kirchhoff's Voltage Law (KVL) The sum of all voltages around a path in a circuit must equal zero.

Kirchhoff's Current Law (KCL) The sum of all currents entering a node must be zero; in other words, any current entering a node must also leave the node.

Node A junction of electronic components.

Notice the difference in the two circuits shown in Figure 2.4 and *Figure 2.9*. Figure 2.4 shows a circuit in which all of the current flowing through the first resistor must flow through the second as well—there is no other path any of the current may pass through. These resistors are in series.

The resistors in Figure 2.9 are tied together at each end. In this case, the voltage across each resistor is exactly the same. These resistors are in **parallel**. Any current which is going to flow into and be split between these two resistors will flow through one or the other, then join together again once it leaves. Resistors in series or parallel must follow these rules exactly. The resistors shown in *Figure 2.10* cannot be considered a simple series or parallel circuit. In this case, we can say that the parallel combination of R2 and R3 is in series with resistor R1, but we cannot say R1 is in series with R2 or R3. We can say that R2 is in parallel with R3.

FIGURE 2.10 *R2 and R3 are in Parallel; the Combination Is in Series with R1*

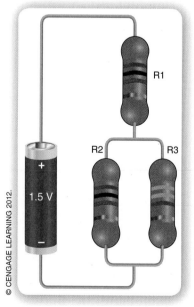

You may take courses in circuit analysis in the future, where you will be able to investigate how different combinations of resistors (and capacitors) can be combined, but we don't have to do so for digital electronics. We will simply use these circuits to better understand current and voltage.

Consider a circuit similar to one we have seen previously. The circuit in *Figure 2.11* has a series of three resistors and one 12-V source. Notice that the resistors are indeed in series: all current flowing through R1 has to flow through R2, and has to flow through R3. We can use Ohm's Law to find the overall current through the resistors:

$$I = V/R = 1.5 \text{ V}/(1{,}000 \ \Omega + 1{,}000 \ \Omega + 3{,}300 \ \Omega) = 1.5 \text{ V}/5{,}300 \ \Omega$$
$$= 0.000283 \text{ A, or } 0.283 \text{ mA}$$

Knowing the current through either 1,000-Ω resistor, we can find the voltage across the resistor:

$$V = R \times I = 1{,}000 \ \Omega \times 0.000283 \text{ A} = 0.283 \text{ V across each 1,000-}\Omega \text{ resistor.}$$

We can also find the voltage across the 3,300-Ω resistor:

$$V = R \times I = 3{,}300\ \Omega \times 0.000283\ A = 0.934\ V$$

What if we add the voltages across each resistor?

$$V_{R1} + V_{R2} + V_{R3} = 0.283\ V + 0.283\ V + 0.934\ V = 1.50\ V$$

If we add the voltages across each resistor, we end up with the original voltage across the series of resistors. This is known as **Kirchhoff's Voltage Law (KVL)**: the voltage from one point in a circuit to another is equal regardless of the path taken to get from that point. In other words, if we add the voltages as we go around a path in the circuit (if we come to a minus sign first, subtract), we end up with zero.

Figure 2.12 shows the same circuit with the voltages labeled. Begin at the + side of the voltage source and go around the circuit. Add the voltage across R1 (+0.283 V) to the voltage across R2 (+0.283 V) to the voltage across R3 (+0.934 V), then continue around the circuit and we reach the negative side of the voltage source (−1.5 V), and we end up with 0.

FIGURE 2.11 R1, R2, and R3 in Series

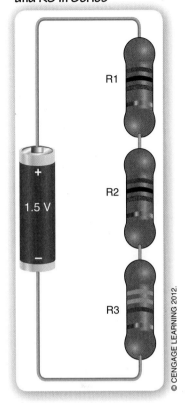

© CENGAGE LEARNING 2012.

Example 2.4

Find the voltages across each resistor shown in *Figure 2.13*, and show that the voltages satisfy Kirchhoff's Voltage Law.

FIGURE 2.13 Example 2.4: Four Resistors in Series

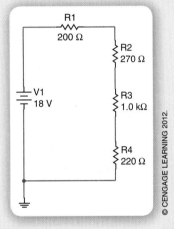

© CENGAGE LEARNING 2012.

Solution

First, we can calculate the current through the series resistors R1, R2, R3, and R4. Using Ohm's Law,

$$I = V/R_{total}$$

We can find the total resistance value R_{total} by adding all of the resistor values,

$$R_{total} = R1 + R2 + R3 + R4 = 200\ \Omega$$
$$+ 270\ \Omega + 1{,}000\ \Omega + 2{,}20\ \Omega = 1{,}690\ \Omega$$

Therefore,

$$I = V/R_{total} = 18\ V/1{,}690\ \Omega = 0.01065\ A$$
$$= 10.65\ mA$$

Now, we can use this value of current to find the voltage across each resistor:

$$V_{R1} = R1 \times I = 200\ \Omega \times 10.65\ mA = 2.130\ V$$
$$V_{R2} = R2 \times I = 270\ \Omega \times 10.65\ mA = 2.876\ V$$
$$V_{R3} = R3 \times I = 1{,}000\ \Omega \times 10.65\ mA = 10.65\ V$$
$$V_{R4} = R4 \times I = 220\ \Omega \times 10.65\ mA = 2.343\ V$$

To prove that this satisfies Kirchhoff's Voltage Law, add the voltages; the sum should be the original voltage course value:

$$V_{R1} + V_{R2} + V_{R3} + V_{R4} = 2.130\ V + 2.876\ V$$
$$+ 10.65\ V + 2.343\ V = 18\ V$$

Final voltages are shown in *Figure 2.14*.

FIGURE 2.12 R1, R2, and R3 in Series with Voltages

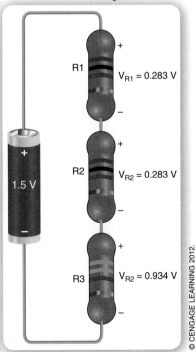

© CENGAGE LEARNING 2012.

FIGURE 2.14 Example 2.4: Four Resistors in Series with Voltages

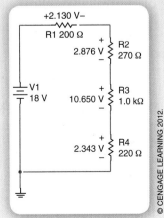

© CENGAGE LEARNING 2012.

FIGURE 2.15 Example 2.5: Three Resistors in Series

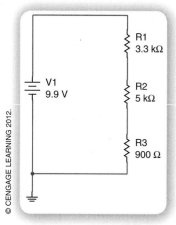

Example 2.5

Multisim Example

Multisim Files: 02.05 KVL.ms11, 02.05 KVL with meters.ms11

Figure 2.15 shows a Multisim circuit with a 9.9-V voltage source and three resistors in series. Open the file 02.05 KVL.ms11 and add multimeters, measure the voltage across each resistor, and verify that the voltages add up to the source voltage.

Solution

Figure 2.16 shows the circuit with multimeters measuring each voltage. Adding the voltages across the resistors gives a total voltage of 9.9 V, proving that KVL applies.

FIGURE 2.16 Example 2.5: Three Resistors in Series with Voltages Shown

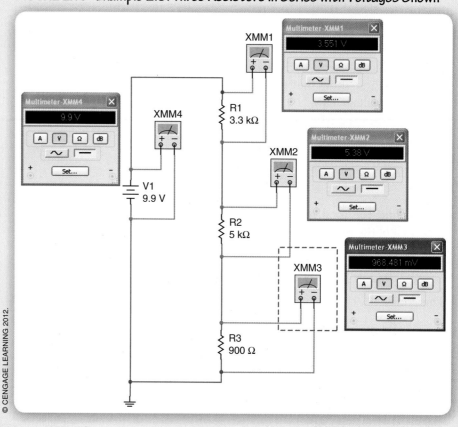

FIGURE 2.17 Series/Parallel Circuit with Current Values

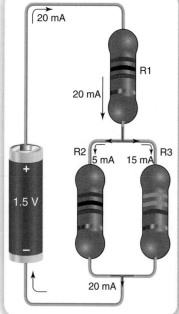

Kirchhoff's Current Law (**KCL**) is seen in *Figure 2.17*. The total current from the voltage supply through resistor R1 is 0.02 A, or 20 mA. The current splits: some current flows through the 1,000-Ω resistor and some goes through the 330-Ω resistor. Because the 330-Ω resistor has less resistance, and is therefore an easier path, more current will flow through resistor R3. Some current will flow through resistor R2. The resistance is higher, so less current will follow this path. We can see in the figure that the total current flowing into the junction—called a **node**—where R1 meets R2 and R3 is 20 mA. From there, 5 mA flows through R2 and

15 mA flows through R3. The total amount of current through R2 plus the current through R3 equals the total current into the node. No current is lost. Recall earlier where we thought of current as water flowing through a pipe. Water enters your home through a large pipe, then splits and goes to the bathroom, kitchen, and anywhere else water is needed in the house. If your pipes don't leak (and they shouldn't), the amount of water you use everywhere in the house equals the amount of water that comes in through the pipe. Kirchhoff's Current Law says that the total amount of current entering a node equals the total current leaving a node.

Example 2.6

Multisim Example

Multisim Files: 02.06 KCL.ms11, 02.06 KCL with meters.ms11

Figure 2.18 shows a Multisim circuit with a 12-V voltage source and three resistors in parallel. Open the file 02.06 KCL.ms11, add multimeters measuring the current through each branch of the circuit, and show that Kirchhoff's Current Law applies.

FIGURE 2.18 Example 2.6: Three Resistors in Parallel

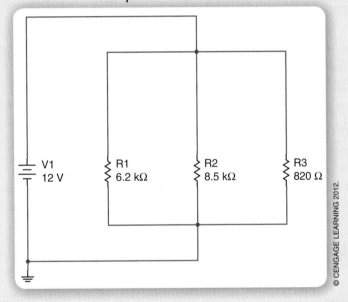

© CENGAGE LEARNING 2012.

Solution

The typical issue when measuring current is that the meter must receive the exact same current as the resistor you intend to measure; this can make it tricky to add the meters. See *Figure 2.19*: notice the meters are directly in the path of the current. Also notice that the "A" button has been pressed on the meters to make them measure current (in amps).

Turn the simulation on until you see a reading. If your circuit is drawn properly, you should see:

$$I_{R1} = 1.936 \text{ mA}$$
$$I_{R2} = 1.412 \text{ mA}$$
$$I_{R3} = 14.635 \text{ mA}$$

continued...

Note . . .

We've discussed current entering a node and current leaving a node: this is one way to define KCL (the sum of the current entering a node must equal the current leaving a node). We can define positive and negative directions for current, which would prove helpful if we need to mathematically solve KCL. We define:

Current entering the node will be considered positive.

Current leaving the node will be considered negative.

FIGURE 2.19   Example 2.6: Three Resistors in Parallel with Current Values

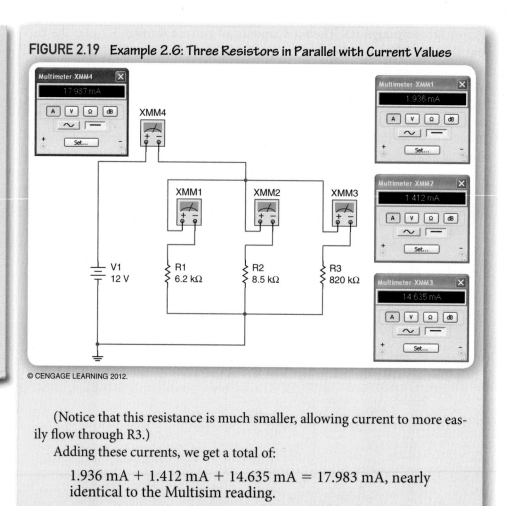

© CENGAGE LEARNING 2012.

(Notice that this resistance is much smaller, allowing current to more easily flow through R3.)

Adding these currents, we get a total of:

1.936 mA + 1.412 mA + 14.635 mA = 17.983 mA, nearly identical to the Multisim reading.

The three laws we have reviewed (Ohm's Law, Kirchhoff's Voltage Law, and Kirchhoff's Current Law) are used in every branch of electronics.

2.6   BREADBOARDING

KEY TERMS

Schematic   A drawing of a circuit being designed that uses special symbols for circuit components.

Trace   A copper line on a circuit board connecting one component to another. A path for current to flow on a circuit board.

Simulation   Entering a schematic, or drawing, of a circuit to test into a computer program which calculates voltages and currents throughout the circuit without physically building the circuit.

Prototype   A circuit built functionally the same as a circuit we wish to test.

Breadboard   A device with hidden interconnections on which we can build a prototype circuit.

Circuit diagrams or **schematics**, such as those in the previous figures, show us the components in a neat and orderly fashion. When we build circuits, we certainly

FIGURE 2.20 Components in a Breadboard

PHOTO: © ISTOCKPHOTO.COM/JEROEN REMANS

FIGURE 2.21 Breadboard

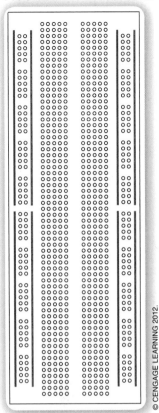

© CENGAGE LEARNING 2012.

don't lay out a battery and resistors on a table and twist the wires together. You could guess this is the case by looking at any circuit board. You can see components and **traces** (lines connecting one component to the next component), usually under a green covering on the circuit board itself. The parts are usually densely packed on the board instead of neatly arranged for us to see. Although we represent circuits using circuit diagrams, which are easy for us to see and analyze (and we will continue to do this), when we need to physically build and test a circuit, we simply need to be sure the components physically *connect* the same way as we drew them in the schematic.

The typical circuit design process involves drawing a schematic followed by **simulating** the diagram. Simulation occurs by entering the schematic into a circuit design package and having the computer calculate values of voltage and current as though we built the circuit—although we didn't physically build the circuit. If all goes well, we typically build a **prototype**. A prototype is built by connecting each component properly and measuring voltages and/or currents. The prototype usually is not built to look like the final product. Instead, we often use a **breadboard** (*Figure 2.20*) to build prototypes.

Figure 2.21 shows a breadboard. A breadboard is covered by patterns of holes, many of which are internally connected to each other. The connection pattern is seen in *Figure 2.22*; holes shown connected by lines are connected to each other internally. Inserting a resistor leg into the hole shown by the arrow in *Figure 2.23A* followed by inserting another in the hole in the same row (*Figure 2.23B*) connects the two leads. This forms a node, shown in *Figure 2.24*. You can see long rows that appear to be "grouped together" along the sides of the breadboard along with the rows of 5 holes connected as previously shown: these are usually used for power. Attaching a wire to the positive (+) side of a battery then to a hole at the top of the board, and connecting one to the negative terminal of the battery (−) and inserting the other end into a row on the other side gives us positive and negative voltage.

FIGURE 2.22 Breadboard —Lines Show Which Holes Are Connected

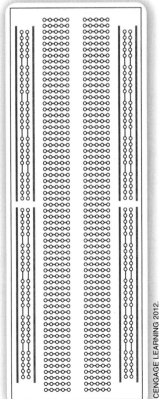

© CENGAGE LEARNING 2012.

FIGURE 2.23A Breadboard with One Hole Selected

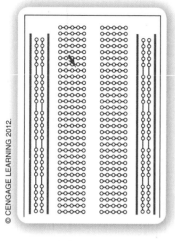

FIGURE 2.23B Breadboard with Two Resistors

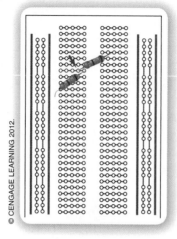

FIGURE 2.24 Resistors Together Form a Node

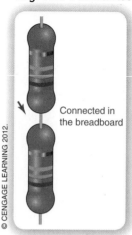

Connected in the breadboard

FIGURE 2.25 Circuit to Be Built on a Breadboard

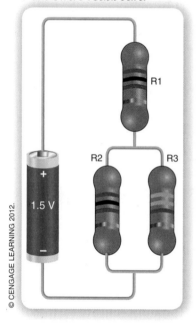

R1

R2   R3

1.5 V

FIGURE 2.26 Breadboard with Circuit from Figure 2.25

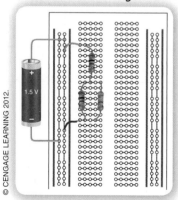

1.5 V

Let's look at an example. The circuit shown in *Figure 2.25* shows

► the positive side of the battery is connected to one leg of R1,
► the other leg of R1 is connected to one leg of R2 and R3, and
► the other legs of R2 and R3 are connected together to the negative side of the battery.

Figure 2.26 shows this circuit built on a breadboard. Notice that each connection shown in the drawing is connected on the breadboard; therefore, the circuits are the same (even though it doesn't necessarily look that way).

Breadboarding is very common when building prototypes. Other methods, such as building inexpensive circuit boards, are also used. In digital electronics, we'll soon find that we aren't usually concerned with voltages, currents, and resistance; instead, we're concerned that a circuit functions correctly. We want to see signals that should be ON are ON, and signals that should be OFF are OFF. For digital circuits, it's often more useful to simulate and download the circuit to a programmable device.

2.7 CIRCUIT DESIGN SOFTWARE

KEY TERMS

Programmable device A device which can be reprogrammed or "rewired" internally, allowing the testing of digital circuits without wiring new components.

Download The process of sending a circuit schematic drawing into a programmable device.

In digital electronics, we follow the same basic procedure as we've described: we draw the circuit using a computer program, then simulate the circuit in the same program, then build a prototype to see if it really worked. The difference between digital electronics and circuits with a voltage source and a bunch of resistors and/ or capacitors is that we want to see if outputs go ON or OFF when we turn inputs

ON or OFF. Fortunately, there are **programmable devices**. These devices allow us to draw a schematic (or a picture of the circuit we want to build), simulate it, then physically build it inside of the device—without needing a breadboard or wires! When we send a diagram, or **download** a program to a device, we can imagine that the computer is breadboarding the circuit inside of the programmable device on the board. There are a few different computer design software products that will allow us to design, simulate, and test digital circuits. You will get to use one or more of these packages in this course.

For example, we can use Multisim to draw a digital circuit, as seen in *Figure 2.27*. This circuit can be drawn, simulated and, if all goes well, downloaded into a programmable device and physically tested, without requiring a breadboard. We'll get to investigate this through the rest of the book.

FIGURE 2.27 *Circuit Drawn Using Multisim Software*

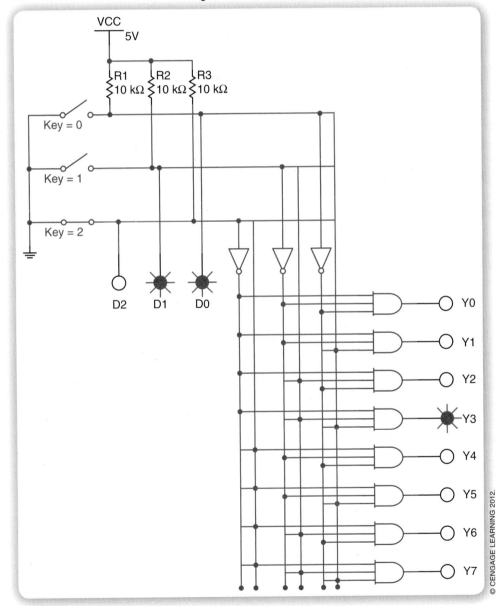

SUMMARY

1. We use units to describe many things in our lives. Using familiar units allows us to easily relate to quantities.

2. The International System of Units (SI units) is the most commonly used, modern form of the metric system.

3. Standard units used in electronics are: amperes, or amps, for current; volts for voltage and ohms for resistance.

4. All material is made from molecules. Molecules are made from the combination of atoms.

5. Atoms have a nucleus at their center made of protons and neutrons, with orbiting electrons.

6. Protons have a positive charge, electrons have a negative charge, and neutrons have no charge (or a neutral charge).

7. Electrons orbit the nucleus in orbital paths, each of which holds a specific number of electrons. Additional electrons must begin to fill the next (further) orbital path.

8. Noble elements are those elements with completely full orbital paths; they do not easily combine with other elements to create molecules.

9. Current is the flow of charge, typically the flow of negatively charged electrons.

10. Voltage is electrical potential energy: it has the potential to produce current if a circuit path exists.

11. Resistance is a characteristic of components that resists the amount of current that can flow.

12. It can be helpful to consider the flow of current similar to water flowing through pipes. In this case, voltage is similar to water stored in a water tank, ready to cause water to flow.

13. Conductors allow current to easily flow. Metals, such as copper, are excellent conductors.

14. Insulators do not allow current to flow. Rubber is an example of an insulator.

15. Ohm's Law states that the voltage across a component equals the current through it multiplied by its resistance:

$$V = I \times R.$$

16. Ohm's Law is useful because, if we know any two quantities (voltage, resistance, or current), we can find the other.

17. Resistors are in series if all current passing through one resistor must pass through the next resistor as well.

18. Resistors are in parallel if they are connected to each other at both ends. Parallel resistors have the same voltage across them, but any current entering a node with both resistors must split between the two resistors, then join back together at the other side.

19. Kirchhoff's Voltage Law (KVL) says that the sum of voltages around a circuit path must be zero.

20. Kirchhoff's Current Law (KCL) says that the sum of all currents entering a node must be zero (in this case, currents leaving a node are considered negative).

21. A node in a circuit is where circuit components connect together.

22. A breadboard is a device with hidden connections allowing us to build a circuit to test its functionality. Building such a circuit is sometimes called breadboarding.

23. A schematic is a diagram of a circuit.

24. Simulation is the process of allowing a computer to calculate voltages and currents in a circuit based on a schematic.

25. A prototype is a circuit built to test the functionality of a circuit design.

26. Schematics can be downloaded into a programmable device; these devices can be reconfigured internally, letting us test and make changes to circuits without wiring or physically building a circuit.

BRING IT HOME

2.1 SI Notation

2.1 Convert 3 kΩ to Ω.

2.2 Convert 2 mA to A.

2.3 Convert 20 mV to V.

2.4 Convert 420,000 μA to mA.

2.5 Convert 0.0067 kV to mV.

2.6 Convert 22.6 km to meters.

2.7 Which of the following units are units in the SI system?
  a. meter
  b. ampere
  c. inch
  d. km
  e. troy ounce
  f. Kelvin
  g. Fahrenheit
  h. ohm

2.8 Convert 46×10^{-6} V to kV.

2.2 Atomic Structure

2.9 Arrange these from smallest to largest: atom, nucleus, proton, molecule.

2.10 Write a sentence describing where neutrons, electrons, and protons are found in an atom.

2.11 Write a sentence describing a noble element.

2.12 If the atomic number of Lithium is 3, how many electrons are in each orbital path?

2.13 If the atomic number of Oxygen is 8, how many electrons are in each orbital path?

2.3 Voltage, Current, and Resistance

2.14 Write a sentence describing voltage.

2.15 Write a sentence describing current.

2.16 Write a sentence describing resistance.

2.17 What units do we use for voltage, current, and resistance?

2.18 How much current is flowing through a battery that is not connected to anything?

2.19 Write a sentence describing an insulator.

2.20 Write a sentence describing a conductor.

2.4 Ohm's Law

2.21 If we know resistance and voltage, we can use Ohm's Law to find _____?

2.22 Find the voltage across a 270-Ω resistor with 1 A of current through it.

2.23 Find the current through a 3,600-Ω resistor with 9.6 V across it.

2.24 Find the resistance of a resistor with 0.03 A of current through it and 0.2 V across it.

2.25 Find the voltage across a 27.0-kΩ resistor with 1.6 μA of current through it.

2.26 Find the current through a 1.25-MΩ resistor with 1.25 V across it.

2.27 Find the resistance of a resistor with 47 mA of current through it and 270 mV across it.

2.28 Find the current through a series combination of a 1,000-Ω and 7.70-kΩ resistor with 4.7 V across the series combination.

2.29 Find the voltage across each resistor of the series combination of a 1,000-Ω and 7.70-kΩ resistor with 4.7 V across the series combination.

2.30 Draw a circuit using symbols for a voltage source and resistor showing that the current through a 30-Ω resistor from a 2.5-V source is 83.3 mA.

2.5 Kirchhoff's Current and Voltage Laws

2.31 Write a sentence describing Kirchhoff's Current Law.

2.32 Write a sentence describing Kirchhoff's Voltage Law.

2.33 Use the results of Problem 2.29 to show Kirchhoff's Voltage Law is true.

2.6 Breadboarding

2.34 Write a sentence describing a breadboard.

2.35 Write a sentence describing the process of creating a schematic, simulating, and prototyping.

2.7 Circuit Design Software

2.36 What is the advantage of using a programmable device for prototyping?

EXTRA MILE

2.2 Atomic Structure

2.37 Describe what would happen if Chlorine came into contact with Sodium.

2.38 How many electrons are in each orbital path of Potassium?

2.4 Ohm's Law

2.39 Which resistors in *Figure 2.28* are in parallel with each other? Which are in series?

FIGURE 2.28: Problems 2.39 and 2.40: Circuit with Five Resistors

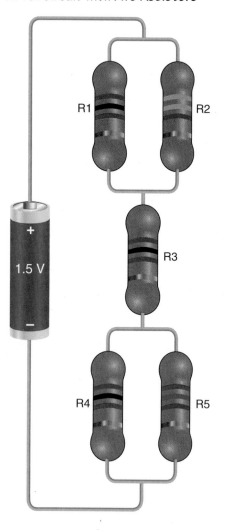

© CENGAGE LEARNING 2012.

2.40 Draw the path of the current through each resistor in *Figure 2.28*.

2.41 Are the resistors in *Figure 2.29* in series with each other? Why or why not?

FIGURE 2.29: Problem 2.41: Circuit with Four Resistors © CENGAGE LEARNING 2012.

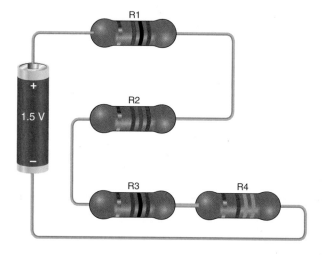

2.42 Are the resistors in *Figure 2.30* in series with each other? Why or why not?

FIGURE 2.30: Problem 2.42: Circuit with Four Resistors

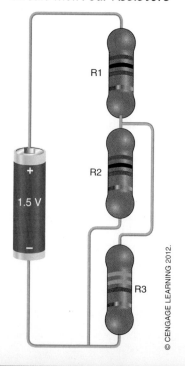

© CENGAGE LEARNING 2012.

CHAPTER 3
Basic Principles of Digital Systems

CHAPTER OBJECTIVES

Upon successful completion of this chapter, you will be able to:

1 Describe some differences between analog and digital electronics.

2 Understand the concept of HIGH and LOW logic levels.

3 Explain the basic principles of a positional notation number system.

4 Translate logic HIGHs and LOWs into binary numbers.

5 Distinguish between the most significant bit and least significant bit of a binary number.

6 Count in binary, decimal, or hexadecimal.

7 Convert a number in binary, decimal, or hexadecimal to any of the other number bases.

8 Describe the difference between periodic, aperiodic, and pulse waveforms.

9 Calculate the frequency, period, and duty cycle of a periodic digital waveform.

10 Calculate the pulse width of a digital pulse.

Binary digits, or bits, 0 and 1, are sufficient to write any number, given enough places.

If you listen to a radio, you may or may not know that it is typically not a digital device. On the other hand, a computer, a DVD player, a video game unit, and a digital clock are all essentially digital devices, or at least have a lot of digital circuitry inside. A radio receives radio waves through the air, converts the waves to different voltages, and then uses these different voltages with magnets to make a speaker vibrate. Digital devices transform the voltage into a digital code of 1's and 0's. For example, your computer stores and processes information in groups of 1's and 0's; each small piece of information can be represented by 8 of these 1's and 0's, and millions or billions of these groups can be processed each second.

Digital electronics is the branch of electronics based on the combination and switching of logic levels. Physically, these logic levels are represented by voltages. Any quantity in the physical world, such as temperature, pressure, or voltage, can be symbolized in a digital circuit by a group of logic levels that, taken together, form a binary number. Logic levels are usually specified as 0 or 1; at times, it may be more convenient to use LOW/HIGH, FALSE/TRUE, or OFF/ON.

Each logic level corresponds to a digit in the binary (base 2) number system. The *binary digits*, or bits, 0 and 1, are sufficient to write any number, given enough places. The hexadecimal (base 16) number system is also important in digital systems. Because every combination of four binary digits can be uniquely represented as a hexadecimal digit, this system is often used as a compact way of writing binary information.

Inputs and outputs in digital circuits are not always static. Often they vary with time. Time-varying digital waveforms can have three forms:

1. Periodic waveforms, which repeat a pattern of logic 1's and 0's. (A special type of waveform called a clock signal is included in this group.)
2. Aperiodic waveforms, which do not repeat.
3. Pulse waveforms, which produce a momentary variation from a constant logic level.

3.1 DIGITAL VERSUS ANALOG ELECTRONICS

> **KEY TERMS**
>
> **Analog** A representation of a physical, continuous quantity. An analog voltage or current can have any value within a defined range.
>
> **Digital** A representation of a physical quantity by a series of binary numbers. A digital representation can have only specific discrete values.
>
> **Continuous** Smoothly connected: an unbroken series of consecutive values with no instantaneous changes.
>
> **Discrete** Separated into distinct segments or pieces. A series of discontinuous values.

Electronic systems and devices are often divided into two areas: **analog** and **digital** electronics. As more electronic systems have been designed using digital technology, devices have become smaller and more powerful. For example, consider telephones and cell phones. These devices had been designed originally as analog devices and were heavy, bulky, and less powerful compared to today's phones. Cell phones that fit into a pocket, can now be used to send text messages, take pictures and videos, hold GPS systems, and even be used as telephones. Music was traditionally available on record albums and cassette tapes, which used analog record players and tape players to play music. The invention of CDs allowed music to be stored and replayed using digital electronics—CDs are smaller, more versatile, and less easily damaged. As digital music players and MP3s came into use, devices as small as a package of gum could now hold hundreds or thousands of songs. Music libraries that used to fill a room and require special care now fit in your pocket! Digital electronics now dominate circuit design, from computers to cars to video-game systems.

The main difference between analog and digital electronics, simply stated, is: analog voltages or currents are **continuous** between values, while digital voltages or currents are **discrete** (or allowed only at distinct levels). Typically, digital signals vary between two discrete values.

Some keywords highlight the differences between digital and analog electronics:

Analog	Digital
Continuously variable	Discrete steps
Amplification	Switching
Voltages	Numbers
AM (Amplitude modulation)	Microprocessors
FM (Frequency modulation)	Binary

© CENGAGE LEARNING 2012.

We can use a common device to better understand the differences between digital and analog signals. A digital signal can be compared to a light switch, whereas an analog signal can be compared to a dimmer switch. The dimmer switch can turn on lights to any level of brightness, and turning the dimmer switch slightly higher gives us slightly brighter light. A typical light switch, however, turns lights on or off. Similar to a digital signal with only two levels, a light controlled by this switch has only two "levels."

If a light switch can only turn lights on and off, what happens if a light switch is put in the middle position? The answer is: the lights may go on, they may go off, they may flicker, or something else may happen. In other words, we just don't know (and we usually don't care) what happens with the light switch between on and off: if the lights are off and we want them on, we push the switch up and they go on. In many ways, if a digital signal is in one state (say it is off), and we want it in the other state (on), we can turn it on. What if it is in-between states? We don't know—and usually don't care—what the value is.

Let's consider digital vs. analog music once again. Music that is stored digitally, such as on CDs and MP3 files on a computer, is very popular because the music is accurate and noise-free when played. This high quality of sound is possible because the music is stored as a coded series of numbers that represent the sound waves, not as a physical copy of the sound waves (as in a vinyl record) or as a magnetic copy (as in an analog tape). If you listed to music played on a record player or from a tape deck, there is some noise (hissing) generated as the needle travels over the surface or the tape travels over the magnetic head. If you listen to a scratched record, you can hear the "pop" as the needle goes over the scratch. Most CD players have built-in circuitry to hide small scratches on CDs, so the music sounds the same even after the CD is played over and over.

Figure 3.1A shows a sample waveform represented as a voltage. This type of waveform could be measured at the output of an amplified microphone. *Figure 3.1B* and *Figure 3.1C* show the voltage as an analog copy and a digital copy of the sound waveform.

The analog copy, shown in Figure 3.1B, may show some distortion with respect to the original. The distortion is usually introduced by both the analog storage and playback processes.

A digital audio system doesn't make a copy of the waveform, but rather stores a code (a series of amplitude numbers) that tells the CD player how to recreate the original sound every time a disc is played. During the recording process, the sound waveform is "sampled" at precise intervals. That is, the voltage of the waveform is measured at certain intervals and each measurement is converted to a representative binary number. A typical encoding scheme might assign the voltage to a value between 0 and 65,535. Such a large number of possible values means the voltage difference between any two consecutive digital numbers is very small. The numbers can thus correspond extremely closely to the actual amplitude of the sound waveform. The signal on a CD is sampled 44,100 times each second; with this very fast sampling and very fine divisions between voltages, the original wave can be reproduced very accurately.

Digital representations of physical quantities are also superior to analog in that they can easily be stored, transferred, and copied without the distortion that accompanies analog processes. Digital values can be stored in a variety of media, such as optical (CD or DVD), magnetic (hard drive on a PC), or semiconductor (flash memory). They can be transmitted over electronic communications systems such as fiber optic, radio, or telephone. As long as the integrity of the digital data is maintained, any copy of the data is as good as any other. Copies can be made from other copies without deterioration between copy generations.

FIGURE 3.1A Analog Signal

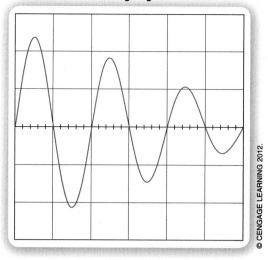

FIGURE 3.1B Digital and Analog Signal Reproduction

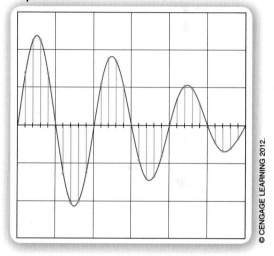

FIGURE 3.1C Digital Signal Reproduction

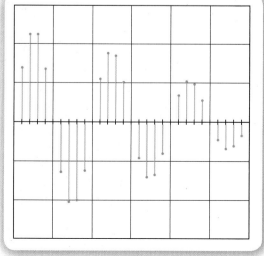

![U-turn sign] **Your Turn**

3.1 What is the basic difference between analog and digital audio reproduction?

3.2 Which would produce a more accurate reproduction of the original analog signal: more samples or fewer samples? Why?

3.2 DIGITAL LOGIC LEVELS

KEY TERMS

Logic level A voltage level that represents a defined digital state in an electronic circuit.

Logic LOW (or **logic 0**) The lower of two voltages in a digital system with two logic levels.

Logic HIGH (or **logic 1**) The higher of two voltages in a digital system with two logic levels.

Positive logic A system in which logic LOW represents binary digit 0 and logic HIGH represents binary digit 1.

Negative logic A system in which logic LOW represents binary digit 1 and logic HIGH represents binary digit 0.

FIGURE 3.2 *Logic Levels Based on +5 V and 0 V*

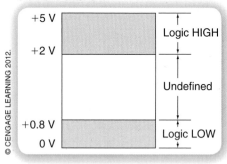

© CENGAGE LEARNING 2012.

Note . . .

For the voltages in Figure 3.2:

+5 V = Logic HIGH
    = HIGH = 1 = ON

0 V = Logic LOW
    = LOW = 0 = OFF

We have described digital signals as similar to light switches: they have only two possible values. Like light switches, we sometimes refer to digital signals as ON or OFF. We can also refer to the two logic levels of a digital signal as HIGH (or logic HIGH) and LOW (or logic LOW), or using the numbers 1 and 0. The binary number system also uses only the two numbers 1 and 0, and we will see shortly how any number can be represented in binary—using only the numbers 1 and 0. Because we are describing a digital quantity electronically, we need to have a system that uses voltages (or currents) to symbolize binary numbers.

For binary systems, digits are used to represent different voltage or logic levels: the lower voltage (usually 0 volts) is called a logic LOW or logic 0 and represents the digit 0. The higher voltage (traditionally 5 V, but in many current systems a different value such as 1.8 V, 2.5 V, or 3.3 V) is called a logic HIGH or logic 1, which represents the digit 1. In reality, there is a range of acceptable values for HIGH and LOW: logic HIGH voltages are at or near the higher voltage, and logic LOW voltages are at or near the lower voltage. Voltages between these values are not defined. *Figure 3.2* shows that voltages between 2 V and 5 V are considered HIGH, and voltages between 0 V and 0.8 V are considered LOW. Voltages between these levels are considered undefined and should not occur in a digital system, much like a light switch should not be between ON and OFF . . . and if it is, the results are undefined.

Assigning the digit 1 to a logic HIGH and digit 0 to logic LOW as described is called positive logic; this is by far the more common system. Throughout the remainder of this text, logic levels will be referred to as HIGH/LOW or 1/0 interchangeably. A complementary system called **negative logic** also exists. If we're using negative logic, we simply change the names of the voltages, where +5 V is named "0" and 0 V is named "1." We'll see some examples much later.

3.3 THE BINARY NUMBER SYSTEM

KEY TERMS

Decimal number system Base-10 number system; the most commonly used number system.

Positional notation A system of writing numbers where the value of a digit depends not only on the digit, but also on its placement within a number.

Binary number system A number system used extensively in digital systems, based on the number 2. It uses two digits, 0 and 1, to write any number.

Bit *Binary digit*. A 0 or a 1.

Positional Notation

The **decimal number system**, or the base-10 number system, is the number system with the numbers you use every day to count, to do math, and so on. We can write any number using only 10 digits, 0 through 9 in the proper columns. For example, when we count, we use the numbers 7 ... 8 ... 9. If we need to count to a number larger than 9, we add a new column to the left—the 10's column. Numbers in the decimal system have a value based on the digits (0 through 9) and their position (1's, 10's, 100's, 1,000's, etc.).

In the decimal number 845, the digit 4 really means 40 (845 = 800 + 40 + 5), whereas in the number 9,426, the digit 4 really means 400 (9,426 = 9,000 + 400 + 20 + 6), as shown in *Figure 3.3*. The value of the digit is determined by *what* the digit is as well as *where* it is.

The decimal system is a **positional notation** system, where the value of a digit depends on its placement within a number. In the decimal system, a digit in the position immediately to the left of the decimal point is multiplied by 1 (10^0, or the 1's column). A digit two positions to the left of the decimal point is multiplied by 10 (10^1, or the 10's column). A digit in the next position left is multiplied by 100 (10^2, or the 100's column), and so on. The positional multipliers, as you move left from the decimal point, are ascending powers of 10.

The **binary number system** is based on the number 2. This means that we can write any number using only two **binary digits** (or **bits**), 0 and 1. The binary system is also a positional notation system; the value of a digit (0 or 1) depends on its placement within a number, like the decimal system. The difference is that the positional multipliers in the binary system are powers of 2 ($2^0 = 1$, $2^1 = 2$, $2^2 = 4$, $2^3 = 8$, $2^4 = 16$, $2^5 = 32$, ... or 1's column, 2's column, 4's column, 8's column, etc.). For example, the binary number 11010 (see *Figure 3.4*) has the decimal equivalent:

$$(1 \times 2^4) + (1 \times 2^3) + (0 \times 2^2) + (1 \times 2^1) + (0 \times 2^0)$$
$$= (1 \times 16) + (1 \times 8) + (0 \times 4) + (1 \times 2) + (0 \times 1)$$
$$= 16 + 8 + 0 + 2 + 0$$
$$= 26$$

FIGURE 3.3 The Number 9,426

9,426				
9	4	2	6	$9 \times 1,000 = 9,000$
1,000s	100s	10s	1s	$4 \times 100 = 400$
10^3	10^2	10^1	10^0	$2 \times 10 = 20$
				$6 \times 1 = 6$
				9,426

© CENGAGE LEARNING 2012.

FIGURE 3.4 The Binary Number 11010

11010					
1	1	0	1	0	$1 \times 16 = 16$
16s	8s	4s	2s	1s	$1 \times 8 = 8$
2^4	2^3	2^2	2^1	2^0	$0 \times 4 = 0$
					$1 \times 2 = 2$
					$0 \times 1 = 0$
					26

© CENGAGE LEARNING 2012.

Note . . .

In cases where there may be some confusion whether the number is binary or decimal (or something else), we can use subscripts to show the base of the number. For example, 101 in binary represents the number 5 in decimal, which is different from 101 in decimal. Thus, 101 in binary can be written as 101_2, and 101 in decimal can be written as 101_{10}. The subscripts are not always shown, but are usually used only if there is some possible confusion on which type of number is written.

Example 3.1

Calculate the decimal equivalents of the binary numbers 1010, 111, and 10010.

Solutions

$$1010 = (1 \times 2^3) + (0 \times 2^2) + (1 \times 2^1) + (0 \times 2^0)$$
$$= (1 \times 8) + (0 \times 4) + (1 \times 2) + (0 \times 1)$$
$$= 8 + 2 = 10$$
$$111 = (1 \times 2^2) + (1 \times 2^1) + (1 \times 2^0)$$
$$= (1 \times 4) + (1 \times 2) + (1 \times 1)$$
$$= 4 + 2 + 1 = 7$$
$$10010 = (1 \times 2^4) + (0 \times 2^3) + (0 \times 2^2) + (1 \times 2^1) + (0 \times 2^0)$$
$$= (1 \times 16) + (0 \times 8) + (0 \times 4) + (1 \times 2) + (0 \times 1)$$
$$= 16 + 2 = 18$$

Binary Inputs

KEY TERMS

Truth table A list of output levels of a circuit corresponding to all different input combinations.

Most significant bit (MSB) The leftmost bit in a binary number. This bit has the number's largest positional multiplier.

Least significant bit (LSB) The rightmost bit of a binary number. This bit has the number's smallest positional multiplier.

A major class of digital circuits, called combinational logic, operates by accepting logic levels at one or more input terminals and producing a logic level at an output. Most of these systems have more than one input, and the signal at each input can be HIGH or LOW. We can use a binary number to represent each input: 1 if the input value is HIGH, 0 if it is LOW. In the analysis and design of these circuits, we will usually find the output logic level of a circuit for all possible combinations of input logic levels.

The digital circuit in the black box in *Figure 3.5* has three inputs. Each input can have two possible states, LOW or HIGH, which can be represented by positive logic as 0 or 1. *Table 3.1* shows a list of all combinations of the input variables both as logic levels and binary numbers, and their decimal equivalents.

The number of possible input combinations is $2^3 = 8$. In general, a circuit with n inputs has $2n$ input combinations, ranging from 0 to $2n - 1$.

FIGURE 3.5 3-Input Digital Circuit

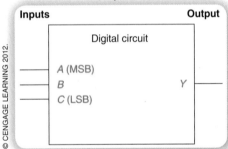

Note . . .

A circuit with n inputs has 2^n input combinations, ranging from 0 to $2^n - 1$.

TABLE 3.1 Possible Input Combinations for a 3-Input Digital Circuit

Logic Level			Binary Value			Decimal Equivalent
A	B	C	A	B	C	
L	L	L	0	0	0	0
L	L	H	0	0	1	1
L	H	L	0	1	0	2
L	H	H	0	1	1	3
H	L	L	1	0	0	4
H	L	H	1	0	1	5
H	H	L	1	1	0	6
H	H	H	1	1	1	7

A list of output logic levels corresponding to *all possible input combinations,* applied in ascending binary order, is called a **truth table**. This is a standard form for showing the function of a digital circuit.

If the input bits on each line of Table 3.1 are read (from left to right) as a series of 3-bit binary numbers, we find that they count from 0 to 7. Because this circuit has 3 inputs, it has $2^3 = 8$ input combinations, ranging from 000_2 to 111_2 (0 to 7).

Bit *A* is called the **most significant bit** (**MSB**), or leftmost bit, and bit *C* is called the **least significant bit** (**LSB**) or rightmost bit. As these terms imply, a change in bit *A* is more significant, because it has the greatest effect on the number of which it is part.

Table 3.2 shows the effect of changing each of these bits in a 3-bit binary number and compares the changed number to the original by showing the difference in magnitude. A change in the MSB of any 3-bit number results in a difference of 4. A change in the LSB of any binary number results in a difference of 1. (Try it with a few different numbers.)

TABLE 3.2 Effect of Changing the LSB and MSB of a Binary Number

	A	B	C	Decimal	
Original	0	1	1	3	
Change MSB	1	1	1	7	Difference = 4
Change LSB	0	1	0	2	Difference = 1

© CENGAGE LEARNING 2012.

Example 3.2

Figure 3.6 shows a 4-input digital circuit. List all the possible binary input combinations to this circuit and their decimal equivalents. What is the value of the MSB?

Solution

Because there are 4 inputs, there will be $2^4 = 16$ possible input combinations, ranging from 0000 to 1111 (0 to 15 in decimal). *Table 3.3* shows the list of all possible input combinations.

TABLE 3.3 Possible Input Combinations for a 4-Input Digital Circuit

A	B	C	D	Decimal
0	0	0	0	0
0	0	0	1	1
0	0	1	0	2
0	0	1	1	3
0	1	0	0	4
0	1	0	1	5
0	1	1	0	6
0	1	1	1	7
1	0	0	0	8
1	0	0	1	9
1	0	1	0	10
1	0	1	1	11
1	1	0	0	12
1	1	0	1	13
1	1	1	0	14
1	1	1	1	15

© CENGAGE LEARNING 2012.

FIGURE 3.6 Example 3.2: 4-Input Digital Circuit © CENGAGE LEARNING 2012.

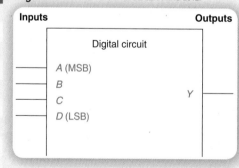

The MSB has a value of 8 (decimal).

Knowing how to construct a binary sequence is a very important skill when working with digital logic systems. Two ways to do this are:

1. *Learn to count in binary.* You should know all the binary numbers from 0000 to 1111 and their decimal equivalents (0 to 15). *Make this your first goal in learning the basics of digital systems.*

   Each binary number is a unique representation of its decimal equivalent. You can work out the decimal value of a binary number by adding the weighted values of all the bits.

   For instance, the binary equivalent of the decimal sequence 0, 1, 2, 3 can be written using two bits: the 1's bit and the 2's bit. The binary count sequence is:

$$00 \ (= 0 + 0)$$
$$01 \ (= 0 + 1)$$
$$10 \ (= 2 + 0)$$
$$11 \ (= 2 + 1)$$

   To count beyond this, you need another bit: the 4's bit. The decimal sequence 4, 5, 6, 7 has the binary equivalents:

$$100 \ (= 4 + 0 + 0)$$
$$101 \ (= 4 + 0 + 1)$$
$$110 \ (= 4 + 2 + 0)$$
$$111 \ (= 4 + 2 + 1)$$

   The two least significant bits of this sequence are the same as the bits in the 0 to 3 sequence; a repeating pattern has been generated.

   The sequence from 8 to 15 requires yet another bit: the 8's bit. The three LSBs of this sequence repeat the 0 to 7 sequence. The binary equivalents of 8 to 15 are:

$$1000 \ (= 8 + 0 + 0 + 0)$$
$$1001 \ (= 8 + 0 + 0 + 1)$$
$$1010 \ (= 8 + 0 + 2 + 0)$$
$$1011 \ (= 8 + 0 + 2 + 1)$$
$$1100 \ (= 8 + 4 + 0 + 0)$$
$$1101 \ (= 8 + 4 + 0 + 1)$$
$$1110 \ (= 8 + 4 + 2 + 0)$$
$$1111 \ (= 8 + 4 + 2 + 1)$$

   Practice writing out the binary sequence, as listed in Table 3.3, until it becomes familiar. In the 0 to 15 sequence, it is standard practice to write each number as a 4-bit value, as in Example 3.2, so that all numbers have the same number of bits. Numbers up to 7 have leading zeros to pad them out to 4 bits.

   When we need to write a binary number using a specified number of bits, we pad the left side with zeros (think of an odometer in a car—a car with 2,309 miles may show 002309).

   While you are still learning to count in binary, you can use a second method.

2. *Follow a simple repetitive pattern.* Look at Table 3.1 and Table 3.3 again. Notice that the least significant bit follows a pattern. The bits alternate with every line, producing the pattern 0, 1, 0, 1 .... The 2's bit alternates every two lines: 0, 0, 1, 1, 0, 0, 1, 1 .... The 4's bit alternates every four lines: 0, 0, 0, 0, 1, 1, 1, 1 .... This pattern can be expanded to cover any number of bits, with the number of lines between alternations doubling with each bit to the left.

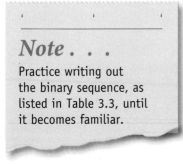

Note . . .

Practice writing out the binary sequence, as listed in Table 3.3, until it becomes familiar.

Decimal-to-Binary Conversion

Two methods are commonly used to convert decimal numbers to binary: sum of powers of 2 and repeated division by 2.

Sum of Powers of 2

You can convert a decimal number to binary by adding up powers of 2 by inspection, adding bits as you need them to fill up the total value of the number (refer to *Figure 3.7*). For example, convert 57_{10} to binary.

$$64_{10} > 57_{10} > 32_{10}$$

▶ We see that 32 $(= 2^5)$ is the largest power of 2 that is smaller than 57. Set the 32's bit to 1 and subtract 32 from the original number, as shown.

$$57 - 32 = 25$$

▶ The largest power of 2 that is less than 25 is 16. Set the 16's bit to 1 and subtract 16 from the accumulated total.

$$25 - 16 = 9$$

▶ 8 is the largest power of 2 that is less than 9. Set the 8's bit to 1 and subtract 8 from the total.

$$9 - 8 = 1$$

▶ 4 is greater than the remaining total. Set the 4's bit to 0.
▶ 2 is greater than the remaining total. Set the 2's bit to 0.
▶ 1 is left over. Set the 1's bit to 1 and subtract 1.

$$1 - 1 = 0$$

▶ Conversion is complete when there is nothing left to subtract. Any remaining bits should be set to 0.

Repeated Division by 2

Any decimal number divided by 2 will leave a remainder of 0 or 1. Repeated division by 2 will leave a string of 0 and 1 remainders that become the binary equivalent of the decimal number. Let us use this method to convert 46_{10} to binary.

1. Divide the decimal number by 2 and note the remainder.

$$46/2 = 23 + \text{remainder } 0 \text{ (LSB)}$$

The remainder is the least significant bit of the binary equivalent of 46.
2. Divide the quotient from the previous division and note the remainder. The remainder is the second LSB.

$$23/2 = 11 + \text{remainder } 1$$

3. Continue this process until the *quotient* is 0. The last remainder is the most significant bit of the binary number.

$$11/2 = 5 + \text{remainder } 1$$
$$5/2 = 2 + \text{remainder } 1$$
$$2/2 = 1 + \text{remainder } 0$$
$$1/2 = 0 + \text{remainder } 1 \text{ (MSB)}$$

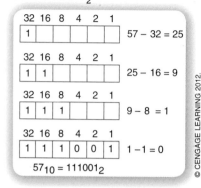

FIGURE 3.7 57_{10} Converted to 111001_2

© CENGAGE LEARNING 2012.

Example 3.3

Convert 92_{10} to binary using the sum-of-powers-of-2 method.

Solution

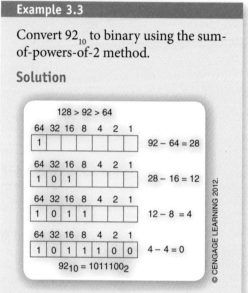

© CENGAGE LEARNING 2012.

To write the binary equivalent of the decimal number, read the remainders from the bottom up.

$$46_{10} = 101110_2$$

Example 3.4

Use repeated division by 2 to convert 115_{10} to a binary number.

Solution

$$115/2 = 57 + \text{remainder } 1 \text{ (LSB)}$$
$$57/2 = 28 + \text{remainder } 1$$
$$28/2 = 14 + \text{remainder } 0$$
$$14/2 = 7 + \text{remainder } 0$$
$$7/2 = 3 + \text{remainder } 1$$
$$3/2 = 1 + \text{remainder } 1$$
$$1/2 = 0 + \text{remainder } 1 \text{ (MSB)}$$

Read the remainders from bottom to top: 1110011.

$$115_{10} = 1110011_2$$

In any decimal-to-binary conversion, the number of bits in the binary number is the exponent of the smallest power of 2 that is larger than the decimal number.

For example, for the number 92_{10}:

$$2^7 = 128 > 92 \quad 7 \text{ bits: } 1011100$$

and 46_{10}:

$$2^6 = 64 > 46 \quad 6 \text{ bits: } 101110$$

Your Turn

3.3 How many different binary numbers can be written with 6 bits?

3.4 How many can be written with 7 bits?

3.5 Write the sequence of 7-bit numbers from 1010000 to 1010111.

3.6 Write the decimal equivalents of the numbers written for the previous problem.

3.4 HEXADECIMAL NUMBERS

KEY TERM

Hexadecimal number system (Hex) Base-16 number system. Hexadecimal numbers are written with 16 digits, 0–9 and A–F, with power-of-16 positional multipliers.

You may have noticed that binary numbers tend to be much longer than decimal numbers: for example, the decimal number 233 is 11101001 binary. The **hexadecimal number system** allows us to work in the binary world (which we need to do in digital electronics), but work with numbers that are shorter and may be easier to work with in many cases. Hex numbers can pack more digital information into fewer digits. The hexadecimal number system is based on powers of 16. After binary numbers, hexadecimal numbers are the most often used in digital applications. Hexadecimal, or **hex**, numbers are primarily used as a shorthand form of binary notation. Because 16 is a power of 2 ($2^4 = 16$), each hexadecimal digit can be easily converted to four binary digits.

Hex numbers have become particularly popular with the increasing use of microprocessors and other computers, which use binary data having 8, 16, 32, or 64 bits. Such data can be represented by 2, 4, 8, or 16 hexadecimal digits, respectively.

Counting in Hexadecimal

The positional multipliers in the hex system are powers of 16: $16^0 = 1$, $16^1 = 16$, $16^2 = 256$, $16^3 = 4096$, and so on.

We need 16 digits to write hex numbers; the decimal digits 0 through 9 are not sufficient. Because we don't have enough digits, we need other symbols to represent 10_{10} through 15_{10}. We will use capital letters A through F to represent 10 through 15. *Table 3.4* shows how hexadecimal digits relate to their decimal and binary equivalents.

Note . . .

Counting rules for hexadecimal numbers:

1. Count in sequence from 0 to F in the least significant digit.
2. Add 1 to the next digit to the left and set the least significant digit to 0.
3. Repeat in all other columns.

TABLE 3.4   Hex Digits and Their Binary and Decimal Equivalents

Hex	Decimal	Binary
0	0	0000
1	1	0001
2	2	0010
3	3	0011
4	4	0100
5	5	0101
6	6	0110
7	7	0111
8	8	1000
9	9	1001
A	10	1010
B	11	1011
C	12	1100
D	13	1101
E	14	1110
F	15	1111

© CENGAGE LEARNING 2012.

For instance, the hex numbers between 19 and 22 are 19, 1A, 1B, 1C, 1D, 1E, 1F, 20, 21, 22. (The decimal equivalents of these numbers are 25_{10} through 34_{10}.)

> *Note . . .*
>
> We have seen subscripts showing the base of the number: 111_2 is a binary number, while 111_{10} is a decimal number. 111_{16} and 111H both represent hex numbers.

Example 3.7

Convert 7C6H to decimal.

Solution

$$7 \times 16^2 = 7_{10} \times 256_{10} = 1792_{10}$$
$$C \times 16^1 = 12_{10} \times 16_{10} = 192_{10}$$
$$6 \times 16^0 = 6_{10} \times 1_{10} = 6_{10}$$
$$\overline{\hspace{3cm}}$$
$$1990_{10}$$

Your Turn

3.7 List the hexadecimal numbers from FA9 to FB0, inclusive.

3.8 List the hexadecimal numbers from 1F9 to 200, inclusive.

Example 3.8

Convert 1FD5H to decimal.

Solution

$$1 \times 16^3 = 1_{10} \times 4096_{10} = 4096_{10}$$
$$F \times 16^2 = 15_{10} \times 256_{10} = 3840_{10}$$
$$D \times 16^1 = 13_{10} \times 16_{10} = 208_{10}$$
$$5 \times 16^0 = 5_{10} \times 1_{10} = 5_{10}$$
$$\overline{\hspace{3cm}}$$
$$8149_{10}$$

Hexadecimal-to-Decimal Conversion

To convert a number from hex to decimal, multiply each digit by its power-of-16 positional multiplier and add the products. In the examples on the left, hexadecimal numbers are indicated by a final "H" (e.g., 1F7H), rather than a "16" subscript.

Your Turn

3.9 Convert the hexadecimal number A30FH to its decimal equivalent.

Decimal-to-Hexadecimal Conversion

Converting decimal numbers to hex numbers is similar to converting decimal numbers to binary numbers. In fact, the conversion from binary to hex is so straightforward, it is often easier to convert from decimal to binary, than from binary to hex (or to use a calculator to do the conversion).

Sum of Weighted Hexadecimal Digits

One method of converting decimal numbers to hex numbers that can be useful for simple conversions (about three digits) is the sum-of-weighted-hex-digits method. The main difficulty we encounter is remembering to convert decimal numbers 10 through 15 into the equivalent hex digits, A through F.

For example, the decimal number 35 is easily converted to the hex value 23.

$$35_{10} = 32_{10} + 3_{10} = (2 \times 16) + (3 \times 1) = 23H$$

Example 3.9

Convert 175_{10} to hexadecimal.

Solution

$$256_{10} > 175_{10} > 16_{10}$$

Because $256 = 16^2$, the hexadecimal number will have two digits.

$$(11 \times 16) > 175 > (10 \times 16)$$

16's	1's	
A		$175 - (A \times 16) = 175 - 160 = 15$

16's	1's	
A	F	$175 - ((A \times 16) + (F \times 1))$

$$= 175 - (160 + 15) = 0$$

Your Turn

3.10 Convert the decimal number 137 to its hexadecimal equivalent.

Conversions between Hexadecimal and Binary

Hexadecimal numbers are used very often because of the ease in converting between binary and hex. In fact, you may find it easier to convert decimal to binary to hex than to convert decimal to hex directly. Table 3.4 shows all 16 hexadecimal digits and their decimal and binary equivalents. Note that for every possible 4-bit binary number, there is a hexadecimal equivalent.

Binary-to-hex and hex-to-binary conversions simply consist of making a conversion between each hex digit and its binary equivalent.

Convert 7EF8H to its binary equivalent.

Solution

Convert each digit individually to its equivalent value:

$$7H = 0111_2$$
$$EH = 1110_2$$
$$FH = 1111_2$$
$$8H = 1000_2$$

The binary number is all of these binary numbers in sequence:

$$7EF8H = 111111011111000_2$$

The leading zero (the MSB of 0111) has been left out.

Sometimes, we might insert spaces every four bits for readability. In this case:

$$7EF8H = 111\ 1110\ 1111\ 1000_2$$

Your Turn

3.11 Convert the hexadecimal number 934BH to binary.

3.12 Convert the binary number 11001000001101001001 to hexadecimal.

3.5 DIGITAL WAVEFORMS

KEY TERMS

Digital waveform A series of logic 1's and 0's plotted as a function of time.

Timing diagram A digital waveform, typically with multiple signals on one plot.

Like a light switch, a digital signal may be ON, may be OFF, or may change every so often. In cases like playing digital music, the digital signals may change quickly—thousands or millions of times per second. The input and output logic levels vary between 0 and 1 with time, creating **digital waveforms**, or plots of the signal vs. time. Digital waveforms of inputs and outputs of a system are an excellent way to demonstrate how a system works; these waveform plots are called **timing diagrams**. There are two types of digital waveforms. *Periodic* waveforms repeat the same pattern of logic levels over a specified period of time. *Aperiodic* waveforms do not repeat. *Pulse* waveforms are a special case where a signal may follow a HIGH-LOW-HIGH or LOW-HIGH-LOW pattern and may be periodic or aperiodic. This is similar to walking into a dark room, turning the light ON for a short time, then turning it back OFF.

Periodic Waveforms

Periodic waveforms repeat the same pattern of HIGHs and LOWs over a specified period of time. The waveform may or may not be symmetrical; that is, it may or may not be HIGH and LOW for equal amounts of time.

Example 3.11

Calculate the **time LOW, time HIGH, period, frequency**, and percent **duty cycle** for each of the periodic waveforms in *Figure 3.8*.

How are the waveforms similar? How do they differ?

FIGURE 3.8 Example 3.11: Periodic Digital Waveforms

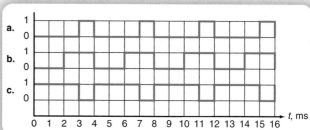

© CENGAGE LEARNING 2012.

Solution

a. Time LOW: $t_l = 3$ ms
   Time HIGH: $t_h = 1$ ms
   Period: $T = t_l + t_h = 3$ ms + 1 ms = 4 ms
   Frequency: $f = 1/T = 1/(4 \text{ ms}) = 0.25$ kHz
   = 250 Hz
   Duty cycle: $\%DC = (t_h/T) \times 100\% = (1 \text{ ms}/4 \text{ ms}) \times 100\% = 25\%$
   (1 ms = 1/1,000 second; 1 kHz = 1,000 Hz)
b. Time LOW: $t_l = 2$ ms
   Time HIGH: $t_h = 2$ ms
   Period: $T = t_l + t_h = 2$ ms + 2 ms = 4 ms
   Frequency: $f = 1/T = 1/(4 \text{ ms}) = 0.25$ kHz = 250 Hz
   Duty cycle: $\%DC = (t_h/T) \times 100\% = (2 \text{ ms}/4 \text{ ms}) \times 100\% = 50\%$
c. Time LOW: $t_l = 1$ ms
   Time HIGH: $t_h = 3$ ms
   Period: $T = t_l + t_h = 1$ ms + 3 ms = 4 ms
   Frequency: $f = 1/T = 1/(4 \text{ ms}) = 0.25$ kHz = 250 Hz
   Duty cycle: $\%DC = (t_h/T) \times 100\% = (3 \text{ ms}/4 \text{ ms}) \infty \times 100\% = 75\%$

The waveforms all have the same period but different duty cycles. A square waveform, shown in Figure 3.8b, has a duty cycle of 50%.

A clock signal, or more simply, a **clock**, is a special case of a symmetrical periodic waveform as shown in Figure 3.8b. Although the duty cycle of a clock doesn't have to be 50%, it typically is close to 50%. Clock signals are useful because we know the frequency, or how often the signal will change. For example, a 1-GHz computer will have an internal clock signal that will change from LOW to HIGH one billion times per second.

Aperiodic Waveforms

KEY TERM

Aperiodic waveform A time-varying sequence of logic HIGHs and LOWs that does not repeat.

FIGURE 3.9 Aperiodic Digital Waveforms

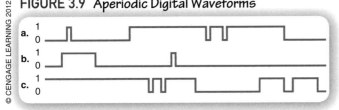

An **aperiodic waveform** does not repeat a pattern of 0's and 1's. Thus, the parameters of time HIGH, time LOW, frequency, period, and duty cycle have no meaning for an aperiodic waveform. Most waveforms of this type are one-of-a-kind specimens.

Figure 3.9 shows some examples of aperiodic waveforms.

Example 3.12

FIGURE 3.10 Example 3.12: Waveforms

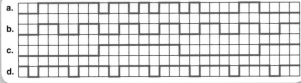

© CENGAGE LEARNING 2012.

A digital circuit generates the following strings of 0's and 1's:

a. 00111111011010110100000110000
b. 0011001100110011001100110011
c. 00000000111111110000000001111
d. 10111011101110111011110111011

The time between two bits is always the same. Sketch the resulting digital waveform for each string of bits. Which waveforms are periodic and which are aperiodic?

Solution

Figure 3.10 shows the waveforms corresponding to the strings of bits just mentioned. The waveforms are easier to draw if you break up the bit strings into smaller groups of, say, 4 bits each. For instance:

a. 0011 1111 0110 1011 0100 0011 0000

All of the waveforms except waveform (a) are periodic.

Pulse Waveforms

KEY TERMS

Pulse A momentary variation of voltage from one logic level to the opposite level and back again.

Rising edge The part of a signal where the logic level is in transition from a LOW to a HIGH. In an ideal pulse, this is instantaneous.

Falling edge (or **trailing edge**) The part of a signal where the logic level is in transition from a HIGH to a LOW.

Amplitude The instantaneous voltage of a waveform. Often used to mean maximum amplitude, or peak voltage, of a pulse.

Pulse width (t_w) (of an ideal pulse) The time from the rising to falling edge of a positive-going pulse, or from the falling to rising edge of a negative-going pulse.

Edge The part of the pulse that represents the transition from one logic level to the other.

In an ideal pulse, the rising edge and the falling edge are vertical. That is, the transitions between logic HIGH and LOW levels are instantaneous. Although there is no such thing as an ideal pulse (i.e., a pulse with absolutely vertical edges) in a real digital circuit, we can usually consider pulses and waveforms to be ideal and not consider the amount of time the transition really takes. Again, like turning on a light, this is the same as saying that a light turns on instantly when the switch is turned to ON, without considering how long the light took to go from OFF to ON. The transition is so fast that we can assume the wave changes instantaneously.

Figure 3.11 shows an ideal **pulse**. The **rising edge** and **falling edge** of an ideal pulse are vertical. That is, the transitions, or **edges**, between logic HIGH and LOW levels are instantaneous.

Pulses can be either positive-going or negative-going, as shown in *Figure 3.12*. In a positive-going pulse, the measured logic level is normally LOW, goes HIGH for the duration of the pulse, and returns to the LOW state. A negative-going pulse acts in the opposite direction.

FIGURE 3.11 Ideal Pulse

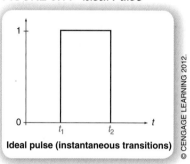

Ideal pulse (instantaneous transitions)

© CENGAGE LEARNING 2012.

FIGURE 3.12 Pulse Edges

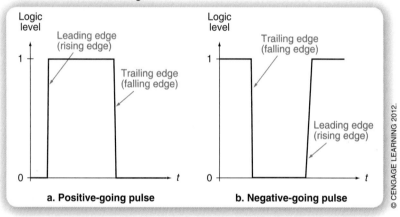

a. Positive-going pulse   b. Negative-going pulse

© CENGAGE LEARNING 2012.

FIGURE 3.13 Pulse Width

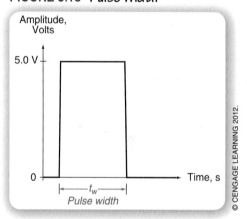

© CENGAGE LEARNING 2012.

The **amplitude** of the pulse is the voltage value of its maximum height; in a digital circuit, this is often 5 volts. The **pulse width**, as shown in *Figure 3.13*, is the time from the rising edge to the falling edge of a positive-going pulse, or the falling to rising edge of a negative-going pulse.

Your Turn

3.13 A digital circuit produces a waveform that can be described by the following periodic bit pattern: 0011001100110011.
   a. What is the duty cycle of the waveform?
   b. Write the bit pattern of a waveform with the same duty cycle and twice the frequency of the original.
   c. Write the bit pattern of a waveform having the same frequency as the original and a duty cycle of 75%.

SUMMARY

1. The two basic areas of electronics are analog and digital electronics. Analog electronics deals with continuously variable quantities; digital electronics represents the world in discrete steps.

2. Digital logic uses defined voltage levels, called logic levels, to represent binary numbers within an electronic system.

3. The higher voltage in a digital system represents the binary digit 1 and is called a logic HIGH or logic 1. The lower voltage in a system represents the binary digit 0 and is called a logic LOW or logic 0.

4. The logic levels of multiple locations in a digital circuit can be combined to represent a multibit binary number.

5. Binary is a positional number system (base 2) with two digits, 0 and 1, and positional multipliers that are powers of 2.

6. The bit with the largest positional weight in a binary number is called the most significant bit (MSB); the bit with the smallest positional weight is called the least significant bit (LSB). The MSB is also the leftmost bit in the number; the LSB is the rightmost bit.

7. A decimal number can be converted to binary by sum of powers of 2 (add place values to get a total) or repeated division by 2 (divide by 2 until the quotient is 0; remainders are the binary value).

8. The positional multipliers in a fractional binary number are negative powers of 2.

9. The hexadecimal number system is based on 16. It uses 16 digits, from 0–9 and A–F, with power-of-16 multipliers.

10. Each hexadecimal digit uniquely corresponds to a 4-bit binary value. Hex digits can thus be used as shorthand for binary.

11. A digital waveform is a sequence of bits over time. A waveform can be periodic (repetitive), aperiodic (nonrepetitive), or pulsed (a single variation and return between logic levels).

12. Periodic waveforms are measured by period (T: time for one cycle), time HIGH (t_h), time LOW (t_l), frequency (f: number of cycles per second), and duty cycle (DC or $\%DC$: fraction of cycle in HIGH state).

13. Pulse waveforms are measured by pulse width (t_w) and amplitude.

BRING IT HOME

3.1 Digital versus Analog Electronics

3.1 Which of the following quantities is analog in nature and which digital? Explain your answers.
   a. Water temperature at the beach
   b. Weight of a bucket of sand
   c. Grains of sand in a bucket
   d. Waves hitting the beach in 1 hour
   e. Height of a wave
   f. People in a square mile

3.2 Which of the following quantities is analog in nature and which digital? Explain your answers.
   a. Number of students in a classroom
   b. Winning score of a basketball game
   c. Height of the tallest player on a team
   d. Speed of a roller coaster
   e. Roller coaster riders per hour

3.2 Digital Logic Levels

3.3 A digital logic system is defined by the voltages 3.3 volts and 0 volts. For a positive logic system, state which voltage corresponds to a logic 0 and which to a logic 1.

3.4 A digital logic system is defined by the voltages 1.8 volts and 0 volts. For a positive logic system, state which voltage corresponds to a logic 0 and which to a logic 1.

3.3 The Binary Number System

3.5 Calculate the decimal values of each of the following binary numbers:

  a. 100
  b. 1000
  c. 11001
  d. 110
  e. 10101
  f. 11101
  g. 100001
  h. 10111001

3.6 Calculate the decimal values of each of the following binary numbers:

  a. 101
  b. 1001
  c. 10110
  d. 111
  e. 11101
  f. 111011
  g. 1010101
  h. 100001

3.7 Translate each of the following combinations of HIGH (H) and LOW (L) logic levels to binary numbers using positive logic:

  a. H H L H
  b. L H L H
  c. H L H L
  d. L L L H

3.8 Translate each of the following combinations of HIGH (H) and LOW (L) logic levels to binary numbers using positive logic:

  a. H L L L
  b. L L L L
  c. H H H L L
  d. H L L H H L

3.9 List the sequence of binary numbers from 101 to 1000.

3.10 List the sequence of binary numbers from 10000 to 11111.

3.11 What is the decimal value of the most significant bit for the numbers in Problem 3.10?

3.12 Convert the following decimal numbers to binary. Use the sum-of-powers-of-2 method for parts a, c, e, and g. Use the repeated-division-by-2 method for parts b, d, f, and h.

  a. 75_{10}
  b. 237_{10}
  c. 198_{10}
  d. 63_{10}
  e. 83_{10}
  f. 64_{10}
  g. 4087_{10}
  h. 8193_{10}

3.13 Convert the following decimal numbers to binary. Use the sum-of-powers-of-2 method.

  a. 65_{10}
  b. 249_{10}
  c. 189_{10}
  d. 98_{10}
  e. 32_{10}
  f. 2177_{10}

3.14 Convert the following decimal numbers to binary. Use the repeated-division-by-2 method.

  a. 35_{10}
  b. 194_{10}
  c. 311_{10}
  d. 89_{10}
  e. 128_{10}
  f. 3247_{10}

3.4 Hexadecimal Numbers

3.15 Write all the hexadecimal numbers in sequence from 308H to 321H inclusive.

3.16 Write all the hexadecimal numbers in sequence from 9F7H to A03H inclusive.

3.17 Convert the following hexadecimal numbers to their decimal equivalents:

  a. 1A0H
  b. 10AH
  c. FFFH
  d. F3C8H
  e. D3B4H
  f. C000H

3.18 Convert the following hexadecimal numbers to their decimal equivalents:

  a. 2BCH
  b. 10FH
  c. 1000H
  d. A38DH
  e. A222H
  f. 30BAFH

continues...

continued...

3.19 Convert the following decimal numbers to their hexadecimal equivalents:
   a. 709_{10}
   b. 1889_{10}
   c. 4225_{10}
   d. 10128_{10}
   e. 32000_{10}
   f. 32768_{10}

3.20 Convert the following decimal numbers to their hexadecimal equivalents:
   a. 907_{10}
   b. 1789_{10}
   c. 4095_{10}
   d. 4096_{10}
   e. 31999_{10}
   f. 33000_{10}

3.21 Convert the following hexadecimal numbers to their binary equivalents:
   a. F3C8H
   b. D3B4H
   c. 8037H
   d. FABDH
   e. 30ACH
   f. 3E7B6H

3.22 Convert the following hexadecimal numbers to their binary equivalents:
   a. 3FFFH
   b. FACEH
   c. A123H
   d. 3214H
   e. 3F36BH
   f. 4952FEH

3.23 Convert the following binary numbers to their hexadecimal equivalents:
   a. 101111010000110_2
   b. 101101101010_2
   c. 110001011011_2
   d. 110101111000100_2
   e. 10101011110000101_2
   f. 11001100010110111_2

3.24 Convert the following binary numbers to their hexadecimal equivalents:
   a. 110110101101_2
   b. 1001010101010_2
   c. 1111101110011_2
   d. 10110011001100101_2
   e. 110000000100001101_2
   f. 101000000000000000_2

3.5 Digital Waveforms

3.25 Calculate the time LOW, time HIGH, period, frequency, and percent duty cycle for the waveforms shown in *Figure 3.14*. How are the waveforms similar? How do they differ?

FIGURE 3.14: Problem 3.25: Periodic Waveforms

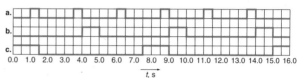

© CENGAGE LEARNING 2012.

3.26 Calculate the time LOW, time HIGH, period, frequency, and percent duty cycle for the waveform shown in *Figure 3.15c*.

FIGURE 3.15: Problems 3.26 and 3.27: Aperiodic and Periodic Waveforms © CENGAGE LEARNING 2012.

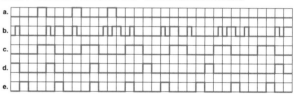

3.27 Which of the waveforms in *Figure 3.15* are periodic and which are aperiodic? Explain your answers.

3.28 Sketch the pulse waveforms represented by the following strings of 0's and 1's. State which waveforms are periodic and which are aperiodic.
   a.   11001111001110110000000110110101
   b.   11100011100011100011100011000111
   c.   11111111000000001111111111111111

3.29 Draw a timing diagram for the signals represented by the following strings of 0's and 1's. State which waveforms are periodic and which are aperiodic.
   a.   01100110011001100110011001100110
   b.   01110110100110100101010011101110
   c.   11111111000000001111111111011111

3.30 Classify each of the waveforms in *Figure 3.16* as aperiodic or periodic. For the periodic waveforms, calculate time HIGH, time LOW, period, frequency, and duty cycle.

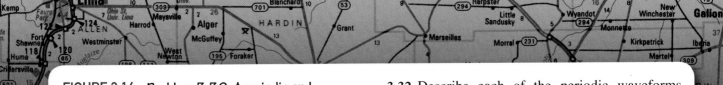

FIGURE 3.16: Problem 3.30: Aperiodic and Periodic Waveforms © CENGAGE LEARNING 2012.

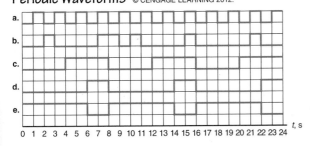

3.31 For each of the periodic waveforms shown in *Figure 3.17*, calculate the period, frequency, time HIGH, time LOW, and percent duty cycle. (The time scale is shown in nanoseconds; 1 ns $= 10^{-9}$ seconds.)

FIGURE 3.17: Problems 3.31 and 3.32: Periodic Waveforms © CENGAGE LEARNING 2012.

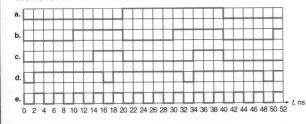

3.32 Describe each of the periodic waveforms shown in Figure 3.17 as a clock signal, specifying its speed.

3.33 Calculate the pulse width of the pulse shown in *Figure 3.18*.

FIGURE 3.18: Problem 3.33: Pulse

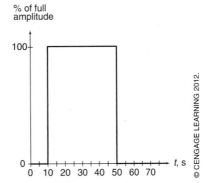

© CENGAGE LEARNING 2012.

CHAPTER 4
Logic Functions and Gates

Menu

START LOCATION	DISTANCE	END LOCATION

CHAPTER OBJECTIVES

Upon successful completion of this chapter, you will be able to:

1 Describe the basic logic functions: AND, OR, and NOT.

2 Draw simple switch circuits to represent AND, OR, and Exclusive OR functions.

3 Describe those logic functions derived from the basic ones: NAND, NOR, Exclusive OR, and Exclusive NOR.

4 Explain the concept of active levels and identify active LOW and HIGH terminals of logic gates.

5 Choose appropriate logic functions to solve simple design problems.

6 Draw the truth table of any logic gate.

7 Draw any logic gate, given its truth table.

8 Draw simple logic switch circuits for single-pole single-throw (SPST) and normally open and normally closed pushbutton switches.

9 Describe the use of light-emitting diodes (LEDs) as indicators of logic HIGH and LOW states.

10 Draw the DeMorgan equivalent form of any logic gate.

11 Determine when a logic gate will pass a digital waveform and when it will block the signal.

12 Describe the behavior of tristate buffers.

13 Describe several types of integrated circuit packaging for digital logic gates.

PHOTO: © ISTOCKPHOTO.COM/INOK.

If you were to look inside of an electronic device (not plugged in, of course), you would most likely see a circuit board covered with electronic components. Many of these components would look exactly the same, or may just have different writing on them. The digital integrated circuits, or chips, found on the board have different functions. Inside the chip, a digital circuit is essentially made from smaller building blocks, all put together in just the right way to perform the correct function.

All digital logic functions can be synthesized by various combinations of the three basic logic functions: AND, OR, and NOT. These so-called Boolean functions are the basis for all further study of combinational logic circuitry: combinational logic circuits are digital circuits whose outputs are functions of their inputs, regardless of the order that the inputs are applied. Standard circuits, called *logic gates,* have been developed for these and for more complex digital logic functions.

Logic gates can be represented in various forms. A standard set of symbols has evolved representing the various functions in a circuit. A useful pair of mathematical theorems, called DeMorgan's theorems, enables us to draw these gate symbols in different ways to represent the same function in two alternative ways.

Simple switches can be configured to apply digital logic levels to a circuit. A single-pole single-throw (SPST) switch and a resistor can be connected to the power supply and ground of a logic circuit to produce logic HIGHs and LOWs in opposite switch positions.

Normally open (NO) and normally closed (NC) pushbuttons can also be used for this purpose.

A light-emitting diode (LED) and a series resistor can be used to indicate the logic level at a particular point in a logic circuit. Depending on the configuration of the LED, it can be used to indicate a logic HIGH or a logic LOW when illuminated.

Logic gates can be used as electronic switches to block or allow passage of digital waveforms. Each logic gate has a different set of properties for enabling (passing) or inhibiting (blocking) digital waveforms.

Data flow can also be controlled by tristate buffers. These devices have three output states: logic HIGH, logic LOW, and high-impedance. When enabled by a control input, the tristate output is either HIGH or LOW. When disabled, the output is in the high-impedance state; just like an open circuit. In this latter state, the gate output is electrically isolated (disconnected) from the rest of the circuit and does not act like a HIGH or a LOW.

Logic gates are available in a variety of packages for use in electronic circuits. For many years, the standard packaging option was the dual in-line package (DIP), with two rows of pins that would be inserted into circuit board holes and soldered to allow connection to the gate inputs and outputs. Packaging options that allow the devices to be mounted directly on the surface of a circuit board are now common. These surface-mount devices are typically smaller in profile than the older DIP varieties and thus can be more densely packed onto a circuit board.

4.1 BASIC LOGIC FUNCTIONS

KEY TERMS

Boolean algebra A system of algebra that operates on Boolean variables. The binary (two-state) nature of Boolean algebra makes it useful for analysis, simplification, and design of combinational logic circuits.

Boolean variable A variable having only two possible values, such as HIGH/LOW, 1/0, On/Off, or True/False.

Boolean expression, Boolean function, or logic function An algebraic expression made up of Boolean variables and operators, such as AND, OR, or NOT.

Logic gate An electronic circuit that performs a Boolean algebraic function.

Note . . .

Boolean variables and constants can have only two possible values: 0 or 1.

At its simplest level, a digital circuit works by accepting logic 1's and 0's at one or more inputs and producing 1's or 0's at one or more outputs. A branch of mathematics known as **Boolean algebra** (named after nineteenth-century mathematician George Boole) describes the relation between inputs and outputs of a digital circuit. We call these input and output values **Boolean variables** and the functions **Boolean expressions, Boolean functions,** or **logic functions.** The distinguishing characteristic of these functions is that they are made up of variables and constants that can have only two possible values: 0 or 1.

All possible operations in Boolean algebra can be created from three basic logic functions: NOT, AND, and OR.[1] Electronic circuits that perform these logic functions are called **logic gates.** When we are analyzing or designing a digital circuit, we usually don't concern ourselves with the actual circuitry of the logic gates, but treat them as "black boxes" that perform specified logic functions. In other words, we don't think about what is inside of a logic gate; we only consider what the gate will *do.* We can think of each variable in a logic function as a circuit input and the whole function as a circuit output.

In addition to gates for the three basic functions, there are also gates for functions that are derived from the basic ones. NAND gates combine the NOT and AND functions in a single circuit. Similarly, NOR gates combine the NOT and OR functions. Gates for more complex functions, such as Exclusive OR and Exclusive NOR, also exist. We will examine all of these devices later in the chapter.

NOT, AND, and OR Functions

KEY TERMS

Truth table A list of all possible input values to a digital circuit, listed in ascending binary order, and the output response for each input combination.

Inverter Also called a NOT gate or an inverting buffer. A logic gate that changes its input logic level to the opposite state.

Distinctive-shape symbols Graphic symbols for logic circuits that show the function of each type of gate by a special shape.

Bubble A small circle indicating logical inversion on a circuit symbol.

Buffer An amplifier that acts as a logic circuit. Its output can be inverting (with the output inverted) or noninverting (where the output is the same as the input).

---

[1] Words in uppercase letters represent either logic functions (AND, OR, NOT) or logic levels (HIGH, LOW). The same words in lowercase letters represent their conventional nontechnical meanings.

NOT Function

The NOT function, the simplest logic function, has one input and one output. The input can be either HIGH or LOW (1 or 0), and the output is always the opposite logic level. We can show these values in a **truth table**, a list of all possible input values and the output resulting from each one. *Table 4.1* shows a truth table for a NOT function, where *A* is the input variable and *Y* is the output.

The NOT function is represented algebraically by the Boolean expression:

$$Y = \overline{A}$$

This is pronounced "*Y* equals NOT *A*" or "*Y* equals *A* bar." We can also say "*Y* is the complement of *A*."

The circuit that produces the NOT function is called the NOT gate or **inverter**. The usual symbol and an alternative for the inverter, both performing the same logic function, are shown in *Figure 4.1*.

Figure 4.1 shows the standard **distinctive-shape symbols** for the inverter. The triangle represents an amplifier circuit, or **buffer**, and the **bubble** (the small circle on the input or output) represents inversion. There are two symbols because, although the inversion typically is shown at the output, it is sometimes convenient to show the inversion at the input. Both symbols represent the same function.

AND Function

> **KEY TERMS**
>
> **Logical product** AND function.
>
> **AND gate** A logic circuit whose output is HIGH when all inputs (e.g., *A* AND *B* AND *C*) are HIGH.

The AND function combines two or more input variables so that the output is HIGH only if all inputs (e.g., *A* and *B*) are HIGH. A sentence that describes the behavior of the AND gate is, "**All** inputs **HIGH** make the output **HIGH**." The partially filled truth table in *Table 4.2A* shows the part of the table for which this condition is true (*A* = 1, *B* = 1, *Y* = 1). Because the gate output can be only 1 or 0, all remaining conditions must have a 0 output, as shown in the complete truth table of *Table 4.2B*.

We can replace the boldface words in the descriptive sentence to describe almost any type of logic gate. We will repeatedly use this systematic "fill-in-the-blanks" method to give us a reliable analytical tool for determining the behavior of logic gates.

Algebraically, the AND function is written:

$$Y = A \cdot B$$

Pronounce this expression "*Y* equals *A* AND *B*." The AND function is similar to multiplication in elementary algebra and thus is sometimes called the **logical product**. The dot between variables may or may not be written, so it is equally correct to write *Y = AB*. The logic circuit symbol for an **AND gate** is shown in *Figure 4.2*.

We can also represent the AND function as a set of switches in series, as shown in *Figure 4.3*. The circuit consists of a voltage source, a lamp, and two series switches. The lamp turns on when switches *A* AND *B* are both closed. For any other condition of the switches, the lamp is off.

TABLE 4.1
NOT Function Truth Table

A	Y
0	1
1	0

© CENGAGE LEARNING 2012.

FIGURE 4.1 Inverter Symbols

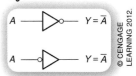

© CENGAGE LEARNING 2012.

TABLE 4.2A Partial Truth Table for a 2-Input AND Gate © CENGAGE LEARNING 2012.

A	B	Y
0	0	
0	1	
1	0	
1	1	1

TABLE 4.2B Complete Truth Table for a 2-Input AND Gate

A	B	Y
0	0	0
0	1	0
1	0	0
1	1	1

© CENGAGE LEARNING 2012.

Note . . .
The output of an AND gate is HIGH only when all inputs are HIGH.

FIGURE 4.2 2-Input AND Gate Symbol

© CENGAGE LEARNING 2012.

FIGURE 4.3 AND Function Represented by Switches in Series

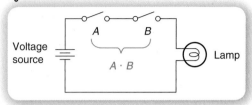

© CENGAGE LEARNING 2012.

FIGURE 4.4 3-Input AND Gate Symbol

© CENGAGE LEARNING 2012.

FIGURE 4.5 3-Input AND Function from 2-Input AND Gates

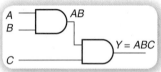

© CENGAGE LEARNING 2012.

TABLE 4.4A Partial Truth Table for a 2-Input OR Gate

A	B	Y
0	0	
0	1	1
1	0	1
1	1	1

© CENGAGE LEARNING 2012.

TABLE 4.4B Complete Truth Table for a 2-Input OR Gate

A	B	Y
0	0	0
0	1	1
1	0	1
1	1	1

© CENGAGE LEARNING 2012.

Note . . .

The OR function has an output Y that is HIGH when any of the inputs is HIGH.

Table 4.3 shows the truth table for a 3-input AND function. Each of the three inputs can have two different values, which means the inputs can be combined in $2^3 = 8$ different ways. In general, n binary (that is, two-valued) variables can be combined in 2^n ways. The condition "**all** inputs **HIGH** make the output **HIGH**" is satisfied only by the last line in the truth table, where $Y = 1$. In all other lines, $Y = 0$.

TABLE 4.3 3-Input AND Function Truth Table

A	B	C	Y
0	0	0	0
0	0	1	0
0	1	0	0
0	1	1	0
1	0	0	0
1	0	1	0
1	1	0	0
1	1	1	1

© CENGAGE LEARNING 2012.

Figure 4.4 shows the logic symbol for the device. The output is HIGH only when all inputs are HIGH.

A 3-input AND gate can be created by using two 2-input AND gates, as shown in *Figure 4.5*. The output of the first gate ($A \cdot B$) is combined with the third variable (C) in the second gate to give the output expression $(A \cdot B) \cdot C = A \cdot B \cdot C$. The circuit in Figure 4.5 is logically equivalent to the gates shown in Figure 4.4.

OR Function

KEY TERMS

Logical sum OR function.

OR gate A logic circuit whose output is HIGH when at least one input (e.g., *A* OR *B* OR *C*) is HIGH.

The 2-input OR function has an output *Y* that is HIGH when either or both of inputs *A* OR *B* are HIGH. Thus we can say, "**At least one** input **HIGH** makes the output **HIGH**." This condition is shown in the partial truth table in *Table 4.4A*. The condition is satisfied for all but one line of the table. Because 0 and 1 are the only possible outputs, the remaining line must have an output value of 0, as shown in the complete truth table of *Table 4.4B*.

The algebraic expression for the OR function is:

$$Y = A + B$$

which is pronounced "*Y* equals *A* OR *B*." This is similar to the arithmetic addition function, but it is not the same. The last line of the truth table tells us that $1 + 1 = 1$ (pronounced "1 OR 1 equals 1"), which is not what we would expect in standard arithmetic. The similarity to the addition function leads to the name **logical sum**. (This is different from the "arithmetic sum," where, of course, $1 + 1$ *does not* equal 1.)

Figure 4.6 shows the logic circuit symbol for an **OR gate**.

The OR function can be represented by a set of switches connected in parallel, as in *Figure 4.7*. The lamp is on when either switch *A* OR switch *B* is closed. (Note that the lamp is also on if *both A and B* are closed. This property makes the OR different from the Exclusive OR function, which we will study later in this chapter.)

Like AND gates, OR gates can have several inputs, such as the 3-input OR gate shown in *Figure 4.8*. *Table 4.5* shows the truth table for this gate. Again, three inputs can be combined in eight different ways. The output is HIGH when at least one input is HIGH. This condition is satisfied by all but the first line in the table.

FIGURE 4.6 2-Input OR Gate Symbol

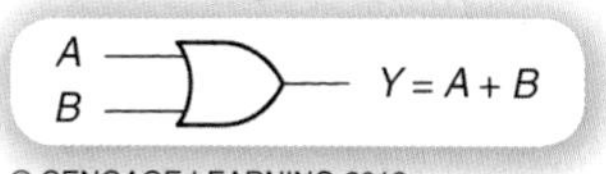

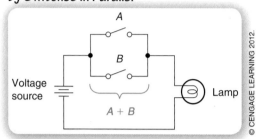

© CENGAGE LEARNING 2012.

FIGURE 4.7 OR Function Represented by Switches in Parallel

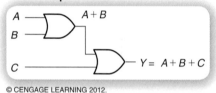

© CENGAGE LEARNING 2012.

TABLE 4.5 3-Input OR Function Truth Table

A	B	C	Y
0	0	0	0
0	0	1	1
0	1	0	1
0	1	1	1
1	0	0	1
1	0	1	1
1	1	0	1
1	1	1	1

© CENGAGE LEARNING 2012.

FIGURE 4.8 3-Input OR Gate Symbol

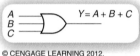

© CENGAGE LEARNING 2012.

FIGURE 4.9 3-Input OR Function from 2-Input OR Gates

© CENGAGE LEARNING 2012.

We can create a 3-input OR function from two 2-input OR gates, as shown in *Figure 4.9*. The first gate combines the inputs *A* and *B* to get (*A* + *B*). This result is combined in the second gate with input *C* to get (*A* + *B*) + *C* = *A* + *B* + *C*. This is the equivalent function to the gates in Figure 4.8. Notice that the output in both cases is HIGH when at least one input is HIGH.

Example 4.1

State which logic function is most suitable for the following operations. Draw a set of switches to represent each function.

1. A manager and one other employee both need a key to open a safe.
2. A light comes on in a storeroom when either (or both) of two doors is open. (Assume the switch closes when the door opens.)
3. For safety, a punch press requires two-handed operation.

Solution

1. Both keys are required, so this is an AND function. *Figure 4.10a* shows a switch representation of the function.
2. One or more switches closed will turn on the lamp. This OR function is shown in *Figure 4.10b*.
3. Two switches are required to activate a punch press, as shown in *Figure 4.10c*. This is an AND function.

continued...

FIGURE 4.10 Example 4.1: Switches Used for AND and OR Functions

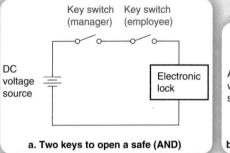

a. Two keys to open a safe (AND)

b. One or more switches turn on a lamp (OR)

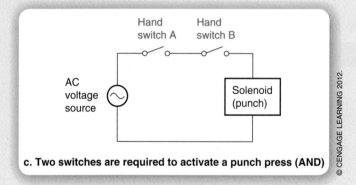

c. Two switches are required to activate a punch press (AND)

© CENGAGE LEARNING 2012.

Example 4.2

FIGURE 4.11A Example 4.2: AND Function in Multisim (LOW Output)

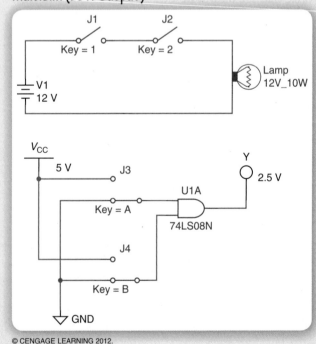

© CENGAGE LEARNING 2012.

Multisim Example

Multisim File: 04.01 AND gate and switches.ms10

Figure 4.11 shows a pair of circuits drawn in Multisim that represent a 2-input AND function, one with a series switch-and-lamp circuit and one with an AND gate. Briefly explain the operation of the circuit.

Solution

The switches J1 and J2 in the top circuit are single-pole single-throw (SPST) switches that must be in the closed position to allow current to flow through the switch. This is shown in Figure 4.11B. If one or both of the switches are open, as shown in Figure 4.11A, current cannot flow through the switches and the lamp is off.

The switches J3 and J4 in the bottom circuit are single-pole double-throw (SPDT) switches. In the upper position, the moveable pole connects to the 5-V power supply, called V_{CC}. In this position, the switch applies a HIGH to the gate input to which it is connected. In the lower position, the moveable pole connects to the circuit ground, which is at zero volts. This applies a logic LOW to the AND gate input to which it is connected.

continued...

Note: In Multisim, the ground component must be the "digital ground," called DGND in the component placement menu, if we are using any digital components in the circuit.

The gate output is monitored by a Multisim component called a digital probe, which goes on if a voltage greater than 2.5 volts is applied to it. The gate output goes HIGH if both inputs are HIGH, as shown in Figure 4.11B. If one or both inputs are LOW, as shown in Figure 4.11A, the digital probe is off, showing a logic LOW.

Interactive Exercise:

Open the Multisim file for this example and run it as a simulation. Test the various switch combinations to see the operation of the SPST switches with the lamp and the SPDT switches with the AND gate. The switches can be operated from the keyboard of the PC running the example circuit. Open and close J1 and J2 by pressing keys 1 and 2 on the PC keyboard. Operate J3 and J4 by pressing keys A and B on the keyboard.

FIGURE 4.11B Example 4.2: AND Function in Multisim (HIGH Output)

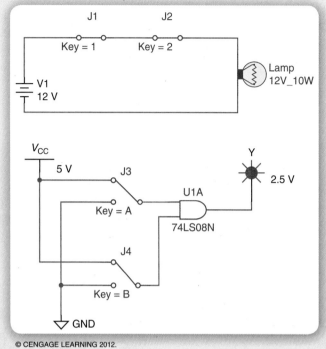

© CENGAGE LEARNING 2012.

a. Does the lamp come on when J1 is open and J2 is closed? Explain.
b. Does the digital probe come on when J3 is in the upper position and J4 is in the lower position? Explain.

Answers to Interactive Exercise:

a. The lamp is off because the current must flow through both switches to turn on the lamp. This cannot happen when J1 is open.
b. The digital probe is off. When J3 is up, it applies a HIGH to the AND gate. When J4 is down it applies a LOW to the gate. Both inputs must be HIGH to make the gate output HIGH and turn on the probe.

Active Levels

KEY TERMS

Active level A logic level defined as the "ON" state for a particular circuit input or output. The active level can be either HIGH or LOW.

Active LOW An active-LOW terminal is considered "ON" when it is in the logic LOW state, indicated by a bubble at the terminal in distinctive-shape symbols.

Active HIGH   An active-HIGH terminal is considered "ON" when it is in the logic HIGH state, indicated by the absence of a bubble at the terminal in distinctive-shape symbols.

An **active level** of a gate input or output is the logic level, either HIGH or LOW, of the terminal when it is performing its designated function. An **active LOW** is shown by a bubble on the affected terminal. If there is no bubble, we assume the terminal is **active HIGH**.

The AND function has active-HIGH inputs and an active-HIGH output. To make the output HIGH, inputs *A* AND *B* must *both* be HIGH. The gate performs its designated function only when *all* inputs are HIGH.

The OR gate requires input *A* OR input *B* (or both) to be HIGH for its output to be HIGH. The HIGH active levels are shown by the absence of bubbles on the terminals.

Your Turn

A 4-input gate has input variables *A*, *B*, *C*, and *D* and output *Y*. Write a descriptive sentence for the active output state(s) if the gate is:

4.1 AND.

4.2 OR.

4.2 DERIVED LOGIC FUNCTIONS

KEY TERMS

NAND gate A logic circuit whose output is LOW when all inputs are HIGH. (A combination of NOT and AND.)

NOR gate A logic circuit whose output is LOW when at least one input is HIGH. (A combination of NOT and OR.)

Exclusive OR (XOR) gate A 2-input logic circuit whose output is HIGH when one input (but not both) is HIGH.

Difference gate An Exclusive OR gate.

Exclusive NOR (XNOR) gate A 2-input logic circuit whose output is the complement of an Exclusive OR gate.

Coincidence gate An Exclusive NOR gate.

The basic logic functions, AND, OR, and NOT, can be combined to make any other logic function. Special logic gates exist for several of the most common of these derived functions. In fact, for reasons that we will discover later, two of these derived-function gates, NAND and NOR, are the most common of all gates, and *each* can be used to create any logic function.

NAND and NOR Functions

The names NAND and NOR are contractions of NOT AND and NOT OR, respectively. The NAND is generated by inverting the output of an AND function. The symbols for the **NAND gate** and its equivalent circuit are shown in *Figure 4.12*.

The algebraic expression for the NAND function is:

$$Y = \overline{A \cdot B}$$

FIGURE 4.12 2-Input NAND Gate Symbols

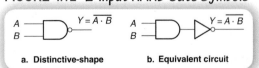

a. Distinctive-shape   b. Equivalent circuit

The NAND gate has active-HIGH inputs and an active-LOW output, shown by the bubble. Because the gate has an AND shape, these conditions lead to the descriptive sentence, "**All** inputs **HIGH** make the output **LOW**." This condition is satisfied only by the last line of the gate's truth table. The partial truth table in *Table 4.6A* shows this condition: when $A = 1$ AND $B = 1$, output $Y = 0$. Because the remaining lines do not satisfy this condition, the output is opposite ($Y = 1$) for all these lines, as shown in the complete truth table in *Table 4.6B*.

TABLE 4.6A Partial Truth Table for a 2-Input NAND Gate

A	B	Y
0	0	
0	1	
1	0	
1	1	0

© CENGAGE LEARNING 2012.

TABLE 4.6B Complete Truth Table for a 2-Input NAND Gate

A	B	Y
0	0	1
0	1	1
1	0	1
1	1	0

© CENGAGE LEARNING 2012.

FIGURE 4.13 2-Input NOR Gate Symbols

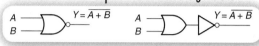

© CENGAGE LEARNING 2012.

Figure 4.13 shows the logic symbols for the **NOR gate**. Because the gate is OR-shaped, with active-HIGH inputs (no bubbles) and an active-LOW output (bubble), it can be described by the sentence, "**At least one** input **HIGH** makes the output **LOW**." *Table 4.7A* shows the lines on the truth table for which this condition is satisfied: at least one input is HIGH in all lines but the first. For each of these lines, $Y = 0$, because the output is active-LOW. The remaining line ($A = 0$, $B = 0$) does not satisfy the condition. Therefore, for this line $Y = 1$, the opposite level from the other lines, as shown in *Table 4.7B*.

TABLE 4.7A Partial Truth Table of a NOR Gate

A	B	Y
0	0	
0	1	0
1	0	0
1	1	0

© CENGAGE LEARNING 2012.

TABLE 4.7B Complete Truth Table of a NOR Gate

A	B	Y
0	0	1
0	1	0
1	0	0
1	1	0

© CENGAGE LEARNING 2012.

The algebraic expression for the NOR function is:

$$Y = \overline{A + B}$$

For both NAND and NOR functions, the inversion covers the entire expression. This is different from inverting each input individually (we will explore this later).

Multiple-Input NAND and NOR Gates

Table 4.8A and *Table 4.8B* show the truth tables of the 3-input NAND and NOR functions. The logic circuit symbols for these gates are shown in *Figure 4.14*.

FIGURE 4.14 3-Input NAND and NOR Gate Symbols

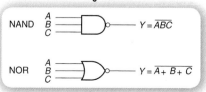

© CENGAGE LEARNING 2012.

TABLE 4.8A   3-Input NAND Truth Table

A	B	C	$\overline{A \cdot B \cdot C}$
0	0	0	1
0	0	1	1
0	1	0	1
0	1	1	1
1	0	0	1
1	0	1	1
1	1	0	1
1	1	1	0

© CENGAGE LEARNING 2012.

TABLE 4.8B   3-Input NOR Truth Table

A	B	C	$\overline{A + B + C}$
0	0	0	1
0	0	1	0
0	1	0	0
0	1	1	0
1	0	0	0
1	0	1	0
1	1	0	0
1	1	1	0

© CENGAGE LEARNING 2012.

The truth tables of these gates can be generated from the active levels of their inputs and outputs, as well as their shape (AND = "all," OR = "at least one"). For the NAND gate, we can say, "**All** inputs **HIGH** make the output **LOW**." This is shown in the last line of the NAND truth table. All other lines have an output with the opposite logic level. For the NOR gate, we can say, "**At least one** input **HIGH** makes the output **LOW**." This condition is met in all lines but the first.

FIGURE 4.15 Expanding a NAND Gate from Two Inputs to Three

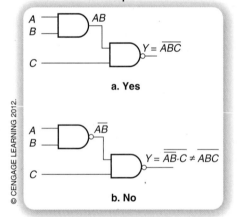

© CENGAGE LEARNING 2012.

Expanding NAND and NOR Gates

Recall that we could use two 2-input AND gates to make a 3-input AND, and two 2-input OR gates to make a 3-input OR. We can also use 2-input gates to make 3-input NAND and NOR gates, but not quite so simply. Remember that a NAND gate combines all of its inputs in an AND function, then inverts the total result. Similarly, a NOR combines all of its inputs in an OR function, then inverts the result. **Therefore, inversion must not be done until the very last step before the output.**

Figure 4.15a shows how a 3-input NAND can be created using a 2-input AND and a 2-input NAND. The AND gate combines A and B. The NAND combines the compound AB with C, then inverts the total result. This is equivalent to the 3-input NAND gate in Figure 4.14. Trying to make the 3-input NAND with two 2-input NANDs, as shown in *Figure 4.15b*, does not work. In this case, we end up inverting a partial result ($\overline{AB}$) before all inputs can be combined in the AND function. The result ($\overline{\overline{AB}C}$) is not equivalent to the 3-input NAND function. To prove the two circuits are different, you can build a truth table for each circuit: if they have different outputs for any input combination, the circuits are different.

Figure 4.16 shows a similar configuration for the 3-input NOR function. Figure 4.16a shows the correct way to get a 3-input NOR from 2-input gates. The OR gate combines A and B. This intermediate result is ORed with C and then the total result is inverted. This is equivalent to the 3-input NOR gate shown in Figure 4.14. Figure 4.16b shows an incorrect connection for a 3-input NOR function. The first NOR combines A OR B, then inverts this partial result ($\overline{A + B}$). When this is combined with C in the second NOR gates, we get $\overline{\overline{A + B} + C}$, which is not equivalent to the 3-input OR function.

FIGURE 4.16 Expanding a NOR Gate from Two Inputs to Three

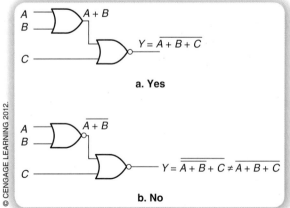

© CENGAGE LEARNING 2012.

NAND and NOR Gates as Inverters

NAND and NOR gates can be used as inverters if we tie their inputs together (short-circuit their inputs), as shown in *Figure 4.17*. The truth tables for the NAND and NOR gates are shown again in *Table 4.9*. If the NAND inputs are shorted, as in Figure 4.16a, then the only lines on the truth table that can be used are the lines where A and B are the same logic level, that is, the first and last lines. As shown in Table 4.9, the input conditions shown in the second and third lines of the NAND truth table cannot occur, so we can cross them out and ignore them. In the first line of the table, both inputs are LOW and the output is HIGH. In the last line, both inputs are HIGH and the output is LOW. This has the effect of inverting the single input that is applied to both inputs of the gate. In a similar way, if we short the NOR inputs, both inputs are the same, yielding the result that if the inputs are LOW, the output is HIGH, and vice versa.

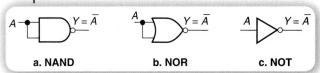

FIGURE 4.17 Three Equivalent Ways of Inverting an Input © CENGAGE LEARNING 2012.

a. NAND **b. NOR** **c. NOT**

> **Note . . .**
> By short-circuiting all the inputs, we can make an inverter from a NAND or NOR gate with any number of inputs.

TABLE 4.9A   Truth Table Showing a NAND Gate as an Inverter

NAND		
A	B	$Y = \overline{A \cdot B}$
0	0	1
0	1	1
1	0	1
1	1	0

© CENGAGE LEARNING 2012.

TABLE 4.9B   Truth Table Showing a NOR Gate as an Inverter

NOR		
A	B	$Y = \overline{A + B}$
0	0	1
0	1	0
1	0	0
1	1	0

© CENGAGE LEARNING 2012.

> **Note . . .**
> The output of a **2-input** XOR gate is **HIGH** when **one and only one** of the inputs is **HIGH**.

Exclusive OR and Exclusive NOR Functions

The Exclusive OR function (abbreviated XOR) is a special case of the OR function. The output of a **2-input** XOR gate is **HIGH** when **one and only one** of the inputs is **HIGH**. (Multiple-input XOR circuits do not expand as simply as other functions. As we will see in Chapter 6 when we study parity circuits, an XOR output is HIGH when an *odd number* of inputs is HIGH.)

A HIGH at both inputs makes the output LOW. (We could say that the case in which both inputs are HIGH is excluded.)

The gate symbol for the **Exclusive OR (XOR) gate** is shown in *Figure 4.18*. *Table 4.10* shows the truth table for the XOR function.

Another way of looking at the Exclusive OR gate is that its output is HIGH when the inputs are *different* and LOW when they are the same. In fact, you may find XOR gates referred to as **difference gates**. This is a useful property in some applications, such as error detection in digital communication systems. (Transmitted data can be compared with received data. If they are the same, no error has been detected.)

The XOR function is expressed algebraically as:

$$Y = A \oplus B$$

The Exclusive NOR (XNOR) function is the complement of the Exclusive OR function and shares some of the same properties. The symbol, shown in

FIGURE 4.18 Exclusive OR (XOR) Gate Symbol

© CENGAGE LEARNING 2012.

TABLE 4.10   Exclusive OR Function Truth Table

A	B	Y
0	0	0
0	1	1
1	0	1
1	1	0

© CENGAGE LEARNING 2012.

FIGURE 4.19 Exclusive
NOR (XNOR) Gate Symbol

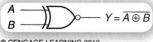

© CENGAGE LEARNING 2012.

Figure 4.19, is an XOR gate with a bubble on the output, implying that the entire function is inverted. *Table 4.11* shows the Exclusive NOR truth table.

The algebraic expression for the Exclusive NOR function is:

$$Y = \overline{A \oplus B}$$

The output of the **Exclusive NOR (XNOR) gate** is HIGH when the inputs are the same and LOW when they are different. For this reason, the XNOR gate is also called a **coincidence gate**. This same/different property is similar to that of the Exclusive OR gate, only opposite in sense. Many of the applications that make use of this property can use either the XOR or the XNOR gate.

TABLE 4.11 Exclusive
NOR Function Truth Table

A	B	Y
0	0	1
0	1	0
1	0	0
1	1	1

© CENGAGE LEARNING 2012.

FIGURE 4.20 Your Turn: Logic Gate Properties

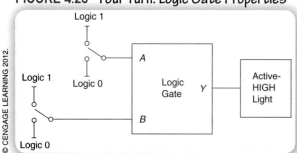

© CENGAGE LEARNING 2012.

Your Turn

A logic gate turns on an active-HIGH light when its output is HIGH. The gate has two inputs, each of which is connected to a logic switch, as shown in *Figure 4.20.*

4.3 What type of gate will turn on the light when the switches are in opposite positions?

4.4 Which gate will turn off the light only when both switches are HIGH?

4.5 What type of gate turns off the light when at least one switch is HIGH?

4.6 Which gate turns on the light when the switches are in the same position?

4.3 DEMORGAN'S THEOREMS AND GATE EQUIVALENCE

TABLE 4.12 NAND Truth Table

A	B	Y
0	0	1
0	1	1
1	0	1
1	1	0

© CENGAGE LEARNING 2012.

KEY TERMS

DeMorgan's equivalent forms Two gate symbols, one AND-shaped and one OR-shaped, that are equivalent according to DeMorgan's theorems.

DeMorgan's theorems Two theorems in Boolean algebra that allow us to transform any gate from an AND-shaped to an OR-shaped gate and vice versa.

Recall the description of a 2-input NAND gate: "**All** inputs **HIGH** make the output **LOW**." This condition is satisfied in the last line of the 2-input NAND truth table, repeated in *Table 4.12*. We could also describe the gate function by saying, "**At least one** input **LOW** makes the output **HIGH**." This condition is satisfied by the first three lines of Table 4.12.

FIGURE 4.21 NAND Gate and DeMorgan Equivalent (Positive and Negative NAND)

© CENGAGE LEARNING 2012.

The gates in *Figure 4.21* represent positive and negative forms of a NAND gate. *Figure 4.22* shows the logic equivalents of these gates. In the first case, we combine the inputs in an AND function, then invert the result. In the second case, we invert the input variables, then combine the inverted inputs in an OR function.

The Boolean function for the AND-shaped gate is given by:

$$Y = \overline{A \cdot B}$$

The Boolean expression for the OR-shaped gate is:

$$Y = \overline{A} + \overline{B}$$

The gates shown in Figure 4.21 are called **DeMorgan equivalent forms**. Both gates have the same truth table, but represent different aspects or ways of looking at the NAND function. We can extend this observation to state that *any* gate (except XOR and XNOR) has two equivalent forms, one AND, one OR.

A gate can be categorized by examining three attributes: *shape, input,* and *output.* A question arises from each attribute:

1. What is its shape (AND/OR)?
   AND: *all*
   OR: *at least one*
2. What active level is at the gate input (HIGH/LOW)?
3. What active level is at the gate output (HIGH/LOW)?

The answers to these questions characterize any gate and allow us to write a descriptive sentence and a truth table for that gate. The DeMorgan equivalent forms of the gate will yield opposite answers to each of these questions.

Thus the gates in Figure 4.21 have the following complementary attributes:

	Basic Gate	DeMorgan Equivalent
Boolean Expression	$\overline{A \cdot B}$	$\overline{A} + \overline{B}$
Shape	AND	OR
Input Active Level	HIGH	LOW
Output Active Level	LOW	HIGH

FIGURE 4.22 Logic Equivalents of Positive and Negative NAND Gates © CENGAGE LEARNING 2012.

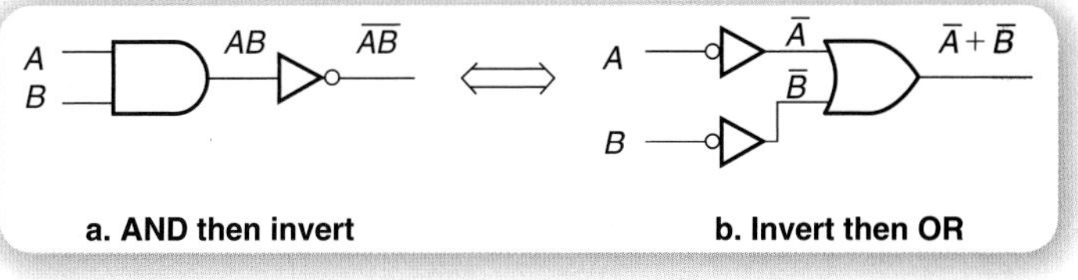

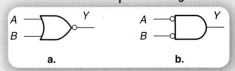

a. AND then invert   b. Invert then OR

Example 4.3

Analyze the shape, input, and output of the gates shown in *Figure 4.23* and write a Boolean expression, a descriptive sentence, and a truth table of each one. Write an asterisk beside the active output level on each truth table. Describe how these gates relate to each other.

FIGURE 4.23 Example 4.3: Logic Gates

a.   b.

© CENGAGE LEARNING 2012.

Solution

a. Boolean expression: $Y = \overline{A + B}$
   Shape: OR *(at least one)*
   Input: HIGH
   Output: LOW
   Descriptive sentence: **At least one** input **HIGH** makes the output **LOW**.
   Truth table: See *Table 4.13*.

b. Boolean expression: $Y = \overline{A} \cdot \overline{B}$
   Shape: AND *(all)*
   Input: LOW

TABLE 4.13 Truth Table of Gate in Figure 4.23a

A	B	Y
0	0	1
0	1	0*
1	0	0*
1	1	0*

© CENGAGE LEARNING 2012.

continued...

TABLE 4.14 Truth Table
of Gate in Figure 4.23b

A	B	Y
0	0	1*
0	1	0
1	0	0
1	1	0

© CENGAGE LEARNING 2012.

Output: HIGH
Descriptive sentence: **All** inputs **LOW** make the output **HIGH**.
Truth table: See *Table 4.14*.

Both gates in this example yield the same truth table. Therefore they are DeMorgan equivalents of one another (positive- and negative-NOR gates).

The gates in Figures 4.21 and 4.23 yield the following algebraic equivalencies:

$$\overline{A \cdot B} = \overline{A} + \overline{B}$$
$$\overline{A + B} = \overline{A} \cdot \overline{B}$$

These equivalencies are known as **DeMorgan's theorems**. (You can remember how to use DeMorgan's theorems by a simple rhyme: "Break the line and change the sign.")

We will look at DeMorgan's theorems more in the next chapter, exploring how we can use these mathematically. For now, we will use these when it is to our advantage to change the shape of the gate in a circuit.

It is tempting to compare the first gate in Figure 4.21 and the second in Figure 4.23 and say that they are the same. Both gates are AND-shaped; both have inversions. However, the comparison is incorrect. The gates have different truth tables, as we have found in Table 4.12 and Table 4.14. Therefore they have different logic functions and are not equivalent. The same is true of the OR-shaped gates in Figures 4.21 and 4.23. The gates may look similar, but because they have different truth tables, they have different logic functions, and are therefore not equivalent.

The confusion arises when, after changing the logic input and output levels, you forget to change the shape of the gate (breaking the line *without* changing the sign). This is a common, but serious, error. These inequalities can be expressed as follows:

$$\overline{A \cdot B} \neq \overline{A} \cdot \overline{B}$$
$$\overline{A + B} \neq \overline{A} + \overline{B}$$

As previously stated, any AND- or OR-shaped gate can be represented in its DeMorgan equivalent form. All we need to do is analyze a gate for its shape, input, and output, then *change everything*.

> ### Note . . .
>
> Use this simple rhyme to remember DeMorgan's theorems: "Break the line and change the sign."

Example 4.4

Analyze the gate in *Figure 4.24* and write a Boolean expression, descriptive sentence, and truth table for the gate. Mark active output levels on the truth table with asterisks. Find the DeMorgan equivalent form of the gate and write its Boolean expression and description.

Solution

Boolean expression: $Y = \overline{\overline{A} + \overline{B} + \overline{C}}$

Shape: OR *(at least one)*

Input: LOW

Output: LOW

FIGURE 4.24 Example
4.4: Logic Gate

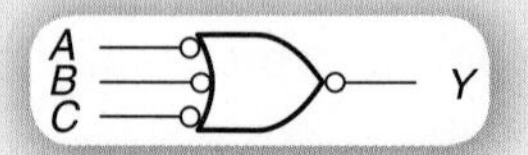

© CENGAGE LEARNING 2012.

continued...

Descriptive sentence: **At least one** input **LOW** makes the output **LOW**.

Truth table: See *Table 4.15*.

TABLE 4.15   Truth Table of Gate in Figure 4.24

A	B	C	Y
0	0	0	0*
0	0	1	0*
0	1	0	0*
0	1	1	0*
1	0	0	0*
1	0	1	0*
1	1	0	0*
1	1	1	1

© CENGAGE LEARNING 2012.

FIGURE 4.25   Example 4.4: DeMorgan Equivalent of Gate Shown in Figure 4.24

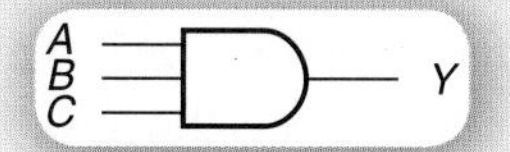

© CENGAGE LEARNING 2012.

Figure 4.25 shows the DeMorgan equivalent form of the gate in Figure 4.24. To create this symbol, we change the shape from OR to AND and invert the logic levels at both input and output. The result is an AND gate.

Boolean expression: $Y = ABC$

Descriptive sentence: **All** inputs **HIGH** make the output **HIGH**.

Your Turn

4.7   The output of a gate is described by the following Boolean expression:

$$Y = \overline{A} + \overline{B} + \overline{C} + \overline{D}$$

Write the Boolean expression for the DeMorgan equivalent form of this gate.

4.4   LOGIC SWITCHES AND LED INDICATORS

Before continuing, we should examine a few simple circuits that can be used for input or output in a digital circuit. Single-pole single-throw (SPST) and pushbutton switches can be used, in combination with resistors, to generate logic voltages for circuit inputs. Light-emitting diodes (LEDs) can be used to monitor outputs of circuits.

Logic Switches

KEY TERMS

V_{cc} The power supply voltage in a transistor-based electronic circuit. The term often refers to the power supply of digital circuits.

Pull-up resistor   A resistor connected from a point in an electronic circuit to the power supply of that circuit.

FIGURE 4.26 *Single-Pole Single-Throw Logic Switch*

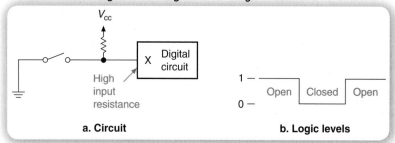

a. Circuit   **b. Logic levels**

Figure 4.26a shows an SPST switch connected as a logic switch. When the switch is closed, point X in the circuit is connected to ground, making it a logic LOW. When the switch is open, point X is connected to the circuit power supply voltage, V_{CC}, via a **pull-up resistor**. This resistor, which typically has a value of 1 kΩ to 10 kΩ, also protects the power supply when the switch is closed by limiting the current from V_{CC} to ground to a few milliamperes or less. *Figure 4.26b* shows the logic levels when the switch is closed and when it is open.

Figure 4.27 shows how pushbuttons can be used as logic inputs. Figure 4.27a shows a normally open pushbutton and a pull-up resistor. The pushbutton has a spring-loaded plunger that makes a connection between two internal contacts when pressed. When released, the spring returns the plunger to the "normal" (open) state. The logic voltage at X is normally HIGH, but LOW when the button is pressed.

Figure 4.27b shows a normally closed pushbutton. The internal spring holds the plunger so that the connection is normally made between the two contacts.

FIGURE 4.27 **Pushbuttons as Logic Switches**

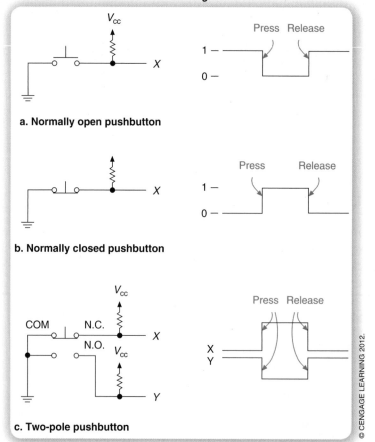

a. Normally open pushbutton

b. Normally closed pushbutton

c. Two-pole pushbutton

© CENGAGE LEARNING 2012.

When the button is pressed, the connection is broken and the resistor pulls up the voltage at X to a logic HIGH. At rest, X is grounded and the voltage at X is LOW.

It is sometimes desirable to have normally HIGH and normally LOW levels available from the same switch. The two-pole pushbutton in Figure 4.27c provides such a function. The switch has a normally open and a normally closed contact. One contact of each switch is connected to the other, in an internal COMMON connection, allowing the switch to have three terminals rather than four. The circuit has two pull-up resistors, one for X and one for Y. Point X is normally HIGH and goes LOW when the switch is pressed. Point Y is opposite.

Example 4.5

Multisim Example

Multisim Files: 04.08 SPST Logic Switch.ms10

Figure 4.28 shows the Multisim design for an SPST switch configured as a logic switch. Open the Multisim file for this example. Run the file as a simulation and operate the space bar on the PC keyboard. Make a table that lists the status of the digital probe, X, and the corresponding logic level for both states of the switch.

FIGURE 4.28 Example 4.5: Single-Pole Single-Throw Logic Switch © CENGAGE LEARNING 2012.

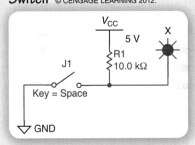

Solution

Table 4.16 shows the status of the switch in terms of its position and logic level.

TABLE 4.16 *Operation of an SPST Logic Switch* © CENGAGE LEARNING 2012.

Switch	Probe	Logic Level
Closed	OFF	LOW
Open	ON	HIGH

Example 4.6

Multisim Example

Multisim Files: 04.09 SPDT PushbuttonLogic Switch.ms10

Figure 4.29 shows a Multisim design for an SPDT pushbutton configured as a two-position logic switch. Open the Multisim file for this example. Run the file as a simulation and operate the pushbutton switch by clicking it with the mouse. To hold the switch in the "pressed" position, click and hold. (The space bar will operate the switch, but will not hold it in place. Thus, it is not the best way to see the operation of the switch.) Make a table that lists the position of the switch, the status of digital probes X and Y, and the logic levels in each case.

FIGURE 4.29 Example 4.6: Single-Pole Double-Throw Pushbutton Logic Switch

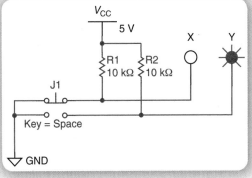

© CENGAGE LEARNING 2012.

Solution

Table 4.17 shows the function of the SPDT switch.

TABLE 4.17 *Operation of an SPDT Pushbutton Switch* © CENGAGE LEARNING 2012.

Switch Position	Probe X	Probe Y	Logic Level X	Logic Level Y
Upper	OFF	ON	LOW	HIGH
Lower	ON	OFF	HIGH	LOW

LED Indicators

FIGURE 4.30 Light-Emitting Diodes (LEDs)

© CENGAGE LEARNING 2012.

FIGURE 4.31 Light-Emitting Diode (LED) © CENGAGE LEARNING 2012.

Anode — Cathode

FIGURE 4.32 Condition for LED Illumination

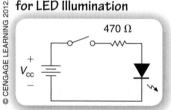

470 Ω

V_{CC}

A device used to indicate the status of a digital output is the **light-emitting diode** or **LED**. This is sometimes pronounced as a word ("led") and sometimes said as separate initials ("ell ee dee"). This device comes in a variety of shapes, sizes, and colors, some of which are shown in the photo in *Figure 4.30*. The circuit symbol, shown in *Figure 4.31*, has two terminals, called the anode (positive) and cathode (negative). The arrow coming from the symbol indicates emitted light.

The electrical requirements for the LED are simple: current flows through the LED if the anode is more positive than the cathode by more than a specified value (about 1.5 to 3 volts, depending on the type of device). If enough current flows, the LED illuminates. If more current flows, the illumination is brighter. (If too much current flows, the LED burns out, so a resistor is used in series with the LED to keep the current in the required range. The series resistor is typically in the range of 180 Ω to 470 Ω. A high-efficiency LED, which requires less current for equal brightness to that of a lower-efficiency LED, could have a higher-valued resistor in series, say about 1 kΩ.) *Figure 4.32* shows a circuit in which an LED illuminates when a switch is closed.

Example 4.7

FIGURE 4.33 Example 4.7: LEDs That Illuminate in Opposite States

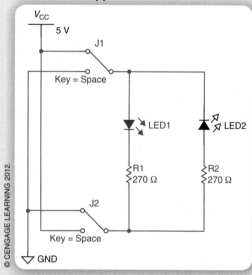

© CENGAGE LEARNING 2012.

Multisim Example

Multisim File: 04.10 LEDs in Opposite States.ms10

Open the Multisim file for this example, shown in *Figure 4.33*. (Ignore the multimeter components off to the side of the circuit in the Multisim file. They will be used in a problem at the end of this chapter.)

Run the file as a simulation and operate the pair of switches by tapping the space bar on the PC keyboard. What happens to each LED in each switch position? Why?

Solution

An LED turns on when current flows from its anode to its cathode, which can only happen when the voltage at the anode is greater than the voltage at the cathode. Voltage configurations where the anode voltage is less than or equal to the cathode voltage result do not allow the LED to turn on, as current will not flow in the direction opposite to the arrow. The LEDs in Figure 4.33 are wired in opposite-conducting directions, so only one will be on at a time.

continued...

When the switches are in the upper position, LED1 is ON because its anode is at 5 volts and its cathode is at 0 volts. LED2 is OFF because its anode is at 0 volts and its cathode is at 5 volts.

When the switches are in the lower position, LED1 is OFF because its anode is at 0 volts and its cathode is at 5 volts. LED2 is ON because its anode is at 5 volts and its cathode is at 0 volts.

Figure 4.34a shows an AND gate driving an LED. The LED is ON when Y is HIGH (5 volts), because the anode of the LED is more positive than the cathode.

In *Figure 4.34b*, the LED is driven by a NAND gate, which has an active-LOW output. The direction of the LED is such that it turns ON when Y is LOW, again because the anode is more positive than the cathode. Note that for either case, the LED is ON when A AND B are both HIGH.

Figure 4.35 shows a circuit in which an LED indicates the status of a logic switch. When the switch is open, the 1 k-Ω pull-up applies a HIGH to the inverter input. The inverter output is LOW, turning on the LED (the anode is more positive than the cathode). When the switch is closed, the inverter input is LOW. The inverter output is HIGH (same value as V_{CC}), making anode and cathode voltages equal. No current flows through the LED, and it is therefore OFF. Thus, the LED is ON for a HIGH state at the switch and OFF for a LOW. Note, however, that the LED is *ON* when the inverter output is *LOW*.

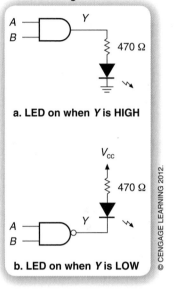

FIGURE 4.34 Logic Gate Driving an LED

a. LED on when Y is HIGH

b. LED on when Y is LOW

© CENGAGE LEARNING 2012.

Your Turn

4.8 A single-pole single-throw switch is connected such that one end is grounded and the other end is connected to a 1 k-Ω pull-up resistor. The other end of the resistor connects to the circuit power supply, V_{CC}. What logic level does the switch provide when it is open? When it is closed?

FIGURE 4.35 LED Indicates Status of Switch

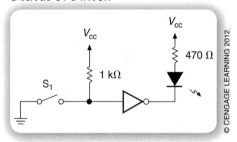

© CENGAGE LEARNING 2012.

4.5 ENABLE AND INHIBIT PROPERTIES OF LOGIC GATES

KEY TERMS

Digital signal (or **pulse waveform**) A series of 0's and 1's plotted over time.

Enable A logic gate is enabled if it allows a digital signal to pass from an input to the output in either true or complement form.

Inhibit (or **disable**) A logic gate is inhibited if it prevents a digital signal from passing from an input to the output.

True form Not inverted.

Complement form Inverted.

In phase Two digital waveforms are in phase if they are always at the same logic level at the same time.

Out of phase Two digital waveforms are out of phase if they are always at opposite logic levels at any given time.

Signal input The input to a logic gate where a digital signal is applied when the gate is used to pass or block the signal.

Control input The input of a logic gate that is used to control whether the digital signal at the signal input will be passed or blocked by the gate.

In Chapter 3, we saw that a **digital signal** is just a string of bits (0's and 1's) generated over time. A major task of digital circuitry is the direction and control of such signals. Logic gates can be used to **enable** (pass) or **inhibit** (block) these signals. (The word "gate" gives a clue to this function; the gate can "open" to allow a signal through or "close" to block its passage.)

AND and OR Gates

The simplest case of the enable and inhibit properties is that of an AND gate used to pass or block a logic signal. *Figure 4.36* shows the output of an AND gate under different conditions of input A when a digital signal (an alternating string of 0's and 1's) is applied to input B.

Recall the properties of an AND gate: both inputs must be HIGH to make the output HIGH. Thus, if input A is LOW, the output will always be LOW, regardless of the state of the other input B. The digital signal applied to B has no effect on the output, and we say that the gate is inhibited or disabled. This is shown in the first half of the timing diagram in Figure 4.36.

If A and B are HIGH, the output is HIGH. When A is HIGH and B is LOW, the output is LOW. Thus, output Y is the same as input B if input A is HIGH; that is, Y and B are **in phase** with each other (or $Y = B$). The input waveform is passed to the output in **true form**, and we say the gate is enabled. The last half of the timing diagram in Figure 4.36 shows this waveform.

It is convenient to define terms for the A and B inputs. Because we apply a digital signal to B, we will call it the **Signal input**. Because input A controls whether or not the signal passes to the output, we will call it the **Control input**. These definitions are illustrated in *Figure 4.37*.

Each type of logic gate has a particular set of enable/inhibit properties that can be predicted by examining the truth table of the gate. Let us examine the truth table of the AND gate to see how the method works.

Divide the truth table in half, as shown in *Table 4.18*. Because we have designated A as the Control input, the top half of the truth table shows the inhibit function ($A = 0$), and the bottom half shows the enable function ($A = 1$). To determine the gate properties, we compare input B (the Signal input) to the output in each half of the table.

Inhibit mode: If $A = 0$ and B is *pulsing* (B is continuously going back and forth between the first and second lines of the truth table), output Y is always 0. Because the Signal input has no effect on the output, we say that the gate is disabled (or inhibited) ($Y = 0$).

Enable mode: If $A = 1$ and B is pulsing (B is going continuously between the third and fourth lines of the truth table), the output is the same as the Signal input. Because the Signal input affects the output, we say that the gate is enabled ($Y = B$).

FIGURE 4.36 **Enable/Inhibit Properties of an AND Gate** © CENGAGE LEARNING 2012.

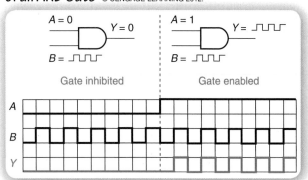

FIGURE 4.37 *Control and Signal Inputs of an AND Gate* © CENGAGE LEARNING 2012.

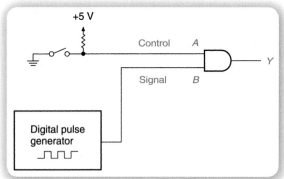

TABLE 4.18 **AND Truth Table Showing Enable/Inhibit Properties** © CENGAGE LEARNING 2012.

A	B	Y	
0	0	0	(Y = 0)
0	1	0	Inhibit
1	0	0	(Y = B)
1	1	1	Enable

Example 4.8

Use the method just described to draw the output waveform of an OR gate if the input waveforms of A and B are the same as in Figure 4.36. Indicate the enable and inhibit portions of the timing diagram.

Solution

Divide the OR gate truth table in half. Designate input A the Control input and input B the Signal input.

As shown in *Table 4.19*, when $A = 0$ and B is pulsing, the output is the same as B and the gate is enabled. When $A = 1$, the output is always HIGH. (At least one input HIGH makes the output HIGH.) Because B has no effect on the output, the gate is inhibited. This is shown in *Figure 4.38* in graphical form.

TABLE 4.19   OR Truth Table Showing Enable/Inhibit Properties © CENGAGE LEARNING 2012.

A	B	Y	
0	0	0	$(Y = B)$
0	1	1	Enable
1	0	1	$(Y = 1)$
1	1	1	Inhibit

FIGURE 4.38   **Example 4.8: OR Gate Enable/Inhibit Waveform**

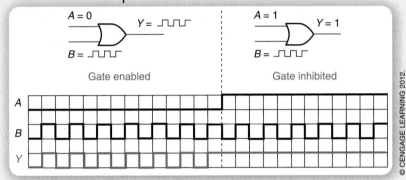

© CENGAGE LEARNING 2012.

Example 4.8 shows that a gate can be in the inhibit state even if its output is HIGH. It is natural to think of the HIGH state as "ON," but this is not always the case. Enable or inhibit states are determined by the effect that the Signal input has on the gate's output. If an input signal does not affect the gate output, the gate is inhibited. If the Signal input does affect the output, the gate is enabled.

Example 4.9

Multisim Example

Multisim Files: 04.12 Enable Inhibit Digital Probe.ms10, 04.13 Enable Inhibit Oscilloscope.ms10

Open the first Multisim file for this example, shown in *Figure 4.39*. This circuit demonstrates the enable and inhibit properties of an AND gate using a digital waveform source, a logic switch, and two digital probes. When the simulation is run, the Signal input probe flashes continuously, but the output probe will flash only when the AND gate is enabled.

The enable/inhibit properties of the gate can also be demonstrated on a virtual oscilloscope in Multisim, as shown in *Figure 4.40*. When this file is open and run as a simulation, the oscilloscope screen can be viewed to show the Signal and Output waveforms, as determined by the state of the Control input of the gate.

FIGURE 4.39   **Example 4.9: Multisim Circuit Showing Enable/Inhibit Properties of an AND Gate with a Digital Probe**

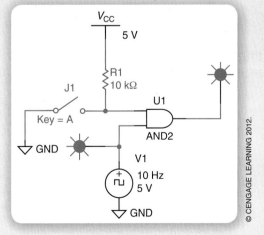

© CENGAGE LEARNING 2012.

continued...

FIGURE 4.40 Example 4.9: Multisim Circuit Showing Enable/Inhibit Properties of an AND Gate with an Oscilloscope © CENGAGE LEARNING 2012.

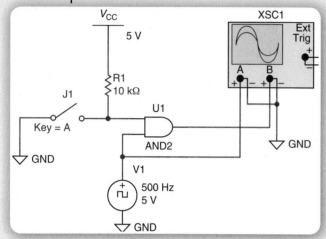

For any given state of the Control input, the output probe can be in only one of four states, each of which corresponds to a state of the output Y. The output can be:

a. always OFF ($Y = 0$);
b. always ON ($Y = 1$);
c. flashing the same as the Signal input ($Y = B$);
d. flashing opposite to the Signal input $Y = \overline{B}$.

Examine the operation of the AND gate, both using the digital probes and the oscilloscope and fill out *Table 4.20* with one of the four possible states of the output for each state of the Control switch.

Solution

When the oscilloscope simulation is run, the result is as shown in *Figure 4.41*.

The AND gate behaves as described in *Table 4.21*.

FIGURE 4.41 Example 4.9: Oscilloscope Traces Showing Signal Input and AND Gate Output Waveforms

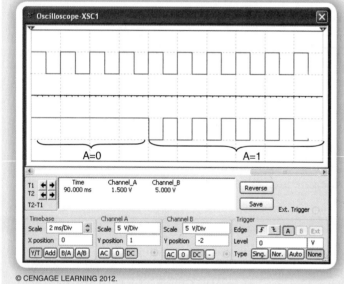

TABLE 4.20 Enable/Inhibit Properties of an AND Gate

Control Switch	A	Output Probe	Y
Closed			
Open			

© CENGAGE LEARNING 2012.

TABLE 4.21 Enable/Inhibit Properties of an AND Gate

Control Switch	A	Output Probe	Y
Closed	0	Always OFF	0
Open	1	Flashing the same as B	B

© CENGAGE LEARNING 2012.

FIGURE 4.42 Enable/Inhibit Properties of a NAND Gate © CENGAGE LEARNING 2012.

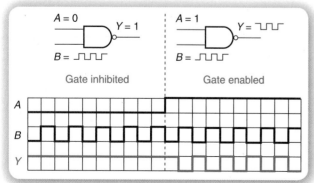

NAND and NOR Gates

When inverting gates, such as NAND and NOR, are enabled, they will invert an input signal before passing it to the gate output. In other words, they transmit the signal in **complement form**. *Figure 4.42* and *Figure 4.43* show the output waveforms of a NAND and a NOR gate when a square waveform is applied to input B and input A acts as a Control input.

The truth table for the NAND gate is shown in *Table 4.22*, divided in half to show the enable and inhibit properties of the gate.

TABLE 4.22 NAND Truth Table Showing Enable/Inhibit Properties © CENGAGE LEARNING 2012.

A	B	Y	
0	0	1	$(Y = 1)$
0	1	1	Inhibit
1	0	1	$(Y = \bar{B})$
1	1	0	Enable

TABLE 4.23 NOR Truth Table Showing Enable/Inhibit Properties © CENGAGE LEARNING 2012.

A	B	Y	
0	0	1	$(Y = \bar{B})$
0	1	0	Enable
1	0	1	$(Y = 0)$
1	1	1	Inhibit

FIGURE 4.43 Enable/inhibit Properties of a NOR Gate © CENGAGE LEARNING 2012.

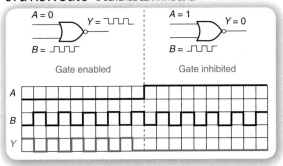

Table 4.23 shows the NOR gate truth table, divided in half to show its enable and inhibit properties.

Figure 4.42 and Figure 4.43 show that when the NAND and NOR gates are enabled, the Signal and output waveforms are opposite to one another; we say that they are **out of phase** or, in this case, $Y = \bar{B}$.

Compare the enable/inhibit waveforms of the AND, OR, NAND, and NOR gates. Gates of the same shape are enabled by the same Control level. The AND and NAND gates are enabled by a HIGH on the Control input and inhibited by a LOW. The OR and NOR are the opposite. A HIGH Control input inhibits the OR/NOR; a LOW Control input enables the gate.

Exclusive OR and Exclusive NOR Gates

Neither the Exclusive OR nor the Exclusive NOR gate has an inhibit state. The Control input on both of these gates acts only to determine whether the output waveform will be in or out of phase with the Signal input. *Figure 4.44* shows the dynamic properties of an XOR gate.

The truth table for the XOR gate, showing the gate's dynamic properties, is given in *Table 4.24*.

Notice that when $A = 0$, the output is in phase with B and when $A = 1$, the output is out of phase with B. A useful application of this property is to use an XOR gate as a programmable inverter. When $A = 1$, the gate is an inverter; when $A = 0$, it is a noninverting buffer.

The XNOR gate has properties similar to the XOR gate. That is, an XNOR has no inhibit state, and the Control input switches the output in and out of phase with the Signal waveform, although not in the same way as an XOR gate does. You will derive these properties in one of the end-of-chapter problems.

Table 4.25 summarizes the enable/inhibit properties of the six gates previously examined.

FIGURE 4.44 Dynamic Properties of an XOR Gate

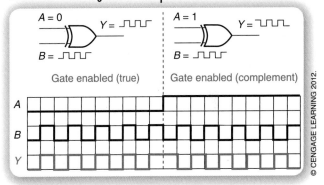

© CENGAGE LEARNING 2012.

TABLE 4.24 XOR Truth Table Showing Dynamic Properties © CENGAGE LEARNING 2012.

A	B	Y	
0	0	0	$(Y = B)$
0	1	1	Enable
1	0	1	$(Y = \bar{B})$
1	1	0	Enable

TABLE 4.25 Summary of Enable/Inhibit Properties © CENGAGE LEARNING 2012.

Control	AND	OR	NAND	NOR	XOR	XNOR
$A = 0$	$Y = 0$	$Y = B$	$Y = 1$	$Y = \bar{B}$	$Y = B$	$Y = \bar{B}$
$A = 1$	$Y = B$	$Y = 1$	$Y = \bar{B}$	$Y = 0$	$Y = \bar{B}$	$Y = B$

4.9 Briefly explain why an AND gate is inhibited by a LOW Control input and an OR gate is inhibited by a HIGH Control input.

Tristate Buffers

> **KEY TERMS**
>
> **High-impedance state** The output state of a tristate buffer that is neither logic HIGH nor logic LOW, but is electrically equivalent to an open circuit; seemingly disconnected from the circuit. (Abbreviation: Hi-Z.)
>
> **Tristate buffer** A gate having three possible output states: logic HIGH, logic LOW, and high-impedance.
>
> **Bus** A common wire or parallel group of wires connecting multiple circuits.

In the previous section, logic gates were used to enable or inhibit signals in digital circuits. For the AND, NAND, NOR, and OR gates, however, in the inhibit state the output was always logic HIGH or LOW. In some cases, it is desirable to have an output state that is neither HIGH nor LOW, but acts to electrically disconnect the gate output from the circuit. This third state is called the **high-impedance state** and is one of three available states in a class of devices known as **tristate buffers**.

Figure 4.45 shows the logic symbols for two tristate buffers, one with a noninverting output and one with an inverting output. The second input, $\overline{OE}$ (Output enable), is an active-LOW signal that enables or disables the buffer output.

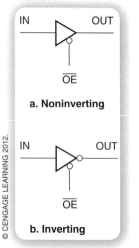

FIGURE 4.45 Tristate Buffers

a. Noninverting

b. Inverting

FIGURE 4.46 Electrical Equivalent of Tristate Operation

IN ▷ OUT = IN
$\overline{OE} = 0$
a. Output enabled

IN ▷ OUT = Hi-Z
$\overline{OE} = 1$
b. Output disabled

When $\overline{OE} = 0$, as shown in *Figure 4.46a*, the noninverting buffer transfers the input value directly to the output as a logic HIGH or LOW. When $\overline{OE} = 1$, as in *Figure 4.46b*, the output is electrically disconnected from any circuit to which it is connected. It appears that there is an open switch at the output of the gate, as if the wire from the output of the device has been cut or pulled out. The open switch in Figure 4.46b does not literally exist. It is shown as a symbolic representation of the electrical disconnection of the output in the high-impedance state.

This type of enable/disable function is particularly useful when digital data are transferred from more than one source to one or more destinations along a common wire (or **bus**), as shown in *Figure 4.47*. (This is the underlying principle in modern computer systems, where multiple components use the same bus to pass data back and forth.)

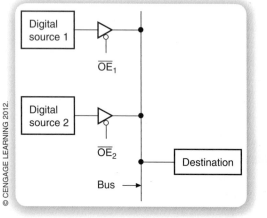

FIGURE 4.47 Using Tristate Buffers to Switch Two Sources to a Single Destination

The destination circuit in Figure 4.47 can receive data from source 1 or source 2. If the source circuits were directly connected to the bus, they could produce contradictory logic levels at the destination. To prevent this, only one source is enabled at a time, with control of this switching left to the two tristate buffers. For example, to transfer data from source 1 to the destination, we make $\overline{OE}_1 = 0$ and $\overline{OE}_2 = 1$. Data is transferred from source 1 to the bus and thus to the destination, whereas source 2 is electrically disconnected from the bus (picture an open switch at the output of the tristate buffer at digital source 2). In this way the data from source 1 and source 2 do not interfere with one another.

Octal Tristate Buffers

Sometimes tristate buffers are packaged in multiples that make it convenient to enable or disable an entire multibit group of signals. The 74LS244 octal tristate buffer, shown in *Figure 4.48*, is such a device. It contains two groups of four noninverting tristate buffers, with each group controlled by a separate $\overline{G}$ (or "gating") input. The gating input has the same function as $\overline{OE}$. ("Octal" means "eight." 74LS244 is an industry standard part number. We shall learn more about such numbers in the next section.)

When $1\overline{G} = 0$, then $1Y = 1A$. Otherwise, $1Y = $ Hi-Z, where Hi-Z is an abbreviation for the high-impedance state. (*1A* and *1Y* are the 4-bit values consisting of *1A1* through *1A4* and *1Y1* through *1Y4*. Thus, a single $\overline{G}$ input controls four *Y* outputs simultaneously.)

Similarly, when $2\overline{G} = 0$, $2Y = 2A$. When $2\overline{G} = 1$, $2Y = $ Hi-Z.

FIGURE 4.48 Octal Tristate Buffer

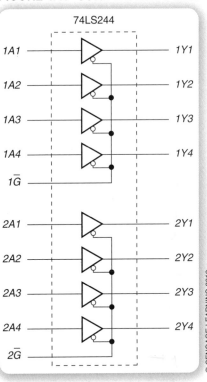

© CENGAGE LEARNING 2012.

Draw a logic circuit showing how a 74LS244 octal tristate buffer can be connected to make a data bus where one of two 4-bit numbers can be transferred to a 4-bit output.

FIGURE 4.49 Example 4.10: Octal Tristate Buffer Connected as a 4-Bit Data Bus Driver

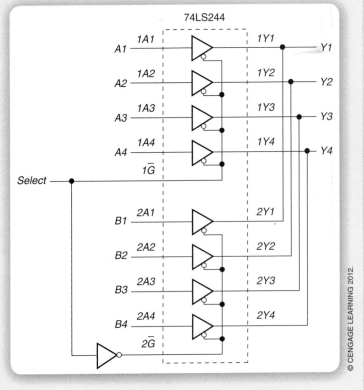

© CENGAGE LEARNING 2012.

Solution

Refer to *Figure 4.49*. The tristate outputs *1Y1* through *1Y4* are connected to outputs *2Y1* through *2Y4*. The inverter connects to the $2\overline{G}$ input to keep it opposite from the $1\overline{G}$ input. This ensures that only one group of four buffers is enabled at any time. When *SELECT* = 0, the *A* inputs connect to *Y*, and *B* is in the Hi-Z state. When *SELECT* = 1, *Y* = *B*, and *A* is in the Hi-Z state.

4.6 INTEGRATED CIRCUIT LOGIC GATES

KEY TERMS

Integrated circuit (IC) An electronic circuit having many gates or other components, such as transistors, diodes, resistors, and capacitors, in a single package.

Small-scale integration (SSI) An integrated circuit having 12 or fewer gates in one package.

Transistor-transistor logic (TTL) A family of digital logic devices whose basic element is the bipolar junction transistor.

Complementary metal-oxide semiconductor (CMOS) A family of digital logic devices whose basic element is the metal-oxide semiconductor field effect transistor (MOSFET).

Chip An integrated circuit. Specifically, a chip of silicon on which an integrated circuit is constructed.

Medium-scale integration (MSI) An integrated circuit having the equivalent of 12 to 100 gates in one package.

Large-scale integration (LSI) An integrated circuit having from 100 to 10,000 equivalent gates.

Very large-scale integration (VLSI) An integrated circuit having more than 10,000 equivalent gates.

Dual in-line package (DIP) A type of IC with two parallel rows of pins for the various circuit inputs and outputs.

Printed circuit board (PCB) A circuit board in which connections between components are made with lines of copper on the surfaces of the circuit board.

Through-hole A means of mounting DIP ICs on a circuit board by inserting the IC leads through holes in the board and soldering them in place.

Breadboard A circuit board for wiring temporary circuits, usually used for prototypes or laboratory work.

Wire-wrap A circuit construction technique in which the connecting wires are wrapped around the posts of a special chip socket or PCB connector, usually used for prototyping or laboratory work.

Quad flat pack (QFP) A square surface-mount IC package with gull-wing leads.

Small outline IC (SOIC) An IC package similar to a DIP, but smaller, which is designed for automatic placement and soldering on the surface of a circuit board. Also called **gull-wing**, for the shape of the package leads.

Thin shrink small outline package (TSSOP) A thinner version of an SOIC package.

Ball grid array (BGA) A square surface-mount IC package with rows and columns of spherical leads underneath the package.

Surface-mount technology (SMT) A system of mounting and soldering integrated circuits on the surface of a circuit board, as opposed to inserting their leads through holes on the board.

Datasheet A printed specification giving details of the pin configuration, electrical properties, and mechanical profile of an electronic device.

Data book A bound collection of datasheets. A digital logic data book usually contains datasheets for a specific logic family or families.

Portable document format (PDF) A format for storing published documents in compressed form.

All the logic gates we have examined so far are available in integrated circuit form. Most of these small-scale integration (SSI) functions are available either in transistor-transistor logic (TTL) or complementary metal-oxide semiconductor (CMOS) technologies. TTL and CMOS devices differ not in their logic functions, but in their construction and electrical characteristics.

TTL and CMOS chips are designated by an industry-standard numbering system, as shown in the following illustration. This system is often referred to as 74-series or 7400-series logic. In the past it was exclusively applied to TTL, but more recently has been used to designate high-speed CMOS devices. Other, more complex, TTL and CMOS devices such as medium-scale integration (MSI) and some large-scale integration (LSI) devices also adopt this numbering system. (An MSI device has between 12 and 100 equivalent gates. An LSI device has between 100 and 10,000 equivalent gates.)

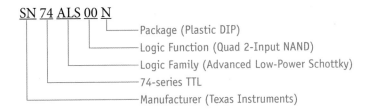

The portions of interest in a part number are those that designate the logic family, which specifies the component's electrical characteristics, and the logic function. For example, in the part number shown, the designation *ALS* indicates that the component belongs to the advanced low-power Schottky TTL family. The digits *00* indicate that the component is a quadruple 2-input NAND gate; that is, a package that contains four NAND gates (indicated by "quadruple"), each with two inputs.

Earlier versions of CMOS had a different set of unrelated numbers of the form 4*NNN*B or 4*NNN*UB where *NNN* was the logic function designator. The suffixes B and UB stand for buffered and unbuffered, respectively. Other, more specialized, very large-scale integration (VLSI) chips have different standard numbering systems (e.g., 27C64 for a 64-kilobit EPROM [a type of memory chip]) or part numbers that are not industry-standard, but relate solely to the products of a particular manufacturer (e.g., XC3S200-4FT256 for a programmable logic device made by Xilinx).

Table 4.26 lists the quadruple 2-input NAND function as implemented in different logic families. All these devices have the same logic function, but different electrical characteristics.

TABLE 4.26   Part Numbers for a Quad 2-Input NAND Gate in Different Logic Families © CENGAGE LEARNING 2012.

Part Number	Logic Family
74LS00	Low-power Schottky TTL
74ALS00	Advanced low-power Schottky TTL
74F00	FAST TTL
74HC00	High-speed CMOS
74HCT00	High-speed CMOS (TTL-compatible inputs)
74LVX00	Low-voltage CMOS
74ABT00	Advanced BiCMOS (TTL/CMOS hybrid)

Table 4.27 lists several logic functions available in the high-speed CMOS family. All these devices have the same electrical characteristics, but different logic functions.

TABLE 4.27 Part Numbers for Different Functions within a Logic Family (High-Speed CMOS) © CENGAGE LEARNING 2012.

Part Number	Function
74HC00	Quadruple 2-input NAND
74HC02	Quadruple 2-input NOR
74HC04	Hex inverter
74HC08	Quadruple 2-input AND
74HC32	Quadruple 2-input OR
74HC86	Quadruple 2-input XOR
74ABT00	Advanced BiCMOS (TTL/CMOS hybrid)

In the past, the most common way to package logic gates was in a plastic or ceramic **dual in-line package**, or **DIP**, which has two parallel rows of pins. The standard spacing between pins in one row is 0.1 inch (or 100 mil). For packages having fewer than 28 pins, the spacing between rows is 0.3 inch (or 300 mil). For larger packages, the rows are spaced by 0.6 inch (600 mil).

This type of package is designed to be inserted in a **printed circuit board** in one of two ways: (a) the pins are inserted through holes in the circuit board and soldered in place; or (b) a socket is soldered to the circuit board and the IC is placed in the socket. Method (a) is referred to as **through-hole** placement. Using a socket, as in method (b), is more expensive, but makes chip replacement much easier. A socket can occasionally cause its own problems by making a poor connection to the pins of the IC.

The DIP is convenient for laboratory and prototype work, as it can be inserted easily into a **breadboard**, a special type of temporary circuit board with internal connections between holes of a standard spacing. It is also convenient for **wire-wrapping**, a technique in which a special tool is used to wrap wires around posts on the underside of special sockets.

The outline of a 14-pin DIP is shown in *Figure 4.50*. There is a notch on one end to show the orientation of the pins. When the IC is oriented as shown and viewed from above, pin 1 is at the top left corner and the pins number counterclockwise from that point.

Figure 4.51 shows the outline of another common IC package, the 240-pin **quad flat pack** (**QFP**). A QFP component is mounted on the surface of a circuit board, rather than soldered as a through-hole component. The package has pins equally distributed on four sides, with pin 1 placed at the top left corner of the package. Pins number counterclockwise from this point. The orientation of the chip is also shown by a cutoff corner, which is at the top left when looking down at the chip from above.

In addition to DIP and QFP, there are numerous other types of packages for digital ICs, including, among others, **small outline IC** (**SOIC**), **thin shrink small outline package** (**TSSOP**), and **ball grid array** (**BGA**) packages. They are used in applications where circuit board space is at a premium and in manufacturing processes relying on **surface-mount technology** (**SMT**). In fact, these devices represent the majority of IC packages found in new designs. Some of these IC packaging options are shown in *Figure 4.52*.

SMT is a sophisticated technology that relies on automatic placement of chips and soldering of pins onto the surface of a circuit board, not through holes in the

FIGURE 4.50 Top View of Pin Numbering on a 14-Pin Dual In-Line Package (DIP)

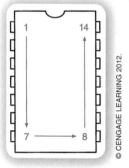

© CENGAGE LEARNING 2012.

FIGURE 4.51 *Top view of Pin Numbering on a 240-Pin Quad Flat Pack (QFP)*

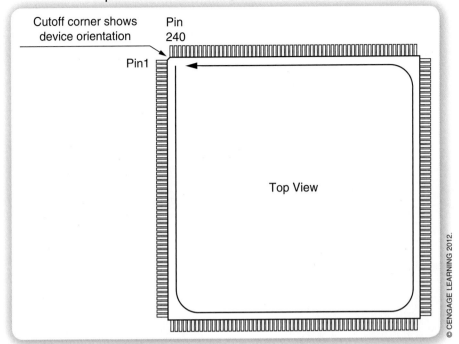

circuit board. This technique allows a manufacturer to mount components on both sides of a circuit board.

Primarily due to the great reduction in board space requirements, many new ICs are available only in the newer surface-mount packages and are not being offered at all in the DIP package. However, we will look at DIP offerings in logic gates because they are inexpensive and easy to use with laboratory breadboards and therefore useful as a learning tool.

Logic gates come in packages containing several gates. Common groupings available in DIP packages are six 1-input gates, four 2-input gates, three 3-input gates, or two 4-input gates, although other arrangements are available. The usual way of stating the number of logic gates in a package is to use the numerical prefixes hex (6), quad or quadruple (4), triple (3), or dual (2).

Some common gate packages are listed in *Table 4.28.*

Information about pin configurations, electrical characteristics, and mechanical specifications of a part is available in a **datasheet** provided by the chip manufacturer. A collection of datasheets for a particular logic family is often bound together in a **data book**. More recently, device manufacturers have been making datasheets available on their corporate Internet sites in **portable**

TABLE 4.28 *Some Common Logic Gate ICs*

Gate	Family	Function
74HC00A	High-speed CMOS	Quad 2-input NAND
74HC02A	High-speed CMOS	Quad 2-input NOR
74ALS04	Advanced low-power Schottky TTL	Hex inverter
74LS11	Low-power Schottky TTL	Triple 3-input AND
74F20	FAST TTL	Dual 4-input NAND
74HC27	High-speed CMOS	Triple 3-input NOR

© CENGAGE LEARNING 2012.

FIGURE 4.52 *Some IC Packaging Options*

SOIC

TSSOP

QFP

DIP

BGA

document format (PDF), readable by a special program such as Adobe Acrobat Reader.

Figure 4.53 shows the internal diagrams of the gates listed in Table 4.28. Notice that the gates can be oriented inside a chip in several ways. That is why it is important to confirm pin connections with a datasheet.

In addition to the gate inputs and outputs there are two more connections to be made on every chip: the power (V_{CC}) and ground connections. In TTL, connect V_{CC} to +5 volts and GND to ground. In CMOS, connect the V_{CC} pin to the supply voltage (+3 V to +6 V) and GND to ground. The gates won't work without these connections.

FIGURE 4.53 Pinouts of ICs Listed in Table 4.28

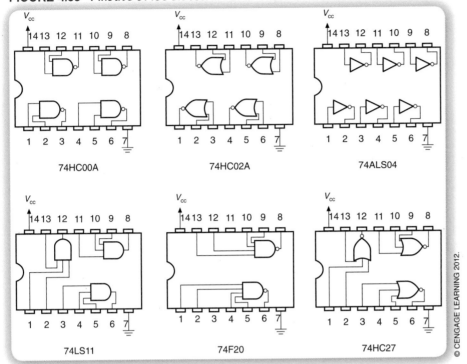

Every chip requires power and ground. This might seem obvious, but it's surprising how often it is forgotten, especially by students who are new to digital electronics. Probably this is because most digital circuit diagrams don't show the power connections, but assume that you know enough to make them.

The only place a chip gets its required power is through the V_{CC} pin. Even if the power supply is connected to a logic input as a logic HIGH, you still need to connect it to the power supply pin.

Even more important is a good ground connection. A circuit with no power connection will not work at all. A circuit without a ground may appear to work, but it will often produce bizarre errors that are very difficult to detect and repair.

In later chapters, we will work primarily with complex ICs in surface-mount packages. The quality of the power and ground connections to these chips are so important that they will not be left to chance; they are provided on a specially designed circuit board. Only input and output pins are accessible for connection by the user.

As digital designs become more complex, it is increasingly necessary to follow good practices in board layout and prototyping procedure to ensure even minimal functionality. Thus, hardware platforms for prototype and laboratory work will need to be at least partially constructed by the board manufacturer to supply the requirements of a stable circuit configuration.

> **Note . . .**
> Most digital circuit diagrams don't show the connections to power or ground, but they are always necessary.

Your Turn

4.10 How are the pins numbered in a DIP?

4.11 How are the pins numbered in a QFP package?

SUMMARY

1. Digital systems can be analyzed and designed using Boolean algebra, a system of mathematics that operates on variables that have one of two possible values.

2. Any Boolean expression can be constructed from the three simplest logic functions: NOT, AND, and OR.

3. A NOT gate, or inverter, has an output state that is in the opposite logic state of the input.

4. The main logic functions are described by the following sentences:

   AND: All inputs **HIGH** make the output **HIGH**.

   OR: At least one input **HIGH** makes the output **HIGH**.

   NAND: All inputs **HIGH** make the output **LOW**.

   NOR: At least one input **HIGH** makes the output **LOW**.

   XOR: Output is **HIGH** if **one** input is **HIGH**, but **not both**. Output is **HIGH** if inputs are **different**.

   XNOR: Output is **LOW** if **one** input is **HIGH**, but **not both**. Output is **HIGH** if inputs are **the same**.

5. The function of a logic gate can be represented by a truth table, a list of all possible inputs in binary order, and the output corresponding to each input state.

6. A 3-input AND function can be made using two 2-input AND gates, where the output of one gate connects to one input of the next gate. The same configuration is possible with OR gates to make a 3-input OR function.

7. A 3-input NAND function can be made using a 2-input AND gate whose output connects to one input of a 2-input NAND gate. A similar connection with an OR and a NOR gate can be used to make a 3-input NOR function. In both cases, the inversion must be the last step in the process. In other words, the AND and NAND are not interchangeable and the OR and NOR are not interchangeable.

8. An inverter can be made from a NAND gate by shorting its inputs together. A NOR gate can also be used this way.

9. DeMorgan's theorems ($\overline{A \cdot B} = \overline{A} + \overline{B}$ and $\overline{A + B} = \overline{A} \cdot \overline{B}$) allow us to represent any gate in either an AND form or an OR form.

10. To change a gate into its DeMorgan equivalent form, change its shape from AND to OR or vice versa and change the active levels of inputs and output ("break the line and change the sign").

11. A logic switch can be created from a single-pole single-throw switch by grounding one end and tying the other end to V_{CC} through a pull-up resistor. The logic level is available on the same side of the switch as the resistor. An open switch is HIGH and a closed switch is LOW. A similar circuit can be made with a pushbutton switch.

12. A light-emitting diode (LED) can be used to indicate logic HIGH or LOW levels. To indicate a HIGH, ground the cathode through a series resistor (about 470 Ω for a 5-volt power supply) and apply the logic level to the anode. To indicate a LOW, tie the anode to V_{CC} through a series resistor and apply the logic level to the cathode.

13. Logic gates can be used to pass or block digital signals. For example, an AND gate will pass a digital signal applied to input B if input A is HIGH ($Y = B$). If input A is LOW, the signal is blocked and the gate output is always LOW

$(Y = 0)$. Similar properties apply to other gates, as summarized in Table 4.25.

14. Tristate buffers have outputs that generate logic HIGH and LOW when enabled and a high-impedance state when disabled. The high-impedance state is electrically equivalent to an open circuit.

15. Logic gates are available as integrated circuits in a variety of packages. Packages that have fewer than 12 gates are called small-scale integration (SSI) devices.

16. Many logic functions have an industry-standard part number of the form 74XXNN, where XX is an alphabetic family designator and NN is a numeric function designator (e.g., 74HC02 = quadruple 2-input NOR gate [02] in the high-speed CMOS [HC] family).

17. Some common IC packages include dual in-line package (DIP), small outline IC (SOIC), thin shrink small outline package (TSSOP), quad flat pack (QFP), and ball grid array (BGA) packages.

18. Most new IC packages are for surface mounting on a printed circuit board. These have largely replaced DIPs in through-hole circuit boards, due to better use of board space.

19. IC pin connections and functional data can be determined from manufacturers' datasheets, available in paper format or electronically via the Internet.

20. All ICs require power and ground, which must be applied to special power supply pins on the chip.

BRING IT HOME

4.1 Basic Logic Functions

4.1 Draw the symbol for the NOT gate (inverter).

4.2 Draw the symbol for a 3-input AND gate.

4.3 Draw the symbol for a 3-input OR gate.

4.4 Write a sentence that describes the operation of a 4-input AND gate that has inputs P, Q, R, and S and output T. Make the truth table of this gate and draw an asterisk beside the line(s) of the truth table indicating when the gate output is in its active state.

4.5 Write a sentence that describes the operation of a 4-input OR gate with inputs J, K, L, and M and output N. Make the truth table of this gate and draw an asterisk beside the line(s) of the truth table indicating when the gate output is in its active state.

4.6 **Multisim Problem**
Multisim File: 04.01 AND gate and switches. ms10
Open the Multisim file for this problem and save it as **04.02 OR gate and switches.ms10**. Replace the 74LS08N 2-input AND gate with a 74LS32N 2-input OR gate. Also rewire the two switches so that they represent a 2-input OR function.

Interactive Exercise:
Test the various switch combinations to see the operation of the SPST switches with the lamp and the SPDT switches with the OR gate.

a. Does the lamp come on when J1 is open and J2 is closed? Explain.

b. Does the digital probe come on when J3 is in the upper position and J4 is in the lower position? Explain.

4.7 **Multisim Problem**
Multisim File: 04.01 AND gate and switches. ms10
Open the Multisim file for this problem and save it as **04.02a 3-in AND gate and switches. ms10**. Replace the 74LS08N 2-input AND gate with a 74LS11N 3-input AND gate and add an SPDT switch for the third input. Add an SPST switch to the lamp circuit and rewire it so that it represents a 3-input AND function. Control the lamp with keys 1, 2, and 3. Control the gate with keys A, B, and C. (To set the key value

continues...

continued...

that controls a switch, double-click the switch symbol to open a dialog box. In the **Value** tab, select the key value from the drop-down box and click **OK**.)

Interactive Exercise:

Test the various switch combinations to see the operation of the SPST switches with the lamp and the SPDT switches with the OR gate.

a. Does the lamp come on when one switch is open and two are closed? Explain.

b. Does the digital probe come on when all three switches are in the upper position? Explain.

4.8 State how four switches must be connected to represent a 4-input OR function. Draw a circuit diagram showing how this function can control a lamp.

4.9 Draw the circuit of a 3-input AND function, made using only 2-input logic gates.

4.10 Draw the circuit of a 3-input OR function, made using only 2-input logic gates.

4.11 Multisim Problem

Multisim File: 04.03 3-in AND gate.ms10

Open the Multisim file for this problem, as shown in *Figure 4.54*. Control the gate inputs with keys *A*, *B*, and *C*. (To set the key value that controls a switch, double-click the switch symbol to open a dialog box. In the **Value** tab, select the key value from the drop-down box and click **OK**.)

FIGURE 4.54 Problem 4.11: 3-Input AND Gates in Multisim © CENGAGE LEARNING 2012.

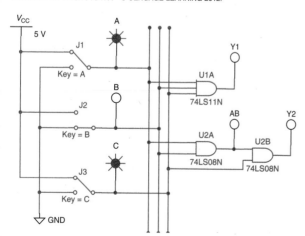

Interactive Exercise:

Test the various switch combinations to see the operation of the SPDT switches with the 3-input AND gate and its equivalent circuit and answer the following questions.

a. Compare the states of *Y1* and *Y2* for all combinations of inputs *A*, *B*, and *C*. What do you observe?

b. What combinations of inputs make the output of gate U2A HIGH?

c. Can *Y2* be HIGH if the output of U2A is LOW? Explain.

4.12 Multisim Problem

Multisim File: 04.03 3-in AND gate.ms10

Open the Multisim file for this problem and save it as **04.04 3-in OR gate.ms10**, as shown in *Figure 4.55*. Replace U1A with a 3-input OR gate called OR3 from the **Miscellaneous Digital** component group. Replace U2A and U2B with 2-input OR gates called OR2 from the **Miscellaneous Digital** component group.

FIGURE 4.55 Problem 4.12: 3-Input OR Gates in Multisim © CENGAGE LEARNING 2012.

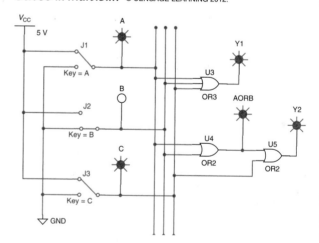

Interactive Exercise:

Test the various switch combinations to see the operation of the SPDT switches with the 3-input OR gate and its equivalent circuit and answer the following questions.

a. Compare the states of *Y1* and *Y2* for all combinations of inputs *A*, *B*, and *C*. What do you observe?

b. What combinations of inputs make the output of gate U2A HIGH?

c. Can Y2 be HIGH if the output of U2A is LOW? Explain.

4.2 Derived Logic Functions

4.13 For a 4-input NAND gate with inputs A, B, C, and D and output Y:

    a. Write the truth table and a descriptive sentence.

    b. Write the Boolean expression.

    c. Draw the logic circuit symbol.

4.14 Repeat Problem 4.13 for a 4-input NOR gate.

4.15 State the active levels of the inputs and outputs of a NAND gate and a NOR gate.

4.16 Write a descriptive sentence of the operation of a 5-input NAND gate with inputs A, B, C, D, and E and output Y. How many lines would the truth table of this gate have?

4.17 Repeat Problem 4.16 for a 5-input NOR gate.

4.18 A pump motor in an industrial plant will start only if the temperature and pressure of liquid in a tank exceed a certain level. The temperature sensor and pressure sensor, shown in *Figure 4.56* each produce a logic HIGH if the measured quantities exceed this value. The logic circuit interface produces a HIGH output to turn on the motor. Draw the symbol and truth table of the gate that corresponds to the action of the logic circuit.

FIGURE 4.56 Problem 4.18: Temperature and Pressure Sensors © CENGAGE LEARNING 2012.

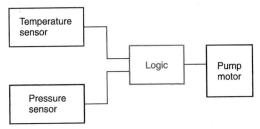

4.19 Repeat Problem 4.18 for the case in which the motor is activated by a logic LOW.

4.20 Multisim Problem

Multisim File: 04.03 3-in AND gate.ms10

Open the Multisim file for this problem, and save it as **04.05 3-in NAND gate.ms10**. Replace the 3-input AND gate with a 74LS10

3-input NAND gate. Replace U2A and U2B with one 2-input gate each, to make an equivalent circuit of a 3-input NAND.

Interactive Exercise:

Test the various switch combinations to see the operation of the SPDT switches with the 3-input NAND gate and its equivalent circuit and answer the following questions.

    a. Compare the states of Y1 and Y2 for all combinations of inputs A, B, and C. What do you observe?

    b. What combinations of inputs make the output of gate U2A LOW?

    c. Can Y2 be LOW if the output of U2A is LOW? Explain.

4.21 Draw the circuit of a 4-input NAND function, made using only 2-input logic gates.

4.22 Multisim Problem

Multisim File: 04.03 3-in AND gate.ms10

Open the Multisim file for this problem and save it as **04.06 3-in NOR gate.ms10**. Replace the 3-input AND gate with a 74LS27 3-input NOR gate. Replace U2A and U2B with one 2-input gate each, to make an equivalent circuit of a 3-input NOR.

Interactive Exercise:

Test the various switch combinations to see the operation of the SPDT switches with the 3-input NOR gate and its equivalent circuit and answer the following questions.

    a. Compare the states of Y1 and Y2 for all combinations of inputs A, B, and C. What do you observe?

    b. What combinations of inputs make the output of gate U2A HIGH?

    c. Can Y2 be LOW if the output of U2A is LOW? Explain.

4.23 Draw the circuit of a 4-input NOR function, made using only 2-input logic gates.

4.24 Find the truth table for the logic circuit shown in *Figure 4.57*.

FIGURE 4.57 Problem 4.24: 3-Input XOR Circuit © CENGAGE LEARNING 2012.

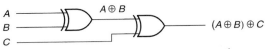

continues...

continued...

4.3 DeMorgan's Theorems and Gate Equivalence

4.25 For each of the gates in *Figure 4.58*:
  a. Write the truth table.
  b. Indicate with an asterisk which lines on the truth table show the gate output in its active state.
  c. Convert the gate to its DeMorgan equivalent form.
  d. Rewrite the truth table and indicate which lines on the truth table show output active states for the DeMorgan equivalent form of the gate.

FIGURE 4.58 Problem 4.25: Logic Gates

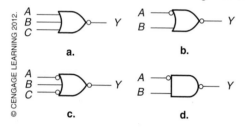

© CENGAGE LEARNING 2012.

4.26 Refer to *Figure 4.59*. State which two gates of the three shown are DeMorgan equivalents of each other. Explain your choice.

FIGURE 4.59 Problem 4.26: Which Gates Are DeMorgan Equivalents? © CENGAGE LEARNING 2012.

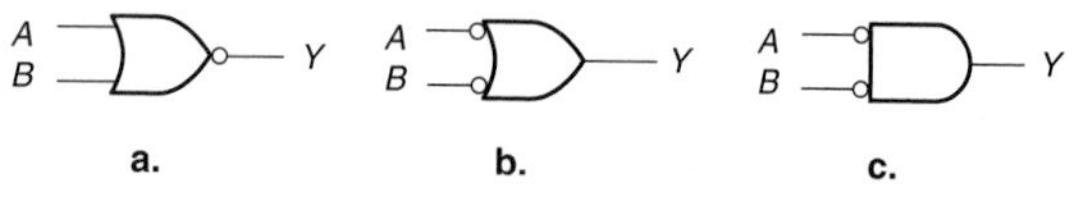

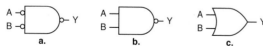

4.27 Refer to *Figure 4.60*. State which two gates of the three shown are DeMorgan equivalents of each other. Explain your choice.

FIGURE 4.60 Problem 4.27: Which Gates Are DeMorgan Equivalents?

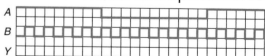

© CENGAGE LEARNING 2012.

4.4 Logic Switches and LED Indicators

4.28 Sketch the circuit of a single-pole single-throw (SPST) switch used as a logic switch. Briefly explain how it works.

4.29 Refer to Figure 4.27 (logic pushbuttons). Should the normally open pushbutton be considered an active HIGH or active LOW device? Briefly explain your choice.

4.30 Should the normally closed pushbutton be considered an active HIGH or active LOW device? Why?

4.31 Briefly state what is required for an LED to illuminate.

4.32 Briefly state the relationship between the brightness of an LED and the current flowing through it. Why is a series resistor required?

4.33 Draw a circuit showing how an OR gate output will illuminate an LED when the gate output is LOW. Assume the required series resistor is $470\ \Omega$.

4.5 Enable and Inhibit Properties of Logic Gates

4.34 Draw the output waveform of the Exclusive NOR gate when a square waveform is applied to one input and
  a. The other input is held LOW.
  b. The other input is held HIGH.
  How does this compare to the waveform that would appear at the output of an Exclusive OR gate under the same conditions?

4.35 Sketch the input waveforms represented by the following 32-bit sequences. (Use 1/4-inch graph paper, 1 square per bit. Spaces are provided for readability only.)
  a. 0000 0000 0000 1111 1111 1111 1111 0000
  b. 1010 0111 0010 1011 0101 0011 1001 1011
  Assume that these waveforms represent inputs to a logic gate. Sketch the waveform for gate output Y if the gate function is:
  a. AND   **d.** NOR
  b. OR   **e.** XOR
  c. NAND   **f.** XNOR

4.36 Repeat Problem 4.35 for the waveforms shown in *Figure 4.61*.

FIGURE 4.61 Problem 4.36: Input Waveforms

A

B

Y

© CENGAGE LEARNING 2012.

4.37 The *A* and *B* waveforms shown in *Figure 4.62* are inputs to an OR gate. Complete the sketch by drawing the waveform for output *Y*.

FIGURE 4.62 Problem 4.37: Input Waveforms

© CENGAGE LEARNING 2012.

4.38 Repeat Problem 4.37 for a NOR gate.

4.39 Make a truth table for the tristate buffers shown in Figure 4.45. Indicate the high-impedance state by the notation "Hi-Z." How do the enable properties of these gates differ from gates such as AND and NAND?

4.6 Integrated Circuit Logic Gates

4.40 Name two logic families used to implement digital logic functions. How do they differ?

4.41 List the industry-standard numbers for a quadruple 2-input NAND gate in low-power Schottky TTL and high-speed CMOS technologies.

4.42 Repeat Problem 4.41 for a quadruple 2-input NOR gate. How does each numbering system differentiate between the NAND and NOR functions?

4.43 List six types of packaging that a logic gate could come in.

EXTRA MILE

4.2 Derived Logic Functions

4.44 *Figure 4.63* shows a circuit for a two-way switch for a stairwell. This is a common circuit that allows you to turn on a light from either the top or the bottom of the stairwell and off at the other end. The circuit also allows anyone coming along after you to do the same thing, no matter which direction they are coming from.

FIGURE 4.63 Problem 4.44: Circuit for Two-Way Switch © CENGAGE LEARNING 2012.

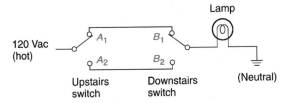

The lamp is ON when the switches are in the same positions and OFF when they are in opposite positions. What logic function does this represent? Draw the truth table of the function and use it to explain your reasoning.

4.45 Recall the description of a 2-input Exclusive OR gate: "Output is HIGH if one input is HIGH, but not both." This is not the best statement of the operation of a multiple-input XOR gate. Look at the truth table derived in Problem 4.24 and write a more accurate description of *n*-input XOR operation.

4.46 **Multisim Problem**
Multisim File: 04.07 Derived Logic Functions. ms10

A circuit showing gates for four derived logic functions is shown in *Figure 4.64*. Enter this circuit in Multisim, using the components listed in *Table 4.29*. Save the file as **04.07 Derived Logic Functions.ms10**.

Interactive Exercise:
Run the Multisim file for this problem as a simulation. Operate switches A and B to make all possible combinations of input logic levels. Write a sentence that describes the operation of each gate.

continues...

continued...

TABLE 4.29 Multisim Components Required for the Circuit in Figure 4.64

Group	Family	Component	Description
SOURCES	POWER_SOURCES	DGND	DIGITAL GROUND
SOURCES	POWER_SOURCES	VCC	TTL SUPPLY
BASIC	SWITCH	SPDT	SINGLE-POLE DOUBLE-THROW SWITCH
INDICATORS	PROBE	PROBE_DIG_RED	RED PROBE—DIGITAL NODE
TTL	74LS	74LS00N	QUADRUPLE 2-INPUT NAND GATE
TTL	74LS	74LS02N	QUADRUPLE 2-INPUT NOR GATE
TTL	74LS	74LS86N	QUADRUPLE 2-INPUT XOR GATE
MISC DIGITAL	TTL	ENOR2	2-INPUT XNOR GATE

© CENGAGE LEARNING 2012.

FIGURE 4.64 Problem 4.46: Derived Logic Functions © CENGAGE LEARNING 2012.

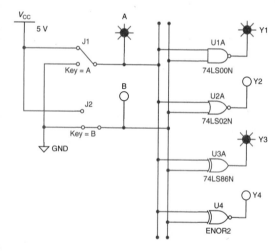

4.4 Logic Switches and LED Indicators

4.47 Multisim Problem

Multisim File: 04.10 LEDs in Opposite States. ms10

Open the Multisim file for this problem and save it as **04.11 LEDs in Opposite States with V and I Meters.ms10**. Modify the circuit by adding four multimeters, as shown in *Figure 4.65*. Meters XMM1 and XMM2 are in series with the LEDs and set as ammeters to measure current through the LEDs. Meters

XMM3 and XMM4 are in parallel with the LEDs and measure the voltage dropped across the LEDs.

FIGURE 4.65 Problem 4.47: LED Circuit with Multimeters © CENGAGE LEARNING 2012.

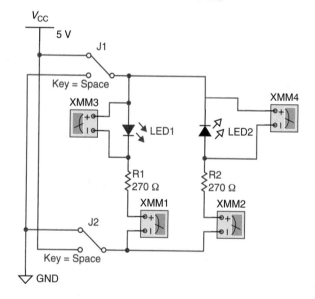

Interactive Exercise:

a. Run the file as a simulation and fill in the following table with the readings on the four multimeters. Also indicate whether each LED is on or off.

Switch Position	LED1 (ON/OFF)	LED2 (ON/OFF)	XMM1	XMM2	XMM3	XMM4
Down						
Up						

b. Write a general statement about the voltage across and current through an LED when it is ON and when it is OFF.

4.5 Enable and Inhibit Properties of Logic Gates

4.48 *Figure 4.66* shows a circuit that will make a lamp flash at 3 Hz when the gasoline level in a car's gas tank drops below a certain point. A float switch in the tank monitors the level of gasoline. What logic level must the float switch produce to make the light flash when the tank is approaching empty? Why?

FIGURE 4.66 *Problem 4.48: Gasoline Level Circuit* © CENGAGE LEARNING 2012.

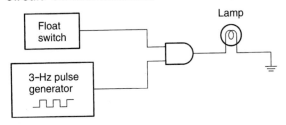

4.49 Repeat Problem 4.48 for the case where the AND gate is replaced by a NOR gate.

4.50 Will the circuit in *Figure 4.66* work properly if the AND gate is replaced by an Exclusive OR gate? Why or why not?

CHAPTER 5
Boolean Algebra and Combinational Logic

CHAPTER OBJECTIVES

Upon successful completion of this chapter, you will be able to:

1 Explain the relationship between the Boolean expression, logic diagram, and truth table of a logic gate network and be able to derive any one from either of the other two.

2 Draw logic gate networks in such a way as to cancel out internal inversions automatically (bubble-to-bubble convention).

3 Write the sum of products (SOP) or product of sums (POS) forms of a Boolean equation.

4 Use rules of Boolean algebra to simplify the Boolean expressions derived from logic diagrams and truth tables.

5 Apply the Karnaugh map method to reduce Boolean expressions and logic circuits to their simplest forms.

6 Use a graphical technique based on DeMorgan equivalent gates to simplify logic diagrams.

7 Apply analysis tools to convert word problems to Boolean equations for the purposes of logic circuit design.

$$a^2 - b = (a+b)(a-b)$$
$$(a+b)^3 = a^3 \pm 3a^2b + 3ab^2 \pm b$$
$$^2 = (a+b)(a-b)$$
$$^3 = (a \pm b)(a^2 \mp ab + b^2)$$

The more digital components, or gates, we use inside of a circuit, the more helpful it might be to simplify the circuit. You have seen this in math class before; if you have a choice between working with a long, complex form of an equation or a simplified equation, you would probably pick the simplified version. We can think of digital circuits in the same way. If we design a big, complicated circuit and then have to build it, it would sure be nice if the circuit could be simplified! We can use rules very similar to algebra to simplify equations, and to simplify circuits.

In Chapter 5, we will examine the basics of combinational logic. A combinational logic circuit is one in which two or more gates are connected together to combine several Boolean inputs. These circuits can be represented several ways: as a logic diagram, truth table, or Boolean expression.

A Boolean expression for a network of logic gates is often not in its simplest form. In such a case, we may be using more components than would be required for the job, so it is of benefit to us if we can simplify the Boolean expression. Several tools are available to us, such as Boolean algebra and a graphical technique known as Karnaugh mapping. We can also simplify the Boolean expression by taking care to draw the logic diagrams so as to automatically eliminate inverting functions within the circuit.

One goal in the design of combinational logic circuits is to translate word problems into gate networks by deriving Boolean expressions for the networks. Once we have a network's Boolean expression, we can apply our tools of simplification and circuit synthesis to complete our design. We will examine a general method of analysis that allows us to derive that initial descriptive equation.

5.1 BOOLEAN EXPRESSIONS, LOGIC DIAGRAMS, AND TRUTH TABLES

KEY TERMS

Logic gate network Two or more logic gates connected together.

Logic diagram A diagram, similar to a schematic, showing the connection of logic gates.

Combinational logic Digital circuitry in which an output is derived from the combination of inputs, independent of the order in which they are applied.

Combinatorial logic Another name for combinational logic.

Sequential logic Digital circuitry in which the output state of the circuit depends not only on the states of the inputs, but also on the sequence in which they reached their present states.

In Chapter 4, we examined the functions of single logic gates. However, most digital circuits require multiple gates. When two or more gates are connected together, they form a **logic gate network**. These networks can be described by a truth table, a **logic diagram** (i.e., a circuit diagram), or a Boolean expression. Any one of these can be derived from any other.

A digital circuit built from gates is called a **combinational** (or **combinatorial**) **logic** circuit. The output of a combinational circuit depends on the *combination* of inputs. The inputs can be applied in any sequence and still produce the same result. For example, an AND gate output will always be HIGH if all inputs are HIGH, regardless of the order in which they became HIGH. This is in contrast to **sequential logic**, in which sequence matters; a sequential logic output may have a different value with two identical sets of inputs if those inputs were applied in a different order. We will study sequential logic in a later chapter.

Boolean Expressions from Logic Diagrams

KEY TERMS

Bubble-to-bubble convention The practice of drawing gates in a logic diagram so that inverting outputs connect to inverting inputs and noninverting outputs connect to noninverting inputs.

Order of precedence The sequence in which Boolean functions are performed, unless otherwise specified by parentheses.

Writing the Boolean expression of a logic gate network is similar to finding the expression for a single gate. The difference is that in a multiple gate network, the inputs will usually not consist of single variables, but compound expressions that represent outputs of previous gates.

These compound expressions are combined according to the same rules as single variables. In an OR gate, with inputs x and y, the output will always be $x + y$ regardless of whether x and y are single variables (e.g., $x = A$, $y = B$, output = $A + B$) or compound expressions (e.g., $x = AB$, $y = AC$, output = $AB + AC$).

Figure 5.1 shows a simple logic gate network, consisting of a single AND and a single OR gate. The AND gate combines inputs A and B to give the output expression AB. The OR combines the AND function and input C to yield the compound expression $AB + C$.

FIGURE 5.1 *Boolean Expression from a Logic Gate* **Network**

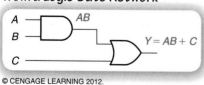

Example 5.1

Derive the Boolean expression of the logic gate network shown in *Figure 5.2a*.

Solution

Figure 5.2b shows the gate network with the output terms indicated for each gate. The AND and NAND functions are combined in an OR function to yield the output expression:

$$Y = AB + \overline{CD}$$

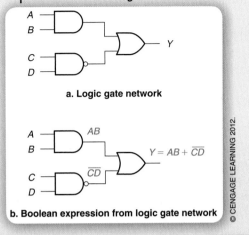

FIGURE 5.2 Example 5.1: Boolean Expression from a Logic Gate Network

a. Logic gate network

b. Boolean expression from logic gate network

© CENGAGE LEARNING 2012.

Note . . .

Recall the basic form of DeMorgan's rule:

$$\overline{A \cdot B} = \overline{A} + \overline{B}$$
$$\overline{A + B} = \overline{A} \cdot \overline{B}$$

The Boolean expression in Example 5.1 includes a NAND function. It is possible to draw the NAND in its DeMorgan equivalent form. If we choose the gate symbols so that outputs with bubbles connect to inputs with bubbles, we will not have bars over groups of variables, except possibly one bar over the entire function. In a circuit with many inverting functions (NANDs and NORs), this results in a cleaner notation and often a clearer idea of the function of the circuit. We will follow this notation, which we will refer to as the **bubble-to-bubble convention**, as much as possible. A more detailed treatment of the bubble-to-bubble convention follows in Section 5.5 of this chapter.

Example 5.2

Redraw the circuit in Figure 5.2 to conform to the bubble-to-bubble convention. Write the Boolean expression of the new logic diagram.

Solution

Figure 5.3 shows the new circuit. The NAND has been converted to its DeMorgan equivalent so that its active-HIGH output drives an active-HIGH input on the OR gate. The new Boolean expression is

$$Y = AB + \overline{C} + \overline{D}$$

FIGURE 5.3 Example 5.2: Using DeMorgan Equivalents to Simplify the Boolean Expression of a Logic Gate Network © CENGAGE LEARNING 2012.

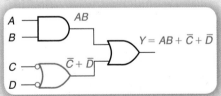

Boolean functions are governed by an **order of precedence.** Unless otherwise specified, AND functions are performed first, followed by ORs. This order results in a form similar to that of algebra, where multiplication is performed before addition, unless otherwise specified.

Figure 5.4 shows two logic diagrams, one whose Boolean expression requires parentheses and one that does not.

The AND functions in Figure 5.4a are evaluated first, eliminating the need for parentheses in the output expression. The expression for Figure 5.4b requires parentheses because the ORs are evaluated first.

FIGURE 5.4 *Order of Precedence of Boolean Operations*

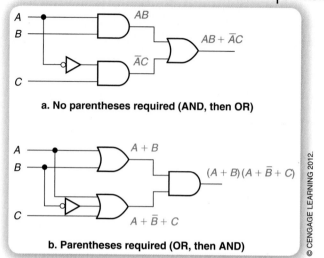

a. No parentheses required (AND, then OR)

b. Parentheses required (OR, then AND)

© CENGAGE LEARNING 2012.

Example 5.3

Write the Boolean expression for the logic diagrams in *Figure 5.5*.

FIGURE 5.5 **Example 5.3: Order of Precedence**

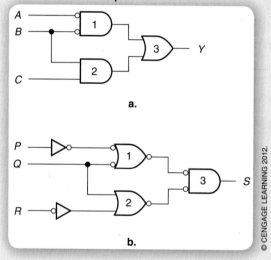

a.

b.

© CENGAGE LEARNING 2012.

Solution

Examine the output of each gate and combine the resultant terms as required.

Figure 5.5a: Gate 1: $\overline{A} \cdot \overline{B}$
                Gate 2: $B \cdot C$
                Gate 3: Y Gate 1 + Gate 2 $= \overline{A} \cdot \overline{B} + B \cdot C$

Figure 5.5b: Gate 1: $\overline{\overline{\overline{P} + Q}} = \overline{P} + \overline{Q}$
                Gate 2: $\overline{Q + \overline{R}}$
                Gate 3: $S = \overline{\overline{\text{Gate 1}} \cdot \overline{\text{Gate 2}}} = \overline{\overline{\overline{P} + \overline{Q}} \cdot \overline{\overline{Q + \overline{R}}}}$
                           $= (P + \overline{Q})(\overline{Q} + \overline{R})$

Note that when two bubbles touch, they cancel out, as in the doubly inverted *P* input or the connection between the outputs of gates 1 and 2 and the inputs of gate 3 in Figure 5.5b. *In the resultant Boolean expression, bars of the same length cancel; bars of unequal length do not.*

Note . . .

In a Boolean expression with two or more overbars, bars of equal length cancel and bars of unequal length do not.

Your Turn

5.1 Write the Boolean expression for the logic diagrams in *Figure 5.6*, paying attention to the order of precedence rules.

Logic Diagrams from Boolean Expressions

FIGURE 5.6 *Logic Gate Networks*

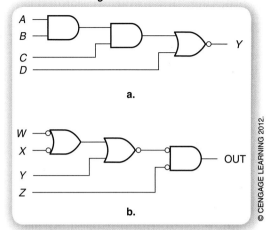

a.

b.

KEY TERMS

Levels of gating The number of gates through which a signal must pass from input to output of a logic gate network.

Double-rail inputs Boolean input variables that are available to a circuit in both true and complement form.

Synthesis The process of creating a logic circuit from a description such as a Boolean equation or truth table.

We can derive a logic diagram from a Boolean expression by applying the order of precedence rules. We examine an expression to create the first **level of gating** from the circuit inputs, then combine the output functions of the first level in the second level gates, and so forth. Input inverters are often not counted as a gating level, as we usually assume that each variable is available in both true (noninverted) and complement (inverted) form. When input variables are available to a circuit in true and complement form, we refer to them as **double-rail inputs**.

The first level usually will be AND gates if no parentheses are present; if OR gates are in the first level, parentheses will usually be necessary. (Not always, however; parentheses merely tell us which functions to synthesize first.) Although we will try to eliminate bars over groups of variables by use of DeMorgan's theorems and the bubble-to-bubble convention, we should recognize that a bar over a group of variables is the same as having those variables in parentheses.

Let us examine the Boolean expression $Y = AC + BD + AD$. Order of precedence tells us that we synthesize the AND functions first. This yields three 2-input AND gates, with outputs AC, BD, and AD, as shown in *Figure 5.7a*. In the next step, we combine these AND functions in a 3-input OR gate, as shown in *Figure 5.7b*.

When the expression has OR functions in parentheses, we synthesize the ORs first, as for the expression $Y = (A + B)(A + C + D)(B + C)$. *Figure 5.8* shows this process. In the first step, we synthesize three OR gates for the terms $(A + B)$, $(A + C + D)$, and $(B + C)$. We then combine these terms in a 3-input AND gate. Again, this is similar to algebra, where multiplication is done before addition unless parentheses are present.

FIGURE 5.7 *Logic Diagram for* $Y = AC + BD + AD$

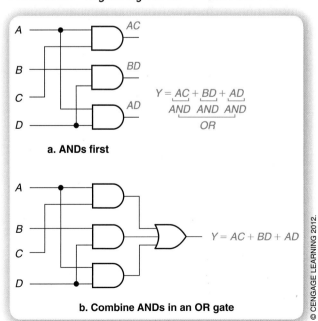

a. ANDs first

$$Y = \underbrace{AC}_{AND} + \underbrace{BD}_{AND} + \underbrace{AD}_{AND}$$
$$OR$$

$Y = AC + BD + AD$

b. Combine ANDs in an OR gate

FIGURE 5.8 Logic Diagram for $Y = (A + B)(A + C + D)(B + C)$

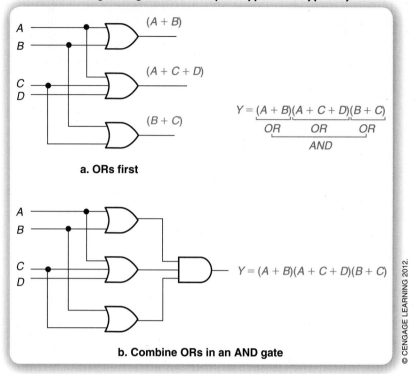

$$Y = \underbrace{(A + B)}_{OR}\underbrace{(A + C + D)}_{OR}\underbrace{(B + C)}_{OR}$$
$$\underbrace{}_{AND}$$

a. ORs first

$Y = (A + B)(A + C + D)(B + C)$

b. Combine ORs in an AND gate

Example 5.4

Synthesize the logic diagrams for the following Boolean expressions:

1. $P = Q\overline{RS} + \overline{S}T$

2. $X = (W + Z + Y)\overline{V} + (\overline{W} + V)\overline{Y}$

FIGURE 5.9 Example 5.4: Logic Diagram for $P = Q\overline{RS} + \overline{S}T$

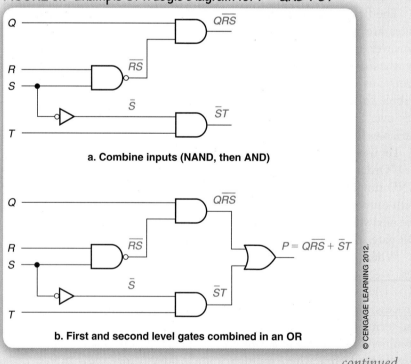

a. Combine inputs (NAND, then AND)

b. First and second level gates combined in an OR

continued...

Solution

1. Recall that a bar over two variables acts like parentheses. Thus, the $Q\overline{RS}$ term is synthesized from a NAND, then an AND, as shown in *Figure 5.9a*. Also shown is the second AND term, $\overline{S}T$.
    Figure 5.9b shows the terms combined in an OR gate.

2. *Figure 5.10* shows the synthesis of the second logic diagram in three stages. Figure 5.10a shows how the circuit inputs are first combined in two OR gates. We do this first because the ORs are in parentheses. In Figure 5.10b, each of these functions is combined in an AND gate, according to the normal order of precedence. The AND outputs are combined in a final OR function, as shown in Figure 5.10c.

FIGURE 5.10 Example 5.4: Logic Diagram for $X = (W + Z + Y)\overline{V} + (\overline{W} + V)\overline{Y}$

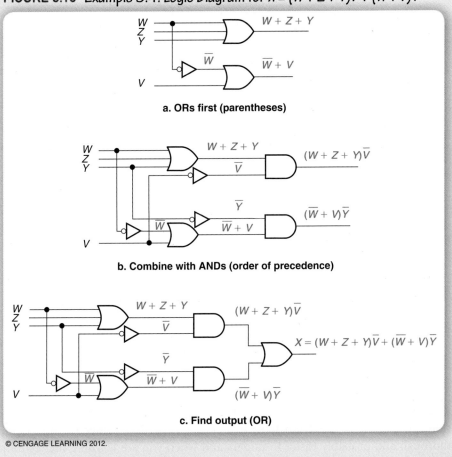

a. ORs first (parentheses)

b. Combine with ANDs (order of precedence)

c. Find output (OR)

Example 5.5

Use DeMorgan's theorem to modify the Boolean equation in part **1** of Example 5.4 so that there is no bar over any group of variables. Redraw Figure 5.9b to reflect the change.

Solution

$$P = Q\overline{RS} + \overline{S}T = Q(\overline{R} + \overline{S}) + \overline{S}T$$

continued...

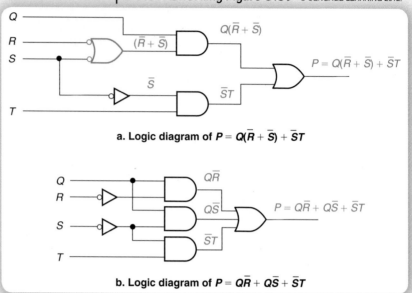

FIGURE 5.11 **Example 5.5: Reworking Figure 5.9b** © CENGAGE LEARNING 2012.

a. Logic diagram of $P = Q(\bar{R} + \bar{S}) + \bar{S}T$

b. Logic diagram of $P = Q\bar{R} + Q\bar{S} + \bar{S}T$

Figure 5.11a shows the modified logic diagram. The levels of gating could be further reduced from three to two (not counting input inverters) by "multiplying through" the parentheses to yield the expression:

$$P = Q\bar{R} + Q\bar{S} + \bar{S}T$$

Figure 5.11b shows the logic diagram for this form. We will examine this simplification procedure more formally in Section 5.3.

FIGURE 5.12 Logic Diagram for $AB + C$

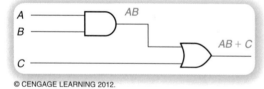

© CENGAGE LEARNING 2012.

TABLE 5.1 Truth Table for Figure 5.12

A	B	C	AB	AB + C
0	0	0	0	0
0	0	1	0	1
0	1	0	0	0
0	1	1	0	1
1	0	0	0	0
1	0	1	0	1
1	1	0	1	1
1	1	1	1	1

© CENGAGE LEARNING 2012.

Truth Tables from Logic Diagrams or Boolean Expressions

There are two basic ways to find a truth table from a logic diagram. We can examine the output of each gate in the circuit and develop its truth table. We then use our knowledge of gate properties to combine these intermediate truth tables into the final output truth table. Alternatively, we can develop a Boolean expression for the logic diagram by examining the expression, then filling in the truth table in a single step. The former method is more thorough and probably easier to understand when you are learning the technique. The latter method is more efficient, but requires some practice and experience. We will look at both.

Examine the logic diagram in *Figure 5.12*. Because there are three binary inputs, there will be eight ways those inputs can be combined. Thus, we start by making an 8-line truth table, as in *Table 5.1*.

The OR gate output will describe the function of the whole circuit. To assess the OR function, we must first evaluate the AND output. We add a column to the truth table for the AND gate and look for the lines in the table where both A AND B equal logic 1 (in this case, the last two rows). For these lines, we write a 1 in the AB column. Next, we look at the values in column C and the AB column. If there is a 1 in either column, we write a 1 in the column for the final output.

Example 5.6

Derive the truth table for the logic diagram shown in *Figure 5.13*.

Solution

The Boolean expression for Figure 5.13 is $(\overline{A} + \overline{B})(A + C)$. We will create a column for each input variable and for each term in parentheses, and a column for the final output. *Table 5.2* shows the result. For the lines where A OR B is 0, we write a 1 in the $(\overline{A} + \overline{B})$ column. Where A OR C is 1, we write a 1 in the $(A + C)$ column. For the lines where there is a 1 in both the $(\overline{A} + \overline{B})$ AND $(A + C)$ columns, we write a 1 in the final output column.

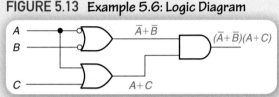

FIGURE 5.13 **Example 5.6: Logic Diagram**

© CENGAGE LEARNING 2012.

TABLE 5.2 **Truth Table for Figure 5.13** © CENGAGE LEARNING 2012.

A	B	C	$(\overline{A} + \overline{B})$	$(A + C)$	$(\overline{A} + \overline{B})(A + C)$
0	0	0	1	0	0
0	0	1	1	1	1
0	1	0	1	0	0
0	1	1	1	1	1
1	0	0	1	1	1
1	0	1	1	1	1
1	1	0	0	1	0
1	1	1	0	1	0

Another approach to finding a truth table is to analyze the Boolean expression of a logic diagram. The logic diagram in *Figure 5.14* can be described by the Boolean expression:

$$Y = \overline{A}\,B\,C + \overline{A}\,\overline{C} + \overline{B}\,\overline{D}$$

We can examine the Boolean expression to determine that the final output of the circuit will be HIGH under one of the following conditions:

$A = 0$ AND $B = 1$ AND $C = 1$
OR,
$A = 0$ AND $C = 0$
OR,
$B = 0$ AND $D = 0$

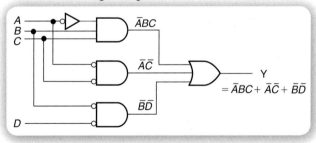

FIGURE 5.14 **Logic Diagram** © CENGAGE LEARNING 2012.

All we have to do is look for these conditions in the truth table and write a 1 in the output column whenever a condition is satisfied. *Table 5.3* shows the result of this analysis with each line indicating which term, or terms, contribute to the HIGH output.

TABLE 5.3 Truth Table for Figure 5.14

A	B	C	D	Y	Terms
0	0	0	0	1	$\overline{A}\,\overline{C}, \overline{B}\,\overline{D}$
0	0	0	1	1	$\overline{A}\,\overline{C}$
0	0	1	0	1	$\overline{B}\,\overline{D}$
0	0	1	1	0	
0	1	0	0	1	$\overline{A}\,\overline{C}$
0	1	0	1	1	$\overline{A}\,\overline{C}$
0	1	1	0	1	$\overline{A}\,BC$
0	1	1	1	1	$\overline{A}\,BC$
1	0	0	0	1	$\overline{B}\,\overline{D}$
1	0	0	1	0	
1	0	1	0	1	$\overline{B}\,\overline{D}$
1	0	1	1	0	
1	1	0	0	0	
1	1	0	1	0	
1	1	1	0	0	
1	1	1	1	0	

FIGURE 5.15 Logic Diagram for Your Turn Problem 5.2 © CENGAGE LEARNING 2012.

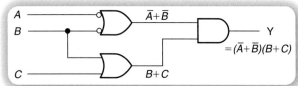

Your Turn

5.2 Find the truth table for the logic diagram shown in *Figure 5.15*.

5.2 SUM-OF-PRODUCTS AND PRODUCT-OF-SUMS FORMS

KEY TERMS

Minterm A product term in a Boolean expression where all possible variables appear once in true or complement form [e.g., $\overline{A}\,\overline{B}\,\overline{C}$; $A\,\overline{B}\,\overline{C}$].

Product term A term in a Boolean expression where one or more true or complement variables are ANDed [e.g., $\overline{A}\,\overline{C}$].

Sum-of-products (SOP) A type of Boolean expression where several product terms are summed (ORed) together [e.g., $\overline{A}\,B\,\overline{C} + \overline{A}\,\overline{B}\,C + A\,B\,C$].

Bus form A way of drawing a logic diagram so that each true and complement input variable is available along a continuous conductor called a bus.

Product-of-sums (POS) A type of Boolean expression where several sum terms are multiplied (ANDed) together [e.g., $(\overline{A} + \overline{B} + C)(A + \overline{B} + \overline{C})(\overline{A} + \overline{B} + \overline{C})$].

Maxterm A sum term in a Boolean expression where all possible variables appear once, in true or complement form [e.g., $\overline{A} + \overline{B} + C$]; [$(A + \overline{B} + C)$].

Sum term A term in a Boolean expression where one or more true or complement variables are ORed [e.g., $\overline{A} + B + \overline{D}$].

Suppose we have an unknown digital circuit, represented by the block in *Figure 5.16*. All we know is which terminals are inputs, which are outputs, and how to connect the power supply (eventually, we will leave power and ground out of these drawings and assume they are present). Given only that information, we can find the Boolean expression of the output.

First, we find the truth table by applying all possible input combinations in binary order and reading the output for each one. Suppose the unknown circuit in Figure 5.16 yields the truth table shown in *Table 5.4*.

FIGURE 5.16 Digital Circuit with Unknown Function

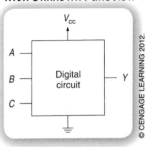

© CENGAGE LEARNING 2012.

TABLE 5.4 Truth Table for Figure 5.16

A	B	C	Y
0	0	0	1
0	0	1	0
0	1	0	0
0	1	1	1
1	0	0	1
1	0	1	0
1	1	0	0
1	1	1	0

FIGURE 5.17 Logic Diagram for
$$Y = \overline{A}\,\overline{B}\,\overline{C} + \overline{A}\,B\,C + A\,\overline{B}\,\overline{C}$$ © CENGAGE LEARNING 2012.

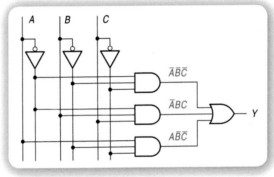

© CENGAGE LEARNING 2012.

The truth table output is HIGH for three conditions:

1. When *A* AND *B* AND *C* are all LOW, OR
2. When *A* is LOW AND *B* AND *C* are HIGH, OR
3. When *A* is HIGH AND *B* AND *C* are LOW.

Each of those conditions represents a **minterm** where the output Boolean expression is HIGH. A minterm is a **product term** (AND term) that includes all variables (*A, B, C*) in true or complement form. The variables in a minterm are written in complement form (with an inversion bar) if the variable is a 0 in the corresponding line of the truth table, and in true form (no bar) if the variable is a 1. The minterms are:

1. $\overline{A}\,\overline{B}\,\overline{C}$
2. $\overline{A}\,B\,C$
3. $A\,\overline{B}\,\overline{C}$

Because condition 1 OR condition 2 OR condition 3 produces a HIGH output from the circuit, the Boolean function *Y* consists of all three minterms summed (ORed) together, as follows:

$$Y = \overline{A}\,\overline{B}\,\overline{C} + \overline{A}\,B\,C + A\,\overline{B}\,\overline{C}$$

This expression is in a standard form called **sum-of-products (SOP)** form. *Figure 5.17* shows the equivalent logic circuit.

Note . . .
We can derive an SOP expression from a truth table as follows:
1. Every line on the truth table that has a HIGH output corresponds to a minterm in the truth table's Boolean expression.
2. Write all truth table variables for every minterm in true or complement form. If a variable is 0, write it in complement form (with a bar over it); if it is 1, write it in true form (no bar).
3. Combine all minterms in an OR function.

The inputs *A*, *B*, and *C* and their complements are shown in **bus form**. Each variable is available, in true or complement form, at any point along a conductor. This is a useful, uncluttered notation for circuits that require several of the input variables more than once. The inverters are shown with the bubbles on their inputs to indicate that the complement line is looking for a LOW input to activate it (note that there is no functional difference in drawing the bubble on the input or the output of an inverter). The true line (no inverter) is looking for a HIGH. Notice that dots indicate connections; no connection exists where lines simply cross each other with no dot.

Example 5.7

Multisim Example

Multisim File: 05.01 SOP Circuit 1.ms11

Figure 5.18 shows a sum-of-products (SOP) circuit drawn in Multisim. The three inputs (*A*, *B*, and *C*) are monitored with logic probes, as are the AND gate outputs (*G1*, *G2*, and *G3*) and the circuit output (*Y*). The circuit is shown with all inputs in the LOW state. Open the file for this example and operate the switches *A*, *B*, and *C* to go through all possible binary input combinations.

   a. Which input states turn on the logic probes for *G1*, *G2*, and *G3*?
   b. How does this correspond to the state of the logic probe for output *Y*?
   c. Write the Boolean expression for the SOP circuit in Figure 5.18.

FIGURE 5.18 **Example 5.7: SOP Circuit with All Inputs LOW**

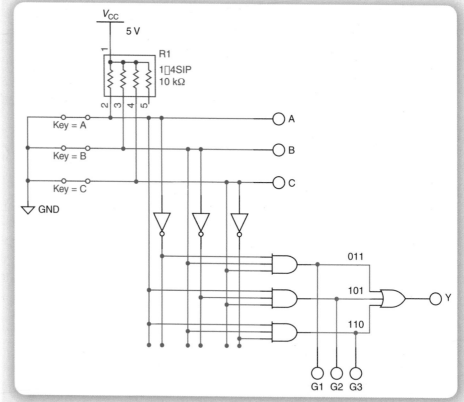

© CENGAGE LEARNING 2012.

Solution

   a. *Figure 5.19A*, *Figure 5.19B*, and *Figure 5.19C* show the three states where one of the AND gate outputs is HIGH.

continued...

FIGURE 5.19A Example 5.7: SOP Circuit with AND Gate 1 Active

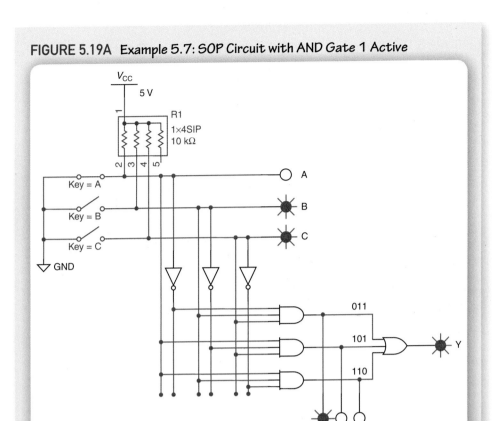

FIGURE 5.19B Example 5.7: SOP Circuit with AND Gate 2 Active

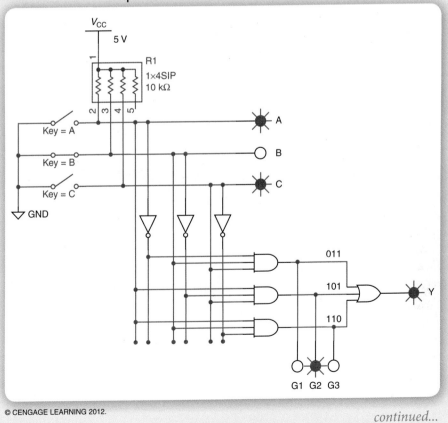

continued...

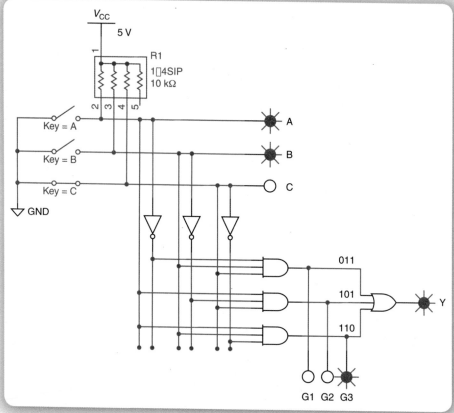

FIGURE 5.19C Example 5.7: SOP Circuit with AND Gate 3 Active

© CENGAGE LEARNING 2012.

Figure 5.19A shows the output of G1 HIGH when $ABC = 011$. In Figure 5.19B, output G2 is HIGH when $ABC = 101$. In Figure 5.19C, G3 is HIGH when $ABC = 110$.

  b. The logic probe for output Y is ON whenever any of the AND gate outputs, G1, G2, or G3 are HIGH.

  c. The Boolean expression for the circuit is determined by the three product terms for the AND gates: $Y = \overline{A}\,B\,C + A\,\overline{B}\,C + A\,B\,\overline{C}$.

Example 5.8

Table 5.5 and *Table 5.6* show the truth tables for the Exclusive OR and the Exclusive NOR functions. Derive the sum-of-products expression for each of these functions and draw the logic diagram for each one.

TABLE 5.5 XOR Truth Table

A	B	$A \oplus B$
0	0	0
0	1	1
1	0	0
1	1	1

© CENGAGE LEARNING 2012.

TABLE 5.6 XOR Truth Table

A	B	$\overline{A \oplus B}$
0	0	1
0	1	0
1	0	0
1	1	1

© CENGAGE LEARNING 2012.

continued...

Solution

XOR: The truth table yields two product terms: $\overline{A}\,B$ and $A\,\overline{B}$. Thus, the SOP form of the XOR function is $A \oplus B = \overline{A}\,B + A\,\overline{B}$. *Figure 5.20* shows the logic diagram for this equation.

XNOR: The product terms for this function are: $\overline{A}\,\overline{B}$ and $A\,B$. The SOP form of the XNOR function is $\overline{A} \oplus \overline{B} = \overline{A}\,\overline{B} + A\,B$. The logic diagram in *Figure 5.21* represents the XNOR function.

FIGURE 5.20 *Example 5.8: SOP Form of* XOR *Function* © CENGAGE LEARNING 2012.

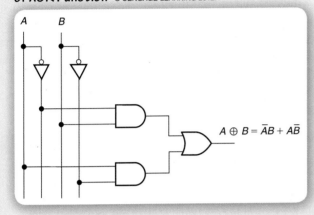

FIGURE 5.21 *Example 5.8: SOP Form of* XNOR *Function* © CENGAGE LEARNING 2012.

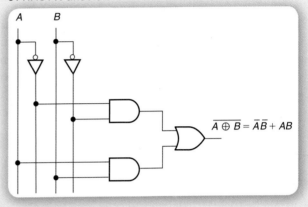

We can also find the Boolean function of a truth table in **product-of-sums** (**POS**) form. The product-of-sums form of a Boolean expression consists of a number of **maxterms** (i.e., **sum terms**, OR terms, containing all variables in true or complement form) that are ANDed together. To find the POS form of Y, we will find the SOP expression for $\overline{Y}$ and apply DeMorgan's theorems.

Recall DeMorgan's theorems:

$$\overline{X + Y + Z} = \overline{X}\,\overline{Y}\,\overline{Z}$$

$$\overline{X\,Y\,Z} = \overline{X} + \overline{Y} + \overline{Z}$$

When the theorems were introduced, they were presented as two-variable theorems, but in fact they are valid for any number of variables.

Let's reexamine Table 5.4. To find the sum-of-products expression for Y, we wrote a minterm for each line where $Y = 1$. To find the SOP expression for $\overline{Y}$, *we must write a minterm for each line where $Y = 0$.* Variables A, B, and C must appear in each minterm, in true or complement form. A variable is in complement form (with a bar over the top) if its value is 0 in that minterm, and it is in true form (no bar) if its value is 1.

We get the following minterms for $\overline{Y}$:

$$\overline{A}\,\overline{B}\,C$$
$$\overline{A}\,B\,\overline{C}$$
$$A\,\overline{B}\,C$$
$$A\,B\,\overline{C}$$
$$A\,B\,C$$

Thus, the SOP form of $\overline{Y}$ is:

$$\overline{Y} = \overline{A}\,\overline{B}\,C + \overline{A}\,B\,\overline{C} + A\,\overline{B}\,C + A\,B\,\overline{C} + A\,B\,C$$

FIGURE 5.22 POS Logic Circuit for Table 5.4

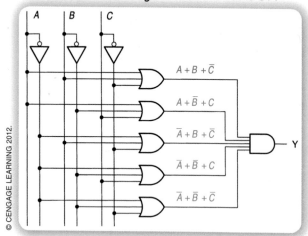

$A + B + \overline{C}$

$A + \overline{B} + C$

$\overline{A} + B + \overline{C}$

$\overline{A} + \overline{B} + C$

$\overline{A} + \overline{B} + \overline{C}$

Y

To get Y in POS form, we must invert both sides of the above expression and apply DeMorgan's theorems to the right-hand side.

$$Y = \overline{\overline{Y}} = \overline{\overline{A}\,\overline{B}\,C + \overline{A}\,B\,\overline{C} + A\,\overline{B}\,C + A\,B\,\overline{C} + A\,B\,C}$$
$$= (\overline{\overline{A}\,\overline{B}\,C})\,(\overline{\overline{A}\,B\,\overline{C}})\,(\overline{A\,\overline{B}\,C})\,(\overline{A\,B\,\overline{C}})\,(\overline{A\,B\,C})$$
$$= (A + B + \overline{C})\,(A + \overline{B} + C)\,(\overline{A} + B + \overline{C})\,(\overline{A} + \overline{B} + C)$$
$$(\overline{A} + \overline{B} + \overline{C})$$

This Boolean expression can be implemented by the logic circuit in *Figure 5.22*.

We don't have to go through the whole process just outlined every time we want to find the POS form of a function. We can find it directly from the truth table, using the procedure summarized in the accompanying Note. Use this procedure to find the POS form of the expression given by Table 5.4. The terms in this expression are the same as those derived by DeMorgan's theorem.

Note . . .

Deriving a POS expression from a truth table:
1. Every line on the truth table that has a LOW output corresponds to a maxterm in the truth table's Boolean expression.
2. Write all truth table variables for every maxterm in true or complement form. If a variable is 1, write it in complement form (with a bar over it); if it is 0, write it in true form (no bar).
3. Combine all maxterms in an AND function.

Note that these steps are all opposite to those used to find the SOP form of the Boolean expression.

Example 5.9

Find the Boolean expression, in both SOP and POS forms, for the logic function represented by *Table 5.7*. Draw the logic circuit for each form.

TABLE 5.7 Truth Table for Example 5.9 (with Minterms and Maxterms)

A	B	C	D	Y	Minterms	Maxterms
0	0	0	0	1	$\overline{A}\,\overline{B}\,\overline{C}\,\overline{D}$	
0	0	0	1	1	$\overline{A}\,\overline{B}\,\overline{C}\,D$	
0	0	1	0	0		$A + B + \overline{C} + D$
0	0	1	1	1	$\overline{A}\,\overline{B}\,C\,D$	
0	1	0	0	0		$A + \overline{B} + C + D$
0	1	0	1	0		$A + \overline{B} + C + \overline{D}$
0	1	1	0	0		$A + \overline{B} + \overline{C} + D$
0	1	1	1	0		$A + \overline{B} + \overline{C} + \overline{D}$
1	0	0	0	1	$A\,\overline{B}\,\overline{C}\,\overline{D}$	
1	0	0	1	0		$\overline{A} + B + C + \overline{D}$
1	0	1	0	1	$A\,\overline{B}\,C\,\overline{D}$	
1	0	1	1	0		$\overline{A} + B + \overline{C} + \overline{D}$
1	1	0	0	1	$A\,B\,\overline{C}\,\overline{D}$	
1	1	0	1	1	$A\,B\,\overline{C}\,D$	
1	1	1	0	1	$A\,B\,C\,\overline{D}$	
1	1	1	1	0		$\overline{A} + \overline{B} + \overline{C} + \overline{D}$

Solution

All minterms (for SOP form) and maxterms (for POS form) are shown in the last two columns of Table 5.7.

continued...

Boolean Expressions

SOP form:

$$Y = \overline{A}\,\overline{B}\,\overline{C}\,\overline{D} + \overline{A}\,\overline{B}\,\overline{C}\,D + \overline{A}\,\overline{B}\,C\,D + A\,\overline{B}\,\overline{C}\,\overline{D} + A\,\overline{B}\,C\,\overline{D} + A\,B\,\overline{C}\,\overline{D}$$
$$+ A\,B\,\overline{C}\,D + A\,B\,C\,\overline{D}$$

POS form:

$$Y = (A + B + \overline{C} + D)(A + \overline{B} + C + D)(A + \overline{B} + C + \overline{D})(A + \overline{B} + \overline{C} + D)$$
$$(A + \overline{B} + \overline{C} + \overline{D})(\overline{A} + B + C + \overline{D})(\overline{A} + B + \overline{C} + \overline{D})\,(\overline{A} + \overline{B} + \overline{C} + \overline{D})$$

The logic circuits are shown in *Figure 5.23* and *Figure 5.24*.

FIGURE 5.23 Example 5.9: SOP Form

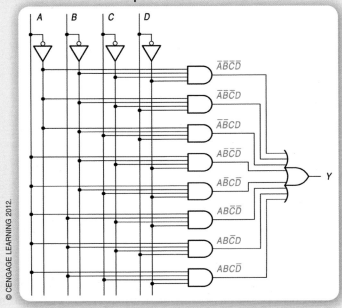

FIGURE 5.24 Example 5.9: POS Form

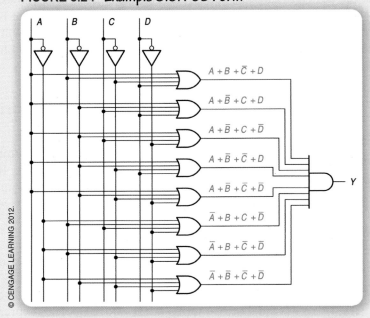

© CENGAGE LEARNING 2012.

Your Turn

5.3 Find the SOP and POS forms of the Boolean functions represented by the following truth tables.

a.

A	B	C	Y
0	0	0	0
0	0	1	0
0	1	0	0
0	1	1	0
1	0	0	1
1	0	1	1
1	1	0	0
1	1	1	0

© CENGAGE LEARNING 2012.

b.

A	B	C	Y
0	0	0	1
0	0	1	0
0	1	0	0
0	1	1	0
1	0	0	1
1	0	1	1
1	1	0	1
1	1	1	0

© CENGAGE LEARNING 2012.

5.3 SIMPLIFYING SOP EXPRESSIONS

> **KEY TERM**
>
> **Maximum SOP simplification** The form of an SOP Boolean expression that cannot be further simplified by canceling variables in the product terms. It may be possible to get a POS form of the expression with fewer terms or variables.

Earlier in this chapter, we discovered that we can generate a Boolean equation from a truth table and express it in sum-of-products (SOP) or product-of-sums (POS) form. From this equation, we can develop a logic circuit diagram. The next step in the design or analysis of a circuit is to simplify its Boolean expression as much as possible, with the ultimate aim of producing a circuit that has fewer physical components than the unsimplified circuit. In this section, we will concentrate on simplifying SOP expressions only. POS expressions can be simplified by a similar technique, but we will not examine it further.

FIGURE 5.25 SOP Circuit from Table 5.8

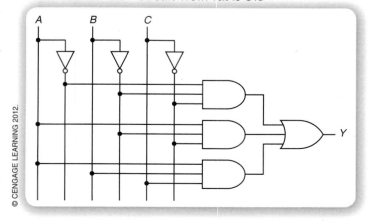

© CENGAGE LEARNING 2012.

Reducing Product Terms by Factoring and Cancellation

Figure 5.25 shows a sum-of-products circuit that is described by the truth table in *Table 5.8*. It is also described by the following Boolean expression:

$$Y = \overline{A}\,\overline{B}\,\overline{C} + A\,\overline{B}\,\overline{C} + A\,B\,C$$

One way to simplify an SOP expression is to look for pairs of product terms that differ by just one variable, then factor out what is common to the term, as you would do in other types of algebra. For example, the first two terms of the above Boolean expression are different in that one term has $\overline{A}$ and the other has A, but both terms have $\overline{B}$ and $\overline{C}$. Thus, we can factor the first two terms, as follows:

$$Y = (\overline{A} + A)\,\overline{B}\,\overline{C} + A\,B\,C$$

We cannot do anything to the last term, as it differs from the original first term by three variables and from the second term by two variables.

The next step in the simplification is to cancel the variables in parentheses. *Figure 5.26* uses logic gates to show how this is done. In Figure 5.26a, a variable is ORed with its complement. Because one input to the OR gate is always HIGH, the OR output is always HIGH, as well. This shows that the OR term in parentheses cancels to a logic 1. This logic 1 is ANDed with the rest of the term, as shown in Figure 5.26b. If the term $\overline{B}\,\overline{C}$ is a 0, the AND output is a 0; if $\overline{B}\,\overline{C}$ is a 1, then the AND output is also a 1. Thus, the output of the AND gate is the same as the input $\overline{B}\,\overline{C}$. The Boolean expression therefore simplifies to:

$$Y = \overline{B}\,\overline{C} + A\,B\,C$$

TABLE 5.8 Truth Table for Figure 5.25

A	B	C	Y
0	0	0	1
0	0	1	0
0	1	0	0
0	1	1	0
1	0	0	1
1	0	1	0
1	1	0	0
1	1	1	1

© CENGAGE LEARNING 2012.

FIGURE 5.26 *Gate Equivalents of Boolean Operations* © CENGAGE LEARNING 2012.

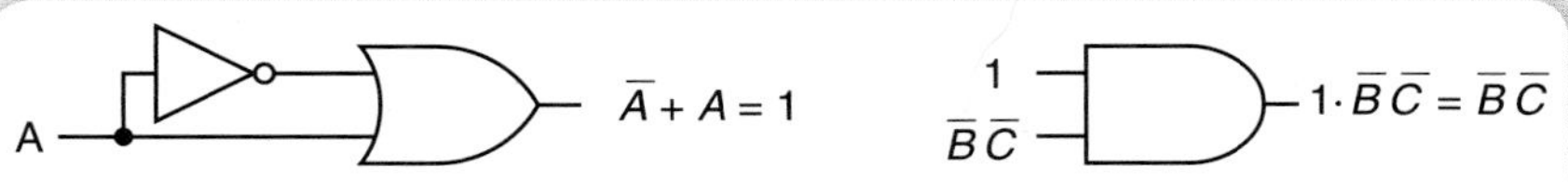

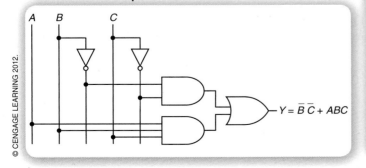

a. Cancelling opposite variables in an OR function

b. Combining a variable group with a logic 1 in an AND function

Note . . .

If two product terms in an SOP expression differ by exactly one variable, the differing variable can be canceled:

$$xyz + xy\overline{z}$$
$$= xy(z + \overline{z}) = xy$$

The simplified SOP circuit is shown in *Figure 5.27*. Because the Boolean expression cannot be simplified any further, we refer to this form as the **maximum SOP simplification**.

Reusing Product Terms

Figure 5.28 shows an SOP circuit with the following Boolean expression:

$$Y = ABC + A\overline{B}C + AB\overline{C}$$

Using the technique from the previous example, we could group the first product term with the second term and cancel the B variables or group the first and third terms and cancel the Cs. Which is better?

FIGURE 5.28 Unsimplified SOP Circuit

FIGURE 5.27 Simplified SOP Circuit

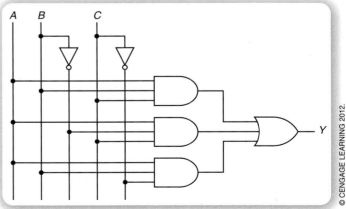

© CENGAGE LEARNING 2012.

© CENGAGE LEARNING 2012.

FIGURE 5.29 *Gate Equivalent of Reusing a Product Term*

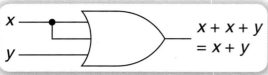

$$x + x + y = x + y$$

It turns out that we can do both. To do so, we have to reuse the first product term (ABC) and group it once with the second term and once with the third term. Is this OK? *Figure 5.29* illustrates why this is possible.

The OR gate in Figure 5.29 has three inputs representing product terms x, x, and y. The output expression of the gate is $x + x + y$. This has the same effect on the output as if we just had one x input, rather than two. (At least one input HIGH makes the OR output HIGH. If y is LOW, it doesn't matter if one or two other inputs are HIGH; the output would be HIGH in either case. If y is HIGH, it doesn't matter what the other inputs are doing; the output will be HIGH anyway. The output will be LOW only if all inputs are LOW.) Therefore, we can write:

$$x + y = x + x + y$$

This implies that a product term (represented by x) in the above expression can be written more than once without having an effect on the output. Applying this principle and the previous cancellation technique, we get the following simplification:

$$Y = ABC + A\overline{B}C + AB\overline{C}$$
$$= ABC + ABC + A\overline{B}C + AB\overline{C}$$
$$= ABC + A\overline{B}C + ABC + AB\overline{C}$$
$$= AC(B + \overline{B}) + AB(C + \overline{C})$$
$$Y = AC + AB$$

Figure 5.30 shows the logic diagram of the simplified Boolean expression.

FIGURE 5.30 *Simplified SOP Circuit*

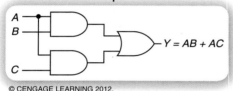

$Y = AB + AC$

© CENGAGE LEARNING 2012.

Example 5.10

Multisim Example

Multisim File: 05.04 SOP Circuit 4a.ms11

Figure 5.31 shows an SOP circuit drawn in Multisim.

  a. Run the circuit simulation and determine the output truth table of the circuit.
  b. From the truth table, find the unsimplified Boolean expression.
  c. Simplify the SOP expression as much as possible.
  d. Modify the circuit in Figure 5.31 to show the circuit for the simplified Boolean expression.

TABLE 5.9 *Truth Table*

A	B	C	Y
0	0	0	0
0	0	1	1
0	1	0	0
0	1	1	0
1	0	0	1
1	0	1	1
1	1	0	0
1	1	1	0

© CENGAGE LEARNING 2012.

Solution

  a. The truth table of the circuit in Figure 5.31 is shown in *Table 5.9*.
  b. The Boolean expression for the circuit is given by: $Y = \overline{A}\,\overline{B}C + A\overline{B}\,\overline{C} + A\overline{B}C$
  c. The Boolean expression can be simplified as follows:

$$Y = \overline{A}\,\overline{B}C + A\overline{B}\,\overline{C} + A\overline{B}C + A\overline{B}C$$
$$= \overline{A}\,\overline{B}C + A\overline{B}C + A\overline{B}\,\overline{C} + A\overline{B}C$$
$$= (\overline{A} + A)\overline{B}C + A\overline{B}(\overline{C} + C)$$
$$Y = \overline{B}C + A\overline{B}$$

continued...

d. The Multisim file for the simplified Boolean expression is shown in
Figure 5.32.

FIGURE 5.31 **Example 5.10: Multisim SOP Circuit** © CENGAGE LEARNING 2012.

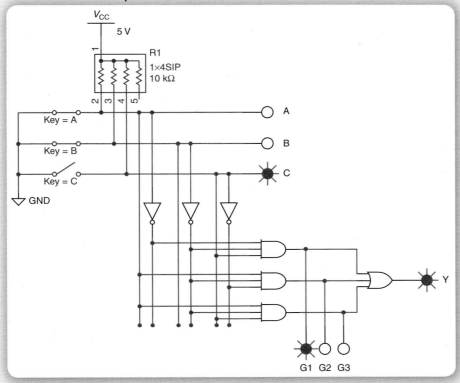

FIGURE 5.32 **Example 5.10: Simplified Multisim SOP Circuit**

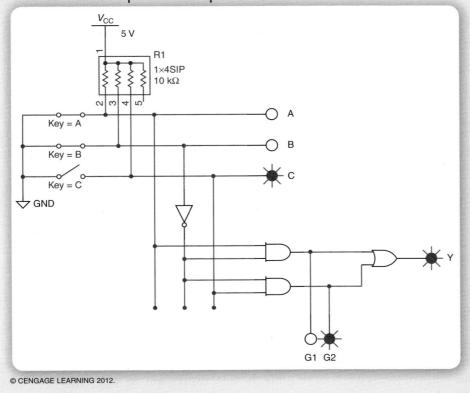

Avoiding Redundant Terms

> **KEY TERM**
>
> **Redundant term** An extra, unneeded product term in a simplified Boolean expression.

Sometimes when we group terms more than once, as we did in the previous two examples, we might end up with more terms than are necessary in a simplified expression. These **redundant terms** make us use more hardware than is required for a particular logic circuit.

For example, the SOP circuit in *Figure 5.33* is represented by the truth table in *Table 5.10*. The unsimplified SOP express on has four product terms, shown in the table. We can refer to the terms by the numbers in parentheses at the far left column of the table. Depending on how we simplify this circuit, we could end up with two terms or with three.

TABLE 5.10   Truth Table for Figure 5.33

	A	B	C	D	Y	
(0)	0	0	0	0	1	$\overline{A}\,\overline{B}\,\overline{C}\,\overline{D}$
(1)	0	0	0	1	1	$\overline{A}\,\overline{B}\,\overline{C}\,D$
(2)	0	0	1	0	0	
(3)	0	0	1	1	0	
(4)	0	1	0	0	0	
(5)	0	1	0	1	1	$\overline{A}\,B\,\overline{C}\,D$
(6)	0	1	1	0	0	
(7)	0	1	1	1	0	
(8)	1	0	0	0	1	$A\,\overline{B}\,\overline{C}\,\overline{D}$
(9)	1	0	0	1	0	
(10)	1	0	1	0	0	
(11)	1	0	1	1	0	
(12)	1	1	0	0	0	
(13)	1	1	0	1	0	
(14)	1	1	1	0	0	
(15)	1	1	1	1	0	

© CENGAGE LEARNING 2012.

FIGURE 5.33   SOP Circuit from Table 5.10

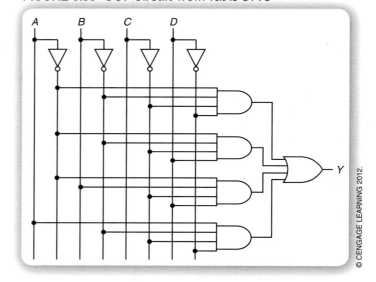

© CENGAGE LEARNING 2012.

The unsimplified Boolean expression for this circuit is $Y = \overline{A}\,\overline{B}\,\overline{C}\,\overline{D} + \overline{A}\,\overline{B}\,\overline{C}\,D + \overline{A}\,B\,\overline{C}\,D + A\,\overline{B}\,\overline{C}\,\overline{D}$, or showing the number value of each product term, rather than the term itself, we can write $Y = (0) + (1) + (5) + (8)$.

Using the earlier method of grouping and simplifying terms, we can make the following groupings:

$$(0)+(1): \overline{A}\,\overline{B}\,\overline{C}\,\overline{D} + \overline{A}\,\overline{B}\,\overline{C}\,D = \overline{A}\,\overline{B}\,\overline{C}\,(\overline{D} + D) = \overline{A}\,\overline{B}\,\overline{C}$$
$$(1)+(5): \overline{A}\,\overline{B}\,\overline{C}\,D + \overline{A}\,B\,\overline{C}\,D = \overline{A}\,\overline{C}\,D\,(\overline{B} + B) = \overline{A}\,\overline{C}\,D$$
$$(0)+(8): \overline{A}\,\overline{B}\,\overline{C}\,\overline{D} + A\,\overline{B}\,\overline{C}\,\overline{D} = (\overline{A} + A)\,\overline{B}\,\overline{C}\,\overline{D} = \overline{B}\,\overline{C}\,\overline{D}$$

The simplified Boolean expression can be written as $Y = \overline{A}\,\overline{B}\,\overline{C} + \overline{A}\,\overline{C}\,D + \overline{B}\,\overline{C}\,\overline{D}$. The logic diagram for this equation is shown in *Figure 5.34a*.

Recall that we found that it is acceptable to use terms more than once if it helps to simplify the final Boolean expression. With that in mind, we grouped terms (0) and (1), both of which overlapped other terms: (0)+(8) and (1)+(5). However, this reuse of terms does not simplify the final expression. In fact, it adds one more term than is required for the maximum Boolean simplification.

Suppose we eliminate this first grouping and use only the last two groups, which then gives us the expression $Y = \overline{A}\,\overline{C}\,D + \overline{B}\,\overline{C}\,\overline{D}$. The logic diagram for this expression is shown in *Figure 5.34b*. This further simplified expression covers all the required product terms of the original expression and also gives us one less AND gate than the other, less simplified form of the Boolean expression.

This points to an important tip to remember when grouping product terms for simplification of an SOP expression. **When grouping terms for simplification, every pair of terms must have at least one term that only belongs to that pair.** Otherwise you will get redundant terms that must be further canceled to get the maximum SOP simplification.

In our example, in the pair (0)+(1), both (0) and (1) were also grouped with other terms. In the other two groups, at least one term belonged only to that group: (5) was only grouped with (1); term (8) was only grouped with (0).

In summary, when simplifying Boolean expressions, the following guidelines apply:

1. Each term must be grouped with another, if possible.
2. When attempting to group all terms, it is permissible to group a term more than once.
3. Each pair of terms should have at least one term that appears only in that pair.

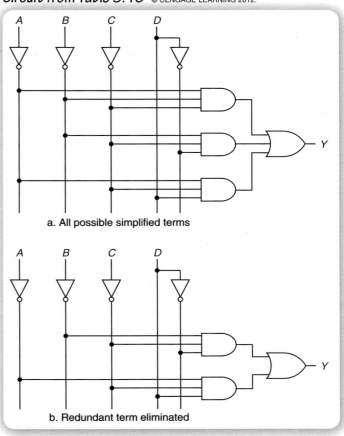

FIGURE 5.34 *Two Possible Simplifications of the SOP Circuit from Table 5.10* © CENGAGE LEARNING 2012.

a. All possible simplified terms

b. Redundant term eliminated

Your Turn

5.4 Find the maximum SOP simplifications for the function represented by *Table 5.11*.

5.4 SIMPLIFICATION BY THE KARNAUGH MAP METHOD

TABLE 5.11 Truth Table for Your Turn Problem 5.4

A	B	C	Y
0	0	0	0
0	0	1	1
0	1	0	1
0	1	1	1
1	0	0	0
1	0	1	0
1	1	0	1
1	1	1	0

© CENGAGE LEARNING 2012.

KEY TERMS

Karnaugh map A graphical tool for finding the maximum SOP or POS simplification of a Boolean expression. A Karnaugh map (or **K-map**) works by arranging the terms of an expression so that variables can be canceled by grouping minterms or maxterms.

Cell The smallest unit of a Karnaugh map, corresponding to one line of a truth table. The input variables are the cell's coordinates, and the output variable is the cell's contents.

Cell coordinate A variable around the edge of a K-map that represents an input variable (e.g., *A, B, C,* or *D*) for the Boolean expression to be simplified.

Cell content A 0 or 1 that represents the value of the output variable (e.g., *Y*) of a Boolean expression for a particular set of coordinates.

Adjacent cell Two cells are adjacent if there is only one variable that is different between the coordinates of the two cells. For example, the cells for minterms ABC and $\overline{A}BC$ are adjacent.

Pair A group of two adjacent cells in a Karnaugh map. A pair cancels one variable in a K-map simplification.

Quad A group of four adjacent cells in a Karnaugh map. A quad cancels two variables in a K-map simplification.

Octet A group of eight adjacent cells in a Karnaugh map. An octet cancels three variables in a K-map simplification.

Reducing circuitry using the laws of Boolean algebra may seem like a complex task. We will explore a more visual method in this section. In the circuit represented by Table 5.10, we derived a sum-of-products Boolean expression from a truth table and simplified the expression by grouping minterms that differed by one variable. We made this task easier by breaking up the truth table into groups of four lines. (It is difficult for the eye to grasp an overall pattern in a group of 16 lines.) We chose groups of four because variables *A* and *B* are the same in any one group and variables *C* and *D* repeat the same binary sequence in each group. This allows us to see more easily when we have terms differing by only one variable.

The **Karnaugh map**, or **K-map**, is a graphical tool for simplifying Boolean expressions that uses a similar idea. A K-map is a square or rectangle divided into smaller squares called **cells**, each of which represents a line in the truth table of the Boolean expression to be mapped. Thus, the number of cells in a K-map is always a power of 2, usually 4, 8, or 16. The **coordinates** ("edge variables") of each cell are the input variables of the truth table. The **cell content** is the value of the output variable on that line of the truth table. *Figure 5.35* shows the formats of Karnaugh maps for Boolean expressions having two, three, and four variables, respectively.

There are two equivalent ways of labeling the cell coordinates: numerically or by true and complement variables. We will use the numerical labeling because it is always the same, regardless of the chosen variables.

The cells in the Karnaugh maps are set up so that the coordinates of any two **adjacent cells** differ by only one variable. By grouping adjacent cells according to specified rules, we can simplify a Boolean expression by canceling variables in their true and complement forms, much as we did algebraically in the previous section.

Two-Variable Map

Table 5.12 shows the truth table of a two-variable Boolean expression.

The Karnaugh map shown in *Figure 5.36* is another way of showing the same information as the truth table. Every line in the truth table corresponds to a cell, or square, in the Karnaugh map.

The coordinates of each cell correspond to a unique combination of input variables (*A, B*). The content of the cell is the output value for that input combination. If the truth table output is 1 for a particular line, the content of the corresponding cell is also 1. If the output is 0, the cell content is 0.

The SOP expression of the truth table is

$$Y = \overline{A}\,\overline{B} + \overline{A}\,B$$

TABLE 5.12 Truth Table for a Two-Variable Boolean Expression

A	B	Y
0	0	1
0	1	1
1	0	0
1	1	0

FIGURE 5.35 Karnaugh Map Formats © CENGAGE LEARNING 2012.

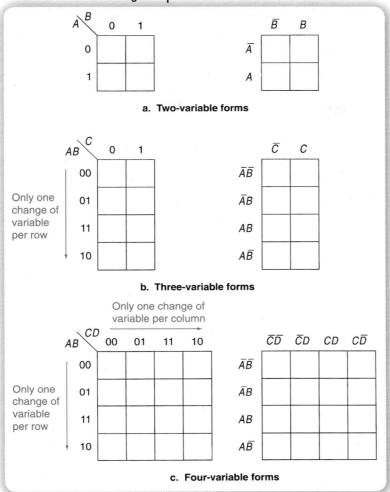

a. Two-variable forms

Only one change of variable per row

b. Three-variable forms

Only one change of variable per column

Only one change of variable per row

c. Four-variable forms

FIGURE 5.36
Karnaugh Map
for Table 5.12

© CENGAGE LEARNING 2012.

FIGURE 5.37
Grouping a Pair
of Adjacent Cells

© CENGAGE LEARNING 2012.

which can be simplified as follows:

$$Y = \overline{A}\,(\overline{B} + B)$$

We can perform the same simplification by grouping the adjacent **pair** of 1's in the Karnaugh map, as shown in *Figure 5.37*.

To find the simplified form of the Boolean expression represented in the K-map, we examine the coordinates of all cells in the circled group. We retain coordinate variables that are the same in all cells and eliminate coordinate variables that are different in different cells.

In this case:

$\overline{A}$ is a coordinate of both cells of the circled pair. (Keep $\overline{A}$.)

$\overline{B}$ is a coordinate of one cell of the circled pair, and B is a coordinate of the other. (Discard B and $\overline{B}$.)

$$Y = \overline{A}\,(\overline{B} + B) = \overline{A}$$

Three- and Four-Variable Maps

Refer to the forms of three- and four-variable Karnaugh maps shown in *Figure 5.35*. Each cell is specified by a unique combination of binary variables. This implies

Note . . .

When we circle a pair of 1's in a K-map, we are grouping the common variable in two minterms, then factoring out and canceling the complements.

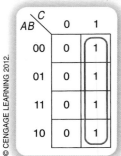

FIGURE 5.38 *Quad*

Note . . .

The number of cells in a group must be a power of 2, such as 1, 2, 4, 8, or 16.

FIGURE 5.39 *Octet*

Note . . .

A Karnaugh map completely filled with 1's implies that all input conditions yield an output of 1. For a Boolean expression Y, $Y = 1$.

that the three-variable map has 8 cells (because $2^3 = 8$) and the four-variable map has 16 cells (because $2^4 = 16$).

The variables specifying the row (both maps) or the column (the four-variable map) do not progress in binary order; they advance such that there is only *one change of variable per row or column.*

A group of four adjacent cells is called a **quad**. *Figure 5.38* shows a Karnaugh map for a Boolean function whose terms can be grouped in a quad. The Boolean expression displayed in the K-map is:

$$Y = \overline{A}\,\overline{B}\,C + \overline{A}\,B\,C + A\,B\,C + A\,\overline{B}\,C$$

A and B are both part of the quad coordinates in true and complement form. (Discard A and B.)

C is a coordinate of *each cell* in the quad. (Keep C.)

$$Y = C$$

Grouping cells in a quad is equivalent to factoring two complementary pairs of variables and canceling them.

$$Y = (\overline{A} + A)(\overline{B} + B)\,C$$

You can verify that this is the same as the original expression by multiplying out the terms.

An **octet** is a group of eight adjacent cells. Figure 5.39 shows the Karnaugh map for the following Boolean expression:

$$Y = \overline{A}\,B\,\overline{C}\,\overline{D} + \overline{A}\,B\,\overline{C}\,D + \overline{A}\,B\,C\,D + \overline{A}\,B\,C\,\overline{D}$$
$$+ A\,B\,\overline{C}\,\overline{D} + A\,B\,\overline{C}\,D + A\,B\,C\,D + A\,B\,C\,\overline{D}$$

Variables A, C, and D are all coordinates of the octet cells in true and complement form. (Discard A, C, and D.)

B is a coordinate of *each* cell. (Keep B.)

$$Y = B$$

The algebraic equivalent of this octet is an expression where three complementary variables are factored out and canceled.

$$Y = (\overline{A} + A)B(\overline{C} + C)\,(\overline{D} + D)$$

Grouping Cells Along Outside Edges

The cells along an outside edge of a three- or four-variable map are adjacent to cells along the opposite edge (only one change of variable). Thus, we can group cells "around the outside" of the map to cancel the variables. In the case of the four-variable map, we can also group the four corner cells as a quad, because they are all adjacent to one another.

Example 5.11

Use Karnaugh maps to simplify the following Boolean expressions:

a. $Y = \overline{A}\,\overline{B}\,\overline{C} + \overline{A}\,B\,C + A\,\overline{B}\,\overline{C} + A\,\overline{B}\,C$

b. $Y = \overline{A}\,\overline{B}\,\overline{C}\,\overline{D} + \overline{A}\,B\,C\,\overline{D} + A\,\overline{B}\,\overline{C}\,\overline{D} + A\,B\,C\,\overline{D}$

Solution

Figure 5.40 shows the Karnaugh maps for the Boolean expressions labeled **a** and **b**. Cells in each map are grouped in a quad.

continued...

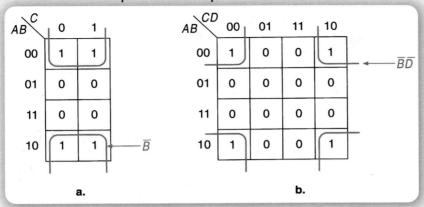

FIGURE 5.40 Example 5.11: K-Maps © CENGAGE LEARNING 2012.

a.

b.

a. *A* and *C* are both coordinates of two cells in true form and two cells in complement form. (Discard *A* and *C*.)

$\overline{B}$ is a coordinate of each cell. (Keep $\overline{B}$.)

$$Y = \overline{B}$$

b. *A* and *C* are both coordinates of two cells in true form and two cells in complement form. (Discard *A* and *C*.)

$\overline{B}$ and $\overline{D}$ are coordinates of each cell. (Keep $\overline{B}$ and $\overline{D}$.)

$$Y = \overline{B}\,\overline{D}$$

> **Note . . .**
> We don't need a Boolean expression to fill a Karnaugh map if we have the function's truth table.

Loading a K-Map from a Truth Table

Figure 5.41 and *Figure 5.42* show truth table and Karnaugh map forms for three- and four-variable Boolean expressions. The numbers in parentheses show the order of terms in binary sequence for both forms.

FIGURE 5.41 Order of Terms (Three-Variable Function)

A	B	C	Y
0	0	0	(0)
0	0	1	(1)
0	1	0	(2)
0	1	1	(3)
1	0	0	(4)
1	0	1	(5)
1	1	0	(6)
1	1	1	(7)

a. Truth table

AB \ C	0	1
00	(0)	(1)
01	(2)	(3)
11	(6)	(7)
10	(4)	(5)

b. K-map

© CENGAGE LEARNING 2012.

FIGURE 5.42 Order of Terms (Four-Variable Function)

A	B	C	D	Y
0	0	0	0	(0)
0	0	0	1	(1)
0	0	1	0	(2)
0	0	1	1	(3)
0	1	0	0	(4)
0	1	0	1	(5)
0	1	1	0	(6)
0	1	1	1	(7)
1	0	0	0	(8)
1	0	0	1	(9)
1	0	1	0	(10)
1	0	1	1	(11)
1	1	0	0	(12)
1	1	0	1	(13)
1	1	1	0	(14)
1	1	1	1	(15)

a. Truth table

AB \ CD	00	01	11	10
00	(0)	(1)	(3)	(2)
01	(4)	(5)	(7)	(6)
11	(12)	(13)	(15)	(14)
10	(8)	(9)	(11)	(10)

b. K-map

© CENGAGE LEARNING 2012.

The Karnaugh map is not laid out in the same order as the truth table. That is, it is not laid out in a binary sequence. This is due to the criterion for cell adjacency: no more than one variable change between rows or columns is permitted.

Filling in a Karnaugh map from a truth table is easy when you understand a system for doing it quickly. For the three-variable map, fill row 1, then row 2, skip to row 4, then go back to row 3. By doing this, you trace through the cells in binary order. Use the phrase "1, 2, skip, back" to help you remember this.

The system for the four-variable map is similar but also must account for the columns. The rows get filled in the same order as the three-variable map, but within each row, fill column 1, then column 2, skip to column 4, then go back to column 3. Again, "1, 2, skip, back."

The four-variable map is easier to fill from the truth table if we split the truth table into groups of four lines, as we have done in Figure 5.42. Each group is one row in the Karnaugh map. Following this system will quickly fill the cells in binary order.

Go back and follow the order of terms on the four-variable map in Figure 5.42, using this system. (Remember, for both rows and columns, "1, 2, skip, back.") Try tracing the order with your finger on the page. See how fast you can do this.

Multiple Groups

Example 5.12

Use the Karnaugh map method to simplify the Boolean function represented by *Table 5.13*.

TABLE 5.13   Truth Table for Example 5.12

A	B	C	D	Y
0	0	0	0	1
0	0	0	1	0
0	0	1	0	1
0	0	1	1	0
0	1	0	0	0
0	1	0	1	1
0	1	1	0	0
0	1	1	1	1
1	0	0	0	0
1	0	0	1	0
1	0	1	0	0
1	0	1	1	0
1	1	0	0	0
1	1	0	1	1
1	1	1	0	0
1	1	1	1	1

© CENGAGE LEARNING 2012.

Solution

Figure 5.43 shows the Karnaugh map for the truth table in Table 5.13. There are two groups of 1, s—a pair and a quad.

Pair:

Variables $\overline{A}$, $\overline{B}$, and $\overline{D}$ are coordinates of both cells. (Keep $\overline{A}$, $\overline{B}$, and $\overline{D}$.) C is a coordinate of one cell and $\overline{C}$ is a coordinate of the other. (Discard C.)

Term: $\overline{A}\,\overline{B}\,\overline{D}$

FIGURE 5.43   Example 5.12: K-Map for Table 5.13

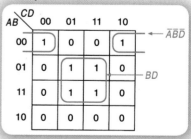

© CENGAGE LEARNING 2012.

Quad:

Both *A* and *C* are coordinates of two cells in true form and two cells in complement form. (Discard *A* and *C*.)

B and *D* are coordinates of all four cells. (Keep *B D*.)

Term: *B D*

Combine the terms in an OR function:

$$Y = \overline{A}\,\overline{B}\,\overline{D} + B\,D$$

Overlapping Groups

Example 5.13

Simplify the function represented by *Table 5.14*.

Solution

The Karnaugh map for the function in Table 5.14 is shown in *Figure 5.44*, with two different groupings of terms.

a. The simplified Boolean expression drawn from the first map has three terms.

$$Y = \overline{A}\,\overline{B} + A\,B + B\,C$$

b. The second map yields an expression with four terms.

$$Y = \overline{A}\,\overline{B} + A\,B + B\,C + \overline{A}\,C$$

One of the last two terms is redundant, as neither of the pairs corresponding to these terms has a cell belonging only to that pair. We could retain either pair of cells and its corresponding term, but not both.

We can show algebraically that the last term is redundant and thus make the expression the same as that in part **a**.

$$
\begin{aligned}
Y &= \overline{A}\,\overline{B} + A\,B + B\,C + \overline{A}\,C \\
&= \overline{A}\,\overline{B}(\overline{C} + C) + A\,B + (\overline{A} + A)B\,C + \overline{A}\,(\overline{B} + B)C \\
&= \overline{A}\,\overline{B}\,\overline{C} + \overline{A}\,\overline{B}\,C + A\,B + \overline{A}\,B\,C + A\,B\,C + \overline{A}\,\overline{B}\,C + \overline{A}\,B\,C \\
&= \overline{A}\,\overline{B}\,\overline{C} + \overline{A}\,\overline{B}\,C + \overline{A}\,\overline{B}\,C + A\,B + \overline{A}\,B\,C + \overline{A}\,B\,C + A\,B\,C \\
&= \overline{A}\,\overline{B}\,\overline{C} + \overline{A}\,\overline{B}\,C + A\,B + \overline{A}\,B\,C + A\,B\,C \\
&= \overline{A}\,\overline{B}\,(\overline{C} + C) + A\,B + (\overline{A} + A)\,B\,C \\
&= \overline{A}\,\overline{B} + A\,B + B\,C
\end{aligned}
$$

FIGURE 5.44 Example 5.13: K-Maps for Table 5.14

a. $Y = \overline{A}\overline{B} + AB + BC$ b. $Y = \overline{A}\overline{B} + AB + BC + \overline{A}C$

One of these is redundant

Only one of these terms is necessary

© CENGAGE LEARNING 2012.

TABLE 5.14 Truth Table for Example 5.13

A	B	C	Y
0	0	0	1
0	0	1	1
0	1	0	0
0	1	1	1
1	0	0	0
1	0	1	0
1	1	0	1
1	1	1	1

© CENGAGE LEARNING 2012.

Conditions for Maximum Simplification

Example 5.14

Find the maximum SOP simplification of the Boolean function represented by *Table 5.15*.

Solution

The values of Table 5.15 are loaded into the three K-maps shown in *Figure 5.45*. Three ways of grouping adjacent cells are shown. One results in maximum simplification; the other two do not.

We get the maximum SOP simplification by grouping the two octets shown in Figure 5.45a. The resulting expression is

$$Y = \overline{B} + D$$

continued...

Note . . .

A cell may be grouped more than once. The only condition is that every group must have at least one cell that does not belong to any other group. Otherwise, redundant terms will result.

TABLE 5.15 Truth Table for Example 5.14

A	B	C	D	Y
0	0	0	0	1
0	0	0	1	1
0	0	1	0	1
0	0	1	1	1
0	1	0	0	0
0	1	0	1	1
0	1	1	0	0
0	1	1	1	1
1	0	0	0	1
1	0	0	1	1
1	0	1	0	1
1	0	1	1	1
1	1	0	0	0
1	1	0	1	1
1	1	1	0	0
1	1	1	1	1

© CENGAGE LEARNING 2012.

Figure 5.45b and Figure 5.45c show two simplifications that are less than the maximum because the chosen cell groups are smaller than they could be. The resulting expressions are:

$$Y = \overline{A}\,\overline{B} + A\,\overline{B} + D$$
$$Y = \overline{B} + B\,D$$

Neither of these expressions is the simplest possible, because both can be reduced by Boolean algebra (or correctly using a K-map) to the form in Figure 5.45a.

FIGURE 5.45 Example 5.14: K-Maps for Table 5.15 © CENGAGE LEARNING 2012.

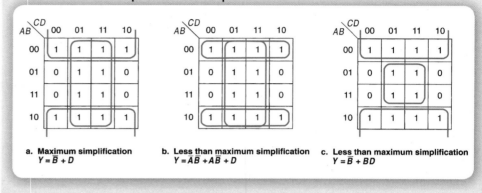

a. **Maximum simplification**
$Y = \overline{B} + D$

b. **Less than maximum simplification**
$Y = \overline{A}\overline{B} + A\overline{B} + D$

c. **Less than maximum simplification**
$Y = \overline{B} + BD$

Don't Care States

KEY TERM

Don't care state An output state that can be regarded as either HIGH or LOW, as is most convenient. A don't care state is the output state of a circuit for a combination of inputs that should never occur under stated design conditions.

Sometimes a digital circuit will be intended to work only for certain combinations of inputs; any other input values will never be applied to the circuit.

In such a case, it may be to our advantage to use so-called **don't care states** to simplify the circuit. A don't care state is shown in a K-map cell as an "X" and can be either a 0 or a 1, depending on which case will yield the maximum simplification.

A common application of the don't care state is a digital circuit designed for binary coded decimal (BCD) inputs. In BCD, a decimal digit (0–9) is encoded as a 4-bit binary number (0000–1001). This leaves six binary states that are never used (1010, 1011, 1100, 1101, 1110, 1111). In any circuit designed for BCD inputs, these states are don't care states.

BCD circuits find a useful application in human-interface devices, such as numeric keypads or digit displays. Research has found that most people prefer to use decimal numbers, rather than hexadecimal, when performing daily tasks, such as programming a cook time into a microwave oven.

In a K-map, all cells containing 1's must be grouped if we are looking for a maximum SOP simplification. (If necessary, a group can contain one cell.) The don't care states can be used to maximize the size of these groups.

Notice that each X must be treated *either* as a 0 or a 1. When we group the X's in a K-map, each circled X becomes a 1; all others must be treated as 0's. We need not group all don't care states, only those that actually contribute to a maximum simplification.

Example 5.15

The circuit in *Figure 5.46* is designed to accept binary-coded decimal inputs. The output is HIGH when the input is the BCD equivalent of 5, 7, or 9. If the BCD equivalent of the input is not 5, 7, or 9, the output is LOW. The output is not defined for input values greater than 9.

Find the maximum SOP simplification of the circuit.

Solution

The Karnaugh map for the circuit is shown in *Figure 5.47a*.

FIGURE 5.47 Example 5.15: Karnaugh Map and Logic Diagram

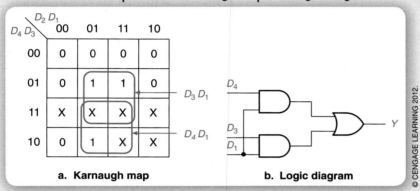

a. Karnaugh map   b. Logic diagram

© CENGAGE LEARNING 2012.

FIGURE 5.46 Example 5.15: BCD Circuit Block

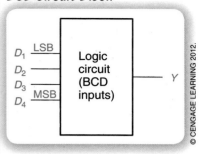

© CENGAGE LEARNING 2012.

We can designate three of the don't care cells as 1's—those corresponding to input states 1011, 1101, and 1111. This allows us to group the 1's into two overlapping quads, which yield the following simplification.

$$Y = D_4 D_1 + D_3 D_1$$

The ungrouped don't care states are treated as 0's. The corresponding circuit is shown in *Figure 5.47b*.

5.5   SIMPLIFICATION BY DEMORGAN EQUIVALENT GATES

In Example 5.2, we showed how a logic gate network could be redrawn to make it conform to the bubble-to-bubble convention. The aim was to cancel the internal inversions within the logic circuit and therefore make the Boolean expression easier to write and to understand.

For example, suppose we have a circuit like the one shown in *Figure 5.48*. We can write the Boolean expression of the circuit as follows:

$$Y = \overline{\overline{A\,B\,\overline{C}} \cdot \overline{\overline{A}\,\overline{B}\,\overline{C}}}$$

This expression is very difficult to interpret in terms of what the Boolean function is actually supposed to do. The levels of inversion bars make it difficult to simplify and to make a truth table. We can use theorems of Boolean algebra, such as DeMorgan's theorems, to simplify the expression as follows:

$$Y = \overline{\overline{A\,B\,\overline{C}}} \cdot \overline{\overline{\overline{A}\,\overline{B}\,\overline{C}}}$$
$$= \overline{A\,B\,\overline{C}} + \overline{\overline{A}\,\overline{B}\,\overline{C}}$$
$$= A\,B\,\overline{C} + \overline{A}\,\overline{B}\,C$$

FIGURE 5.48 NAND Circuit   © CENGAGE LEARNING 2012.

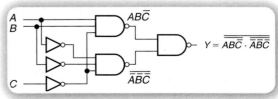

FIGURE 5.49 NAND Circuit Simplified Using a DeMorgan Equivalent Gate © CENGAGE LEARNING 2012.

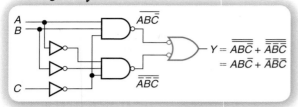

Alternatively, you could redraw the last NAND gate (the one driving output Y), in its DeMorgan equivalent form, as shown in *Figure 5.49*. The inputs of this gate are now active-LOW, as are the outputs of the two gates driving them. These two inversions automatically cancel, giving the simplified output expression without having to do the algebra.

Apply the following guidelines when making a circuit conform to the bubble-to-bubble convention:

1. Start at the output and work back. There are always two possible choices for the output gate: a positive- or negative-logic form. One of these will be AND-shaped and one will be OR-shaped. Choose the OR-shaped gate for an SOP result and the AND-shaped gate for a POS result. If the active level of the output is important, choose the desired level, converting the output gate if necessary.
2. Go back toward the circuit inputs to the next level of gating. Match the output levels of these gates to the input levels of the gate described in Step 1, converting the next level gates to their DeMorgan equivalent if necessary.
3. Repeat Step 2 until you reach the circuit inputs.

There are always two different correct results for this procedure, one with an active-HIGH output and one with an active-LOW.

Example 5.16

a. Write the Boolean expression for the logic diagram shown in *Figure 5.50*.
b. Redraw the circuit in Figure 5.50 to make it conform to the bubble-to-bubble convention. Show two different solutions.
c. Write the Boolean expressions of the redrawn circuits.

FIGURE 5.50 Example 5.16: Original Circuit

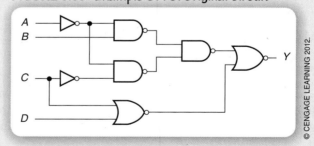

© CENGAGE LEARNING 2012.

Solution

a. The Boolean expression for the logic diagram in Figure 5.50 is:

$$Y = \overline{\overline{\overline{A\,B} \cdot \overline{A\,C}} + \overline{C + D}}$$

b. *Figure 5.51* shows the logic diagram with the output gate converted to its DeMorgan equivalent form and the remaining gates changed to conform to the bubble-to-bubble convention. The changed gates are shown in color.

Figure 5.52 shows the second solution. The output gate is left in its original form and the two gates at the next level, shown in color, are changed
continued...

FIGURE 5.51 Example 5.16: DeMorgan Equivalent Circuit

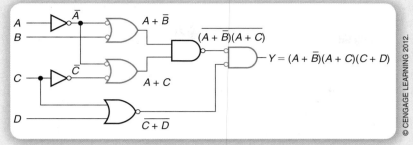

FIGURE 5.52 Example 5.16: DeMorgan Equivalent Circuit

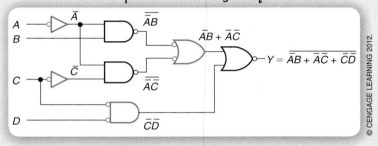

to suit. The two inverters are also redrawn. Even though their function does not change, it indicates that the inputs driving the inverters are active-LOW.

c. The Boolean expressions for the redrawn circuits are:

Figure 5.51: $Y = (A + \overline{B})(A + C)(C + D)$

Figure 5.52: $Y = \overline{\overline{\overline{A} B} + \overline{\overline{A} \overline{C}} + \overline{\overline{C} \overline{D}}}$ OR $\overline{Y} = \overline{A} B + \overline{A} \overline{C} + \overline{C} \overline{D}$

Your Turn

5.5 Redraw the logic diagram in *Figure 5.53* to make it conform to the bubble-to-bubble convention. Write the Boolean expression of the redrawn network.

FIGURE 5.53 Your Turn OR in Your Turn Problem 5.5 (see Fig 4.20) Unsimplified Logic Circuit

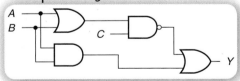

5.6 A GENERAL APPROACH TO LOGIC CIRCUIT DESIGN

In previous sections of this chapter, we have seen several example design problems, in which we start with a verbal description of a problem and translate it into a combinational logic circuit. Each of these examples has been used to explain a problem under study at the time, but in doing so, we have not followed any consistent method of design. In this section, we will show how to design a logic circuit in a systematic way.

When designing a combinational logic circuit, ultimately we are looking for a Boolean equation. Any Boolean combinational function can be written using

AND, OR, and NOT functions, which can be arranged in sum-of-products or product-of-sums form, as derived from a truth table and simplified using Boolean algebra or a Karnaugh map. This is all just mechanics; the tricky bit is knowing how to translate the words of a design idea into that required Boolean equation.

The first step in the design process is to have *an accurate description of the problem*. Generally, this consists of understanding the effect of all inputs, singly and in combination, on the circuit output or outputs. These should be listed systematically and examined to make sure that all possible input combinations have been accounted for. Active levels of inputs and outputs should be properly specified, as should any special constraints on inputs (e.g., binary-coded decimal values only).

Once the problem has been stated, *each output of the circuit should be described, either verbally or with a truth table*. If using a verbal description, look for keywords, such as AND, OR, NOT that can be translated into a Boolean expression. If using a truth table, enter the requirements for each circuit output, as it relates to the circuit inputs.

Using Boolean algebra or Karnaugh maps, *simplify the Boolean equations* of the circuit.

Example 5.17

A sealed tank in a chemical factory, shown in *Figure 5.54*, is used in an industrial process with specific constraints on the volume of liquid in the tank, its pressure, and its temperature. A level sensor determines if the volume of liquid in the tank is too large. Pressure is monitored by a pressure sensor and temperature by a temperature sensor. If any of these parameters is exceeded, the corresponding sensor generates a logic HIGH.

The three sensors are monitored by a safety system, shown in *Figure 5.55*, which controls the behavior of three active-HIGH indicator lights. A green light, designated "OK," indicates that the system is functioning normally. This is defined as the condition where no more than one sensor indicates an out-of-range condition. A yellow light, called the "ALERT" light, illuminates when two or more sensors indicate an out-of-range condition. The red light, designated as the "DANGER" condition, comes on when all sensors are out of range. (Note that in the final condition, both the yellow and red lights are on.)

FIGURE 5.54 **Example 5.17: Level, Temperature, and Pressure Sensors in a Tank**

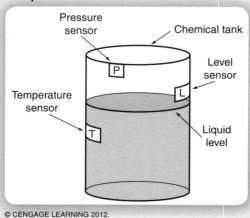

© CENGAGE LEARNING 2012.

FIGURE 5.55 **Example 5.17: Active-HIGH Controller for Status Lamps**

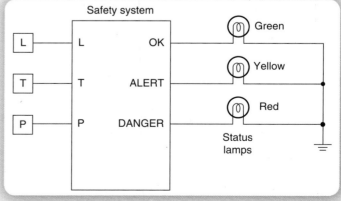

© CENGAGE LEARNING 2012.

continued...

Design a circuit for the system in Figure 5.55 that will fulfill the requirements for the three indicator lights.

Solution

Let's summarize the input and output requirements of the circuit:

1. There are three active-HIGH sensors for out-of-range parameters: level (L), temperature (T), and pressure (P).
2. There are three active-HIGH status lamps: green (OK), yellow (ALERT), and red (DANGER).
3. *Table 5.16* shows the relationship between lamps and sensors.

Because both sensors and lamps are active-HIGH, an ON means a logic 1 in a truth table. *Table 5.17* shows a truth table for this circuit. At this point, we are ready to use the mechanics of Boolean simplification to draw the logic diagram of the indicator lamp control circuit.

Figure 5.56 shows Karnaugh maps for the OK and ALERT expressions. The DANGER function can be derived directly from the truth table.

$$OK = \overline{L}\,\overline{T} + \overline{L}\,\overline{P} + \overline{T}\,\overline{P}$$
$$ALERT = LT + LP + TP$$
$$DANGER = LTP$$

These expressions can be read as follows:

OK: The active-HIGH green lamp is ON when any two sensors are OFF.

ALERT: The active-HIGH yellow lamp is ON when any two sensors are ON.

DANGER: The active-HIGH red lamp is ON when all three sensors are ON.

Note that the outputs for the OK and ALERT lights are exactly opposite from one another. Therefore, an alternative implementation for the OK light would be:

$$OK = \overline{LT + LP + TP}$$

However, we will stick to the original equation, as its output is active-HIGH and thus more accurately represents the behavior of the green status lamp. The completed circuit is shown in *Figure 5.57*.

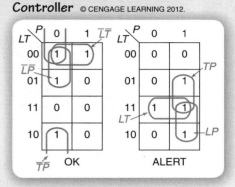

FIGURE 5.56 Example 5.17: K-Maps for Active-HIGH Lamp Controller © CENGAGE LEARNING 2012.

TABLE 5.16 Lamps and Sensors for Figure 5.55

State	Number of Sensors ON	Green Lamp	Yellow Lamp	Red Lamp
OK	None	ON	OFF	OFF
OK	One	ON	OFF	OFF
ALERT	Two	OFF	ON	OFF
DANGER	Three	OFF	ON	ON

© CENGAGE LEARNING 2012.

TABLE 5.17 Truth Table for the Circuit in Figure 5.55

L	T	P	OK	ALERT	DANGER
0	0	0	1	0	0
0	0	1	1	0	0
0	1	0	1	0	0
0	1	1	0	1	0
1	0	0	1	0	0
1	0	1	0	1	0
1	1	0	0	1	0
1	1	1	0	1	1

© CENGAGE LEARNING 2012.

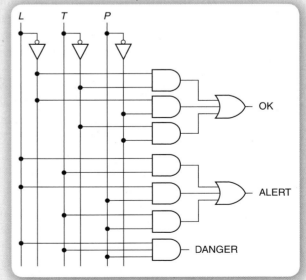

FIGURE 5.57 Example 5.17: Logic Diagram for Active-HIGH Lamp Controller

© CENGAGE LEARNING 2012.

SUMMARY

1. Two or more gates connected together form a logic gate network or combinational logic circuit, which can be described by a truth table, a logic diagram, or a Boolean expression.

2. The output of a combinational logic circuit is always the same with the same combination of inputs, regardless of the order in which they are applied.

3. The order of precedence in a logic gate network is AND, then OR, unless otherwise indicated by parentheses.

4. DeMorgan's theorems:

$$\overline{x \cdot y} = \overline{x} + \overline{y}$$
$$\overline{x + y} = \overline{x} \cdot \overline{y}$$

5. Inequalities:

$$\overline{x \cdot y} \neq \overline{x} \cdot \overline{y}$$
$$\overline{x + y} \neq \overline{x} + \overline{y}$$

6. A logic gate network can be drawn to simplify its Boolean expression by ensuring that bubbled (active-LOW) outputs drive bubbled inputs and outputs with no bubble (active-HIGH) drive inputs with no bubble. Some gates might need to be drawn in their DeMorgan equivalent form to achieve this.

7. In Boolean expressions, logic inversion bars of equal lengths cancel; bars of unequal lengths do not. Bars of equal lengths represent bubble-to-bubble connections.

   (e.g., $\overline{\overline{x + y + z}} = x + y + z$,
   but $\overline{\overline{x} + \overline{y} + \overline{z}} \neq x + y + z$)

8. A logic diagram can be derived from a Boolean expression by order of precedence rules: synthesize ANDs before ORs, unless parentheses indicate otherwise. Inversion bars act as parentheses for a group of variables.

9. A truth table can be derived from a logic gate network either by finding truth tables for intermediate points in the network and combining them by the laws of Boolean algebra, or by simplifying the Boolean expression into a form that can be directly written into a truth table.

10. A sum-of-products (SOP) network combines inputs in AND gates to yield a group of product terms that are combined in an OR gate (logical sum) output.

11. A product-of-sums (POS) network combines inputs in OR gates to yield a group of sum terms that are combined in an AND gate (logical product) output.

12. An SOP Boolean expression can be derived from the lines in a truth table where the output is at logic 1. Each product term contains all inputs in true or complement form, where inputs at logic 0 have a bar and inputs at logic 1 do not.

13. Rules similar to those of algebra are helpful in simplifying Boolean expressions.

14. SOP networks can be simplified by grouping pairs of product terms and applying the Boolean identity $x\,y\overline{z} + xyz = xy(\overline{z} + z) = xy$.

15. To achieve maximum simplification of an SOP or POS network, each product or sum term should be grouped with another if possible. A product or sum term can be grouped more than once, as long as each group has a term that is only in that group.

16. A Karnaugh map can be used to graphically reduce a Boolean expression to its simplest form by grouping adjacent cells containing 1's. One cell is equivalent to one line of a truth table. A group of adjacent cells that contain 1's represents a simplified product term.

17. Adjacent cells in a K-map differ by only one variable. Cells around the outside of the map

are considered adjacent. For example, a cell in the top row of the map is adjacent to a cell in the bottom row, provided they are in the same column. A cell in the far-left column of the map is adjacent to a cell in the far-right column, provided they are in the same row. The four corners of the map are adjacent to one another, as there is a single difference of input variable in any vertical or horizontal direction.

18. A group in a K-map must be a power of 2 in size: 1, 2, 4, 8, or 16. A group of two is called a pair, a group of four is a quad, and a group of eight is an octet.

19. A pair cancels one variable. A quad cancels two variables. An octet cancels three variables.

20. A K-map can have multiple groups. Each group represents one simplified product term in a sum-of-products expression.

21. Groups in K-maps can overlap as long as each group has one or more cells that appear only in that group.

22. Groups in a K-map should be as large as possible for maximum SOP simplification.

23. Don't care states represent output states of input combinations that should never occur in a circuit. They are represented by X's in a truth table or K-map and can be used as 0's or 1's, whichever is more advantageous for the simplification of the circuit.

24. When using DeMorgan equivalent gates to simplify a logic circuit, redraw the logic diagram by starting at the output and working back to the inputs. Use DeMorgan's theorems to modify gates as required to ensure that connections conform to the bubble-to-bubble convention.

25. Combinational logic circuit design largely consists of translating word problems to Boolean equations, then applying tools of simplification and synthesis to make the best circuit.

26. Boolean equations required by a combinational logic design process can be derived from verbal descriptions by creating an accurate description of the problem that analyzes the effects of each of the inputs on each circuit output. These descriptions can then be translated into Boolean expressions either directly or by creating a truth table.

BRING IT HOME

5.1 Boolean Expressions, Logic Diagrams, and Truth Tables

5.1 Write the unsimplified Boolean expression for each of the logic gate networks shown in *Figure 5.58*.

5.2 The circuit in *Figure 5.59* is called a majority vote circuit. It will turn on an active-HIGH indicator lamp only if a majority of inputs (at least two out of three) are HIGH. Write the Boolean expression for the circuit.

FIGURE 5.58 *Problem 5.1: Logic Circuits*

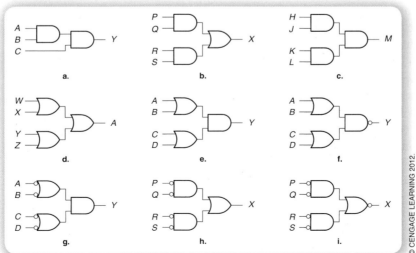

© CENGAGE LEARNING 2012.

continues...

continued...

FIGURE 5.59 Problem 5.2: Majority Vote Circuit

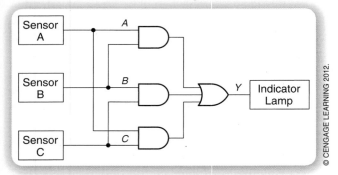

© CENGAGE LEARNING 2012.

5.3 Suppose you wish to design a circuit that indicates when at least three out of four inputs are HIGH. The circuit has four inputs, A, B, C, and D and an active-HIGH output, Y. Write the Boolean expression for the circuit and draw the logic circuit.

5.4 Draw the logic circuit for each of the following Boolean expressions:
- **a.** $Y = AB + BC$
- **b.** $Y = ACD + BCD$
- **c.** $Y = (A + B)(C + D)$
- **d.** $Y = A + BC + D$

5.5 Write the truth tables for the logic diagrams in *Figure 5.58*, parts **b, e, f,** and **g**.

5.2 Sum-of-Products and Product-of-Sums Forms

5.6 Multisim Problem
Multisim File: 05.01 SOP Circuit 1.ms11
Open the Multisim File for this problem and save it as **05.02 SOP Circuit 2.ms11**. The three inputs (A, B, and C) are monitored with logic probes, as are the AND gate outputs (G1, G2, and G3) and the circuit output (Y). Modify the circuit so that G1 goes ON when $ABC = 001$, G2 goes on when $ABC = 100$, and G3 goes ON when $ABC = 110$.

Operate the switches A, B, and C to go through all possible binary input combinations.
- **a.** How does the state of the logic probe for output Y correspond to the states of G1, G2, and G3?
- **b.** Write the Boolean expression for the SOP circuit created in this problem.

5.7 Find the Boolean expression, in both sum-of-products (SOP) and product-of-sums (POS) forms, for the logic function represented by the following truth table. Draw the logic diagram for each form.

A	B	C	Y
0	0	0	1
0	0	1	1
0	1	0	1
0	1	1	1
1	0	0	0
1	0	1	0
1	1	0	0
1	1	1	0

© CENGAGE LEARNING 2012.

5.8 Find the Boolean expression, in both sum-of-products (SOP) and product-of-sums (POS) forms, for the logic function represented by the following truth table. Draw the logic diagram for the SOP form only.

A	B	C	Y
0	0	0	0
0	0	1	1
0	1	0	0
0	1	1	0
1	0	0	1
1	0	1	0
1	1	0	1
1	1	1	0

© CENGAGE LEARNING 2012.

5.9 Find the Boolean expression, in both sum-of-products (SOP) and product-of-sums (POS) forms, for the logic function represented by the following truth table. Draw the logic diagram for the POS form only.

A	B	C	Y
0	0	0	0
0	0	1	1
0	1	0	1
0	1	1	0
1	0	0	0
1	0	1	1
1	1	0	1
1	1	1	1

© CENGAGE LEARNING 2012.

5.10 Find the Boolean expression, in both sum-of-products (SOP) and product-of-sums (POS) forms, for the logic function represented by the following truth table. Draw the logic diagram for the SOP form only.

A	B	C	D	Y
0	0	0	0	0
0	0	0	1	0
0	0	1	0	0
0	0	1	1	0
0	1	0	0	1
0	1	0	1	0
0	1	1	0	1
0	1	1	1	0
1	0	0	0	1
1	0	0	1	0
1	0	1	0	0
1	0	1	1	0
1	1	0	0	0
1	1	0	1	1
1	1	1	0	0
1	1	1	1	0

5.11 Use the truth table of a 2-input XOR gate to write its Boolean expression in POS form. Draw the logic diagram of the POS form of the XOR function.

5.12 Use the truth table of a 2-input XNOR gate to write its Boolean expression in POS form. Draw the logic diagram of the POS form of the XNOR function.

5.3 Simplifying SOP Expressions

5.13 Use the rules of Boolean algebra to find the maximum SOP simplification of the function represented by the following truth table.

A	B	C	Y
0	0	0	0
0	0	1	1
0	1	0	0
0	1	1	1
1	0	0	0
1	0	1	1
1	1	0	0
1	1	1	1

5.14 Use the rules of Boolean algebra to find the maximum SOP simplification of the function represented by the following truth table.

A	B	C	Y
0	0	0	1
0	0	1	0
0	1	0	1
0	1	1	0
1	0	0	0
1	0	1	0
1	1	0	1
1	1	1	0

5.15 Use the rules of Boolean algebra to find the maximum SOP simplification of the function represented by the following truth table.

A	B	C	Y
0	0	0	0
0	0	1	1
0	1	0	0
0	1	1	1
1	0	0	0
1	0	1	1
1	1	0	0
1	1	1	0

5.16 Use the rules of Boolean algebra to find the maximum SOP simplification of the function represented by the following truth table.

A	B	C	D	Y
0	0	0	0	0
0	0	0	1	0
0	0	1	0	0
0	0	1	1	0
0	1	0	0	1
0	1	0	1	1
0	1	1	0	0
0	1	1	1	0
1	0	0	0	0
1	0	0	1	1
1	0	1	0	0
1	0	1	1	1
1	1	0	0	1
1	1	0	1	1
1	1	1	0	0
1	1	1	1	1

continues...

continued...

5.17 Use the rules of Boolean algebra to find the maximum SOP simplification of the function represented by the following truth table.

A	B	C	D	Y
0	0	0	0	1
0	0	0	1	0
0	0	1	0	1
0	0	1	1	0
0	1	0	0	1
0	1	0	1	0
0	1	1	0	0
0	1	1	1	0
1	0	0	0	0
1	0	0	1	0
1	0	1	0	0
1	0	1	1	0
1	1	0	0	1
1	1	0	1	1
1	1	1	0	0
1	1	1	1	0

© CENGAGE LEARNING 2012.

5.18 Multisim Problem
Multisim File: 05.06 SOP Circuit 5a.ms11
Figure 5.60 shows an SOP circuit drawn in Multisim.

a. Run the circuit simulation and determine the output truth table of the circuit.
b. From the truth table, find the unsimplified Boolean expression.
c. Simplify the SOP expression as much as possible.
d. Modify the circuit in Figure 5.60 to show the circuit for the simplified Boolean expression.

5.4 Simplification by the Karnaugh Map Method

5.19 Use the Karnaugh map method to find the maximum SOP simplification of the logic diagram in Figure 5.23.

5.20 Use the Karnaugh map method to reduce the following Boolean expressions to their maximum SOP simplifications:

a. $Y = \overline{A}\,\overline{B}\,C + \overline{A}\,B\,C + A\,B\,C$
b. $Y = \overline{A}\,\overline{B}\,C + \overline{A}\,B\,C + A\,\overline{B}\,C + A\,B\,\overline{C} + A\,B\,C$

FIGURE 5.60 **Problem 5.18: Multisim SOP Circuit** © CENGAGE LEARNING 2012.

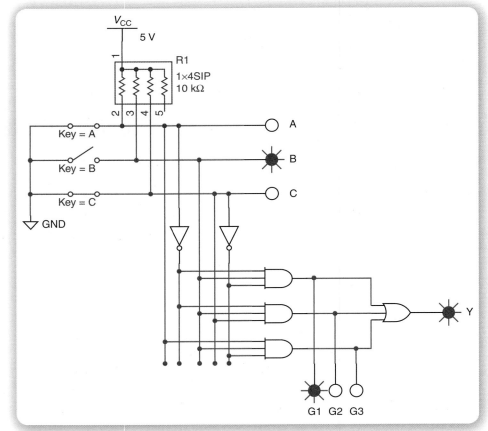

5.21 Use the Karnaugh map method to reduce the Boolean expression represented by the following truth table to its simplest SOP form.

A	B	C	D	Y
0	0	0	0	0
0	0	0	1	0
0	0	1	0	0
0	0	1	1	1
0	1	0	0	1
0	1	0	1	1
0	1	1	0	1
0	1	1	1	1
1	0	0	0	0
1	0	0	1	0
1	0	1	0	0
1	0	1	1	1
1	1	0	0	0
1	1	0	1	0
1	1	1	0	0
1	1	1	1	1

© CENGAGE LEARNING 2012.

5.23 Use the Karnaugh map method to reduce the Boolean expression represented by the following truth table to its simplest SOP form.

A	B	C	D	Y
0	0	0	0	0
0	0	0	1	0
0	0	1	0	1
0	0	1	1	1
0	1	0	0	0
0	1	0	1	0
0	1	1	0	0
0	1	1	1	0
1	0	0	0	0
1	0	0	1	1
1	0	1	0	X
1	0	1	1	X
1	1	0	0	X
1	1	0	1	X
1	1	1	0	X
1	1	1	1	X

© CENGAGE LEARNING 2012.

5.22 Use the Karnaugh map method to reduce the Boolean expression represented by the following truth table to its simplest SOP form.

A	B	C	D	Y
0	0	0	0	0
0	0	0	1	1
0	0	1	0	0
0	0	1	1	1
0	1	0	0	0
0	1	0	1	1
0	1	1	0	0
0	1	1	1	1
1	0	0	0	1
1	0	0	1	0
1	0	1	0	0
1	0	1	1	0
1	1	0	0	0
1	1	0	1	1
1	1	1	0	0
1	1	1	1	1

© CENGAGE LEARNING 2012.

5.24 Use the Karnaugh map method to reduce the Boolean expression represented by the following truth table to its simplest SOP form.

A	B	C	D	Y
0	0	0	0	0
0	0	0	1	1
0	0	1	0	0
0	0	1	1	1
0	1	0	0	0
0	1	0	1	1
0	1	1	0	0
0	1	1	1	1
1	0	0	0	1
1	0	0	1	0
1	0	1	0	0
1	0	1	1	1
1	1	0	0	0
1	1	0	1	0
1	1	1	0	0
1	1	1	1	1

© CENGAGE LEARNING 2012.

continues...

continued...

5.25 Use the Karnaugh map method to reduce the Boolean expression represented by the following truth table to its simplest SOP form.

A	B	C	D	Y
0	0	0	0	1
0	0	0	1	1
0	0	1	0	0
0	0	1	1	0
0	1	0	0	1
0	1	0	1	1
0	1	1	0	0
0	1	1	1	1
1	0	0	0	0
1	0	0	1	0
1	0	1	0	1
1	0	1	1	1
1	1	0	0	0
1	1	0	1	1
1	1	1	0	1
1	1	1	1	1

© CENGAGE LEARNING 2012.

5.26 Use the Karnaugh map method to reduce the Boolean expression represented by the following truth table to its simplest SOP form.

A	B	C	D	Y
0	0	0	0	0
0	0	0	1	1
0	0	1	0	0
0	0	1	1	1
0	1	0	0	0
0	1	0	1	1
0	1	1	0	0
0	1	1	1	1
1	0	0	0	0
1	0	0	1	1
1	0	1	0	0
1	0	1	1	1
1	1	0	0	0
1	1	0	1	1
1	1	1	0	0
1	1	1	1	1

© CENGAGE LEARNING 2012.

5.5 Simplification by DeMorgan Equivalent Gates

5.27 a. Write the unsimplified Boolean expression for the logic circuit shown in *Figure 5.61*.

FIGURE 5.61 Problem 5.27: Logic Circuit

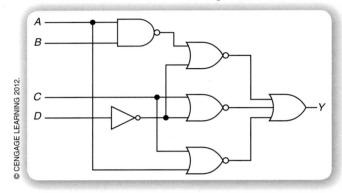

© CENGAGE LEARNING 2012.

b. Redraw the circuit in Figure 5.61 to conform to the bubble-to-bubble convention. Write the Boolean expression of the redrawn circuit.

c. Redraw the circuit in Figure 5.61 so that it conforms to the bubble-to-bubble convention, but is drawn differently than the circuit in part **b**. Write the Boolean expression of the second redrawn circuit.

5.28 a. Write the unsimplified Boolean expression for the logic circuit shown in *Figure 5.62*.

FIGURE 5.62 Problem 5.28: Logic Circuit

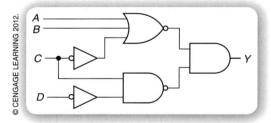

© CENGAGE LEARNING 2012.

b. Redraw the circuit in Figure 5.62 to conform to the bubble-to-bubble convention. Write the Boolean expression of the redrawn circuit.

c. Redraw the circuit in Figure 5.62 so that it conforms to the bubble-to-bubble convention, but is drawn differently than the circuit in part **b**. Write the Boolean expression of the second redrawn circuit.

5.6 A General Approach to Logic Circuit Design

5.29 Multisim Problem

Multisim File: 05.10 3-bit Bar Graph.ms11

A digital circuit is required to display a digital signal on a bar-graph LED, similar to a sound meter on a stereo. You know that the input to your circuit has values from 0 to 3 coded in binary format. You need to indicate the value on a bar graph where progressively more LEDs turn on as the input number gets higher, as shown in *Figure 5.63*. When the highest number is reached, all LEDs are on. A Multisim file for the required circuit is shown in *Figure 5.64*. The bar-graph LED shown has active-HIGH LEDs.

FIGURE 5.63 Problem 5.29: LED Bar-Graph Indicator

© CENGAGE LEARNING 2012.

a. Open the Multisim file for this problem and go through all the combinations of A and B. Write the truth table for the circuit required to drive the LEDs.

b. Write and simplify the Boolean equations for the three outputs to show that the circuit design is correct. (Note that Y_2 is not directly connected to input A. The electrical properties of the bar-graph component would cause A to be stuck LOW if this connection were to be made. In order to prevent this, we must insert a **buffer**, found in the **Miscellaneous Digital** component group, between Y_2 and A. The buffer has no logic function, but provides electrical isolation between A and Y_2.)

5.30 Multisim Problem

Multisim File: 05.11 7-bit Bar Graph.ms11

Repeat Problem 5.29 for a 7-position bar-graph LED, as shown in *Figure 5.65*. The Multisim version of the circuit is shown in *Figure 5.66*. Use one 3-variable Karnaugh map per output to simplify the seven Boolean equations for the circuit.

FIGURE 5.64 Problem 5.29: Multisim Circuit for a 3-Bit Bar-Graph Decoder

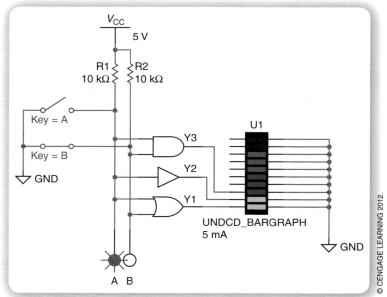

© CENGAGE LEARNING 2012.

FIGURE 5.65 Problem 5.30: LED Bar-Graph Indicator

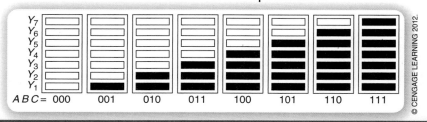

© CENGAGE LEARNING 2012.

continues...

continued...

FIGURE 5.66 Problem 5.30: Multisim Circuit for a 7-Bit Bar-Graph Decoder

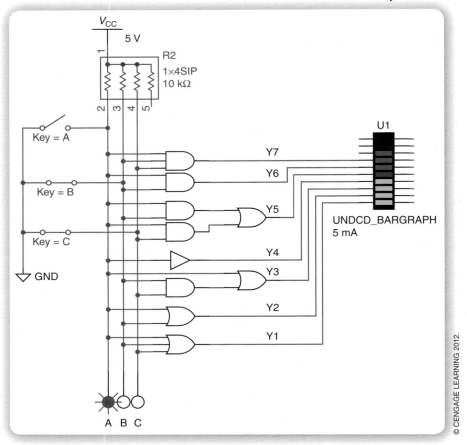

EXTRA MILE

5.1 Boolean Expressions, Logic Diagrams, and Truth Tables

5.31 Write the unsimplified Boolean expression for each of the logic gate networks shown in *Figure 5.67*.

5.32 Redraw the logic diagrams of the gate networks shown in Figure 5.67a, e, f, h, i, and j so that they conform to the bubble-to-bubble convention. Rewrite the Boolean expression of each of the redrawn circuits.

5.33 a. Write the unsimplified Boolean equation for each of the logic diagrams in *Figure 5.68*.

   b. Redraw each of the logic diagrams in Figure 5.68 so that they conform to the bubble-to-bubble convention. Rewrite the Boolean expression for each of the redrawn circuits.

5.34 a. Write the unsimplified Boolean equation for the logic diagram in *Figure 5.69*.

   b. Redraw the logic diagram in Figure 5.69 so that it conforms to the bubble-to-bubble convention. Rewrite the Boolean expression of the redrawn circuit.

5.35 Draw the logic circuit for each of the following Boolean expressions:

   a. $Y = \overline{AC} + \overline{B} + C$
   b. $Y = \overline{\overline{AC} + B} + C$
   c. $Y = \overline{\overline{ABD} + \overline{BC}} + \overline{A} + C$
   d. $Y = \overline{\overline{AB} + \overline{AC}} + \overline{BC}$
   e. $Y = \overline{\overline{AB} + \overline{AC}} + \overline{BC}$

FIGURE 5.67 Problem 5.31: Logic Diagrams

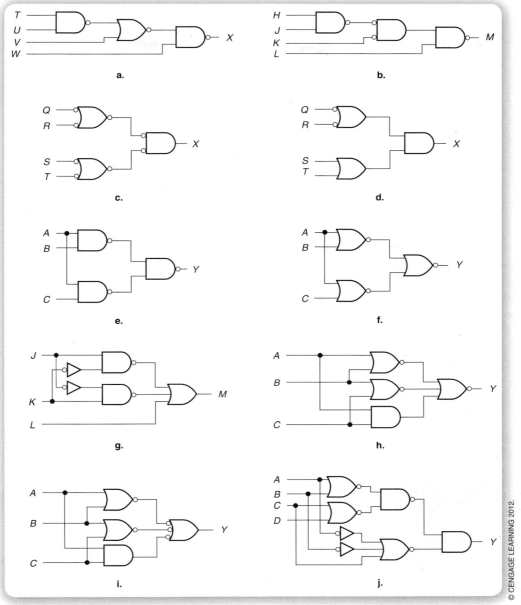

a.

b.

c.

d.

e.

f.

g.

h.

i.

j.

FIGURE 5.68 Problem 5.33: Logic Diagrams

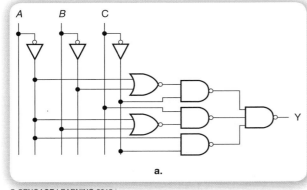

a.

b.

continues...

continued...

FIGURE 5.69 Problem 5.34: Logic Diagrams

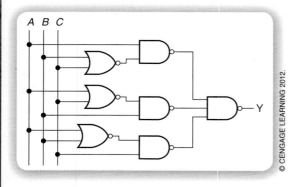

5.36 Use DeMorgan's theorems to modify the Boolean equations in Problem 5.35 so that there is no bar over any group of variables. Redraw the logic diagrams of the circuits to reflect the changes. (The final circuit versions should conform to the bubble-to-bubble convention.)

5.37 Write the truth tables for the logic diagrams in Figure 5.67, parts a, h, i, and j.

5.38 Write the truth tables for the Boolean expressions in Problem 5.35.

5.2 Sum-of-Products and Product-of-Sums Forms

5.39 Multisim Problem

Multisim File: 05.01 SOP Circuit 1.ms11

Open the Multisim file for this problem and save it as **05.03 SOP Circuit 3.ms11**. Modify the circuit so that it has four inputs, A, B, C, and D. Make the SOP circuit with five 4-input AND gates (found in the Miscellaneous Digital component group in Multisim) and label them G1, G2, G3, G4, and G5. Connect them so that they behave as follows:

> G1 is HIGH when ABCD = 0001
> G2 is HIGH when ABCD = 0010
> G3 is HIGH when ABCD = 1000
> G4 is HIGH when ABCD = 1011
> G5 is HIGH when ABCD = 1101

Connect the outputs of the AND gates to the inputs of a 5-input OR gate.

Operate the switches A, B, C, and D to go through all possible binary input combinations.

a. How does the state of the logic probe for output Y correspond to the states of G1, G2, G3, G4, and G5?

b. Write the Boolean expression for the SOP circuit created in this problem.

5.3 Simplifying SOP Expressions

5.40 Use the rules of Boolean algebra to find the maximum SOP simplification of the function represented by the following truth table.

A	B	C	D	Y
0	0	0	0	0
0	0	0	1	1
0	0	1	0	0
0	0	1	1	1
0	1	0	0	0
0	1	0	1	1
0	1	1	0	1
0	1	1	1	1
1	0	0	0	0
1	0	0	1	1
1	0	1	0	0
1	0	1	1	0
1	1	0	0	0
1	1	0	1	1
1	1	1	0	1
1	1	1	1	0

5.41 Use the rules of Boolean algebra to find the maximum SOP simplification of the function represented by the following truth table.

A	B	C	D	Y
0	0	0	0	1
0	0	0	1	0
0	0	1	0	0
0	0	1	1	1
0	1	0	0	1
0	1	0	1	0
0	1	1	0	0
0	1	1	1	0
1	0	0	0	1
1	0	0	1	0
1	0	1	0	1
1	0	1	1	1
1	1	0	0	1
1	1	0	1	0
1	1	1	0	1
1	1	1	1	1

5.42 Use the rules of Boolean algebra to find the maximum SOP simplification of the function represented by the following truth table.

A	B	C	D	Y
0	0	0	0	0
0	0	0	1	1
0	0	1	0	0
0	0	1	1	0
0	1	0	0	0
0	1	0	1	0
0	1	1	0	1
0	1	1	1	1
1	0	0	0	1
1	0	0	1	0
1	0	1	0	0
1	0	1	1	0
1	1	0	0	0
1	1	0	1	0
1	1	1	0	1
1	1	1	1	1

5.43 Use the rules of Boolean algebra to find the maximum SOP simplification of the function represented by the following truth table.

A	B	C	D	Y
0	0	0	0	0
0	0	0	1	0
0	0	1	0	0
0	0	1	1	1
0	1	0	0	0
0	1	0	1	1
0	1	1	0	0
0	1	1	1	1
1	0	0	0	0
1	0	0	1	0
1	0	1	0	0
1	0	1	1	1
1	1	0	0	0
1	1	0	1	1
1	1	1	0	0
1	1	1	1	1

5.44 Multisim Problem
Multisim File: 05.08 SOP Circuit 6a.ms11
Figure 5.70 shows an SOP circuit drawn in Multisim.

a. Run the circuit simulation and determine the output truth table of the circuit.
b. From the truth table, find the unsimplified Boolean expression.
c. Simplify the SOP expression as much as possible.
d. Modify the circuit in Figure 5.70 to show the circuit for the simplified Boolean expression.

FIGURE 5.70 Problem 5.44: Multisim Circuit © CENGAGE LEARNING 2012.

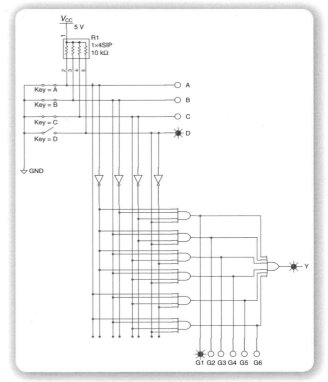

5.4 Simplification by the Karnaugh Map Method

5.45 Use the Karnaugh map method to reduce the following Boolean expression to its maximum SOP simplification:

a. $Y = \overline{A}\,\overline{B}\,\overline{C} + \overline{A}\,B\,C + A\,\overline{B}\,C + A\,B\,C$

continues...

continued...

5.46 Use the Karnaugh map method to reduce the Boolean expression represented by the following truth table to its simplest SOP form.

A	B	C	D	Y
0	0	0	0	1
0	0	0	1	1
0	0	1	0	1
0	0	1	1	1
0	1	0	0	0
0	1	0	1	0
0	1	1	0	1
0	1	1	1	0
1	0	0	0	1
1	0	0	1	0
1	0	1	0	1
1	0	1	1	0
1	1	0	0	0
1	1	0	1	1
1	1	1	0	1
1	1	1	1	0

5.48 Use the Karnaugh map method to reduce the Boolean expression represented by the following truth table to its simplest SOP form.

A	B	C	D	Y
0	0	0	0	1
0	0	0	1	0
0	0	1	0	1
0	0	1	1	1
0	1	0	0	0
0	1	0	1	0
0	1	1	0	0
0	1	1	1	1
1	0	0	0	0
1	0	0	1	0
1	0	1	0	1
1	0	1	1	1
1	1	0	0	0
1	1	0	1	0
1	1	1	0	0
1	1	1	1	1

5.47 Use the Karnaugh map method to reduce the Boolean expression represented by the following truth table to its simplest SOP form.

A	B	C	D	Y
0	0	0	0	0
0	0	0	1	0
0	0	1	0	0
0	0	1	1	0
0	1	0	0	1
0	1	0	1	1
0	1	1	0	1
0	1	1	1	1
1	0	0	0	0
1	0	0	1	1
1	0	1	0	0
1	0	1	1	0
1	1	0	0	1
1	1	0	1	0
1	1	1	0	1
1	1	1	1	0

5.49 Use the Karnaugh map method to reduce the Boolean expression represented by the following truth table to its simplest SOP form.

A	B	C	D	Y
0	0	0	0	1
0	0	0	1	1
0	0	1	0	1
0	0	1	1	1
0	1	0	0	0
0	1	0	1	0
0	1	1	0	1
0	1	1	1	1
1	0	0	0	1
1	0	0	1	1
1	0	1	0	1
1	0	1	1	1
1	1	0	0	0
1	1	0	1	0
1	1	1	0	0
1	1	1	1	0

5.50 Use the Karnaugh map method to reduce the Boolean expression represented by the following truth table to its simplest SOP form.

A	B	C	D	Y
0	0	0	0	0
0	0	0	1	0
0	0	1	0	0
0	0	1	1	0
0	1	0	0	0
0	1	0	1	0
0	1	1	0	0
0	1	1	1	1
1	0	0	0	1
1	0	0	1	1
1	0	1	0	1
1	0	1	1	1
1	1	0	0	1
1	1	0	1	1
1	1	1	0	0
1	1	1	1	1

© CENGAGE LEARNING 2012.

5.5 Simplification by DeMorgan Equivalent Gates

5.51 a. Write the unsimplified Boolean expression for the logic circuit shown in *Figure 5.71*.

FIGURE 5.71 Problem 5.51: Logic Diagram

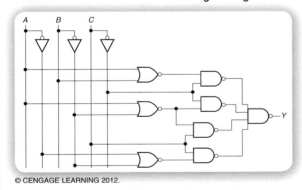

© CENGAGE LEARNING 2012.

b. Redraw the circuit in Figure 5.71 to conform to the bubble-to-bubble convention. Write the Boolean expression of the redrawn circuit.

c. Redraw the circuit in Figure 5.71 so that it conforms to the bubble-to-bubble convention, but is drawn differently than the circuit in part **b**. Write the Boolean expression of the second redrawn circuit.

5.6 A General Approach to Logic Circuit Design

5.52 The Multisim circuit, shown in *Figure 5.72*, drives a 9-bit LED bar-graph display, much like the 7-bit bar-graph circuit in Problem 5.30. The logic gates for the circuit are hidden inside the rectangle on the diagram. The gate inputs are connected to four logic switches with pull-up resistors. Their outputs are connected to nine segments of a bar-graph LED display. It is your job to design the circuit in the box.

The binary inputs *ABCD* are indicated on the logic probes at the bottom of the diagram. This binary input value is equivalent to the number of active-HIGH bars that are turned on. For example, *ABCD* = 0000 means no bars are on. *ABCD* = 0101 turns on five bars, since 0101 (binary) is the same as 5 (decimal). Values of *ABCD* greater than 1001 are to be treated as don't care states.

a. Make the truth table for the circuit, showing all nine outputs, *Y1* to *Y9*, for every combination of *ABCD*.

b. Use the Karnaugh map method to simplify the Boolean expressions for the nine outputs of the circuit.

c. Draw a Multisim circuit for the 9-bit bar graph. (You could start with the Multisim file for the 7-bit bar graph [**05.11 7-bit Bar Graph.ms11**] and modify it accordingly.)

FIGURE 5.72 Problem 5.52: Multisim Circuit for a 9-Bit Bar-Graph © CENGAGE LEARNING 2012.

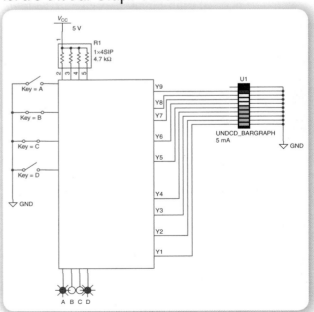

CHAPTER 6
Combinational Logic Functions

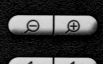

START LOCATION	DISTANCE	END LOCATION

Menu

Chapter Objectives

Upon successful completion of this chapter, you will be able to:

1 Design binary decoders using logic gates.

2 Create decoder designs in Multisim.

3 Interpret simulation files to verify the operation of combinational circuits.

4 Design BCD-to-seven-segment and hexadecimal-to-seven-segment decoders, including special features such as ripple blanking.

5 Use Multisim to generate the design for a 3-bit binary and a BCD priority encoder.

6 Describe the circuit and operation of a simple multiplexer.

7 Draw logic circuits for multiplexer applications, such as single-channel data selection and multibit data selection.

8 Describe demultiplexer circuits.

9 Explain how the same device can be used as a decoder or demultiplexer.

10 Define the operation of a CMOS analog switch and its use in multiplexers and demultiplexers.

11 Define the operation of a magnitude comparator.

12 Explain the use of parity as an error-checking system and draw simple parity generation and checking circuits.

PHOTO: ©ISTOCKPHOTO.COM/TEBNED.

You have probably seen a wide variety of digital devices in your home or school. Some devices such as a digital thermostat, calculator, or some kinds of alarms, let you control something by pressing a button or two, no matter how fast or slow you press. Many other devices, like a digital clock, depend on some element of time. The devices that just give an output (like the temperature) after you give an input (like pressing a button) use combinational circuits.

We have seen circuit designs that perform particular tasks, and other circuits that weren't designed for anything other than to match a K-map, a truth table, or equation. Many standard combinational logic functions have been developed that represent many of the useful tasks that can be performed with digital circuits. Rather than redesigning or analyzing the same circuits many times, we can think of a circuit based on

what we want it to *do*; in other words, we can think of each circuit's *function* instead of analyzing it at the gate level each time.

Decoders detect the presence of particular binary states and can activate other circuits based on their input values or can convert an input code to a different output code. Encoders generate a binary or binary coded decimal (BCD) code corresponding to an active input.

Multiplexers and demultiplexers are used for data routing. They select a specific path for incoming or outgoing data, based on a selection made by a set of binary-related inputs.

Magnitude comparators determine whether one binary number is less than, greater than, or equal to another binary number.

Parity generators and checkers are used to implement a system of checking for errors in groups of data.

6.1 DECODERS

The general function of a **decoder** is to activate one or more outputs when a specific set of values is applied to the circuit's inputs. The simplest decoder is a single logic gate, such as a NAND or AND, whose output activates when *all* its inputs are HIGH. When combined with one or more inverters, a NAND or AND can detect any unique combination of binary input values.

An extension of this type of decoder is a device containing several such gates, each of which responds to a different input state. Usually, for an *n*-bit input, there are 2^n logic gates, each of which decodes a different combination of input variables. For example, a decoder with a 3-bit input has $2^3 = 8$ separate gates, each of which responds to a different input combination. A variation is a binary coded decimal (BCD) device with 4 input variables and 10 outputs, each of which activates for a different BCD input.

Some types of decoders translate binary inputs to other forms, such as the decoders that drive seven-segment numerical displays, those familiar figure-8 arrangements of LED or LCD outputs ("segments"). The decoder has one output for every segment in the display. These segments illuminate in unique combinations for each input code.

Single-Gate Decoders

The simplest decoder is a single gate, sometimes in combination with one or more inverters, used to detect the presence of one particular binary value. *Figure 6.1* shows two such decoders, both of which detect an input $D_3D_2D_1D_0 = 1111$.

The decoder in Figure 6.1a generates a logic HIGH when its input is 1111. The decoder in Figure 6.1b responds to the same input, but makes the output LOW instead.

In Figure 6.1, we designate D_3 as the most significant bit of the input and D_0 the least significant bit. We will continue this convention for multibit inputs.

In Boolean expressions, we will indicate the active levels of inputs and outputs separately. For example, in Figure 6.1, the inputs to both gates are the same, so we write $D_3D_2D_1D_0$ for the inputs of both gates. The gates in Figures 6.1a and Figure 6.1b have outputs with opposite active levels, so we write the output variables as complements (Y and $\overline{Y}$).

> **Note . . .**
>
> The simplest type of decoder is an AND or NAND gate with input inverters.

> **Note . . .**
>
> For multibit inputs, such as *D*, designate D_3 as the most significant bit of the input and D_0 the least significant bit.

FIGURE 6.1 *Single-Gate Decoders* © CENGAGE LEARNING 2012.

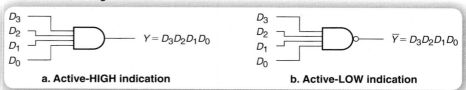

a. Active-HIGH indication   b. Active-LOW indication

Example 6.1

Figure 6.2 shows three single-gate decoders. For each one, state the output active level and the input code that activates the decoder. Also write the Boolean expression of each output.

FIGURE 6.2 Example 6.1: Single-Gate Decoders © CENGAGE LEARNING 2012.

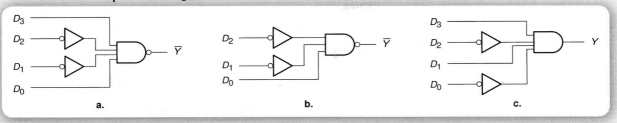

Solution

Each decoder is a NAND or AND gate. For each of these gates, the output is *active* when *all inputs are HIGH*. Because of the inverters, each circuit has a different code that fulfills this requirement.

Figure 6.2a: Output: Active LOW
Input code: $D_3D_2D_1D_0 = 1001$

$$\bar{Y} = D_3\bar{D}_2\bar{D}_1D_0$$

Figure 6.2b: Output: Active LOW
Input code: $D_2D_1D_0 = 001$

$$\bar{Y} = \bar{D}_2\bar{D}_1D_0$$

Figure 6.2c: Output: Active HIGH
Input code: $D_3D_2D_1D_0 = 1010$

$$\bar{Y} = D_3\bar{D}_2D_1\bar{D}_0$$

Example 6.2

Multisim Example

Multisim File: 06.01 Single Gate Decoder 1010 HIGH.ms11

Figure 6.3 shows a circuit of a single-gate decoder drawn using Multisim. The inputs are supplied by logic switches controlled on a PC keyboard by keys 0 to 3. Run the Multisim design in Simulation mode and try typing 0, 1, 2, and 3 to see the switches change state. The input and output state are monitored by logic probe components, which light when HIGHs are applied to their monitored points.

What is the input code that is decoded by this circuit? Write the Boolean expression of the circuit output. Write the truth table of the single-gate decoder. Is the output active-HIGH or active-LOW? Why?

FIGURE 6.3 Example 6.2: Single-Gate Active-HIGH Decoder © CENGAGE LEARNING 2012.

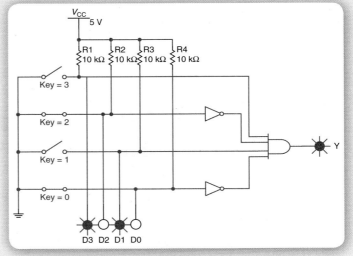

continued...

Solution

The input code decoded by this decoder is $D_3D_2D_1D_0 = 1010$.
Boolean expression: $Y = D_3\bar{D}_2D_1\bar{D}_0$

Truth table:

D_3	D_2	D_1	D_0	Y
0	0	0	0	0
0	0	0	1	0
0	0	1	0	0
0	0	1	1	0
0	1	0	0	0
0	1	0	1	0
0	1	1	0	0
0	1	1	1	0
1	0	0	0	0
1	0	0	1	0
1	0	1	0	1
1	0	1	1	0
1	1	0	0	0
1	1	0	1	0
1	1	1	0	0
1	1	1	1	0

© CENGAGE LEARNING 2012.

The output is active-HIGH because the output goes HIGH when the selected code 1010 is applied to the inputs.

Interactive Exercise:

Open the Multisim example for the single-gate decoder of Figure 6.3. Run it as a simulation and work through all combinations of inputs. Verify that the output corresponds to the truth table written in the solution for this example.

Example 6.3

Multisim Example

Multisim File: 06.02 Single Gate Decoder 1010 LOW.ms11

Figure 6.4 shows a Multisim circuit of a single-gate decoder.
What is the input code that is decoded by this circuit? Write the Boolean expression of the circuit output. Write the truth table of the single-gate decoder. Is the output active-HIGH or active-LOW? Why?

continued...

FIGURE 6.4 Example 6.3: Single-Gate Active-LOW Decoder

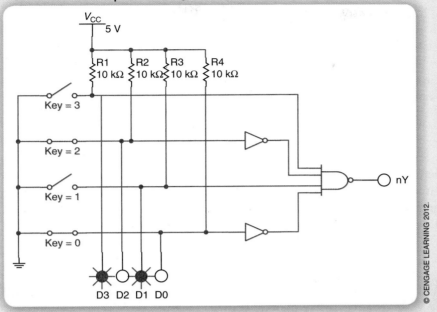

Solution

The input code decoded by this decoder is $D_3D_2D_1D_0 = 1010$.

Boolean expression: $\overline{Y} = D_3\overline{D}_2D_1\overline{D}_0$

The output variable in the Multisim circuit is shown as nY because it is *not* possible to write a logic inversion overbar over the variable name in Multisim.

Truth table:

D_3	D_2	D_1	D_0	Y
0	0	0	0	1
0	0	0	1	1
0	0	1	0	1
0	0	1	1	1
0	1	0	0	1
0	1	0	1	1
0	1	1	0	1
0	1	1	1	1
1	0	0	0	1
1	0	0	1	1
1	0	1	0	0
1	0	1	1	1
1	1	0	0	1
1	1	0	1	1
1	1	1	0	1
1	1	1	1	1

continued...

The output is active-LOW because the output goes LOW only when the selected code 1010 is applied to the inputs.

Interactive Exercise:

Open the Multisim example for the single-gate decoder of Figure 6.4. Run it as a simulation and work through all combinations of inputs. Verify that the output corresponds to the truth table written in the solution for this example.

Single-gate decoders are often used to activate other circuits under various operating conditions, particularly if there is a choice of circuits to activate. For example, single-gate decoders can be used to enable the different lights in a traffic signal. A combination of binary values specifies a unique set of conditions to turn on a particular light.

Example 6.4

Multisim Example

Multisim File: 06.03 Traffic Light Decoder.ms11

Figure 6.5A, *Figure 6.5B*, and *Figure 6.5C* show the states of the three lights in a traffic signal (RED, YELLOW, GREEN), controlled by a set of three single-gate decoders.

Based on the circuit diagram and the states of the inputs and outputs shown in Figure 6.5, write a Boolean expression for each gate output. Also write a truth table for the decoders. Are there any input states where no traffic light is on?

FIGURE 6.5A Example 6.4: Traffic Light Decoder (Red)

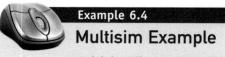

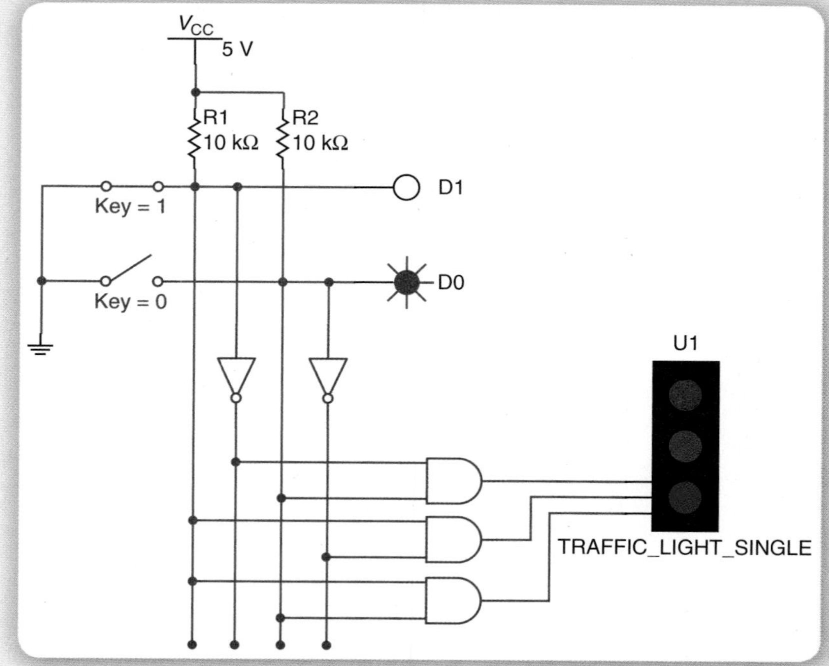

© CENGAGE LEARNING 2012.

continued...

158   Digital Electronics

FIGURE 6.5B Example 6.4: Traffic Light Decoder (Yellow)

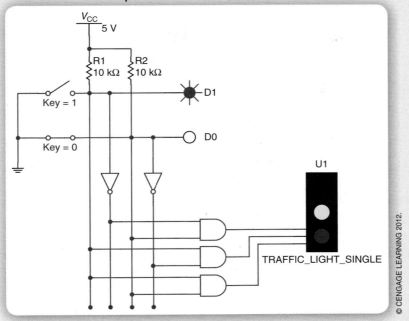

FIGURE 6.5C Example 6.4: Traffic Light Decoder (Green)

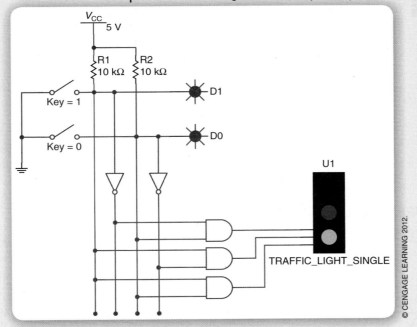

Solution

By examining the logic gate circuits and the input/output states of the decoders, we see that the red light is turned on when $D_1 D_0 = 01$, the yellow light when $D_1 D_0 = 10$, and the green light when $D_1 D_0 = 11$. Thus, the Boolean expressions for the gates are given by:

$$RED = \bar{D}_1 D_0$$
$$YELLOW = D_1 \bar{D}_0$$
$$GREEN = D_1 D_0$$

continued...

Note . . .

Why would we ever choose $D_1D_0 = 00$? We probably don't need to design this into a real traffic light, but because we have two input variables, we have four states. We needed only three states (RED, YELLOW, and GREEN). We often end up with extra states we don't really "need," and in a final design, we would make sure our system wouldn't go to these unwanted states.

Note . . .

A decoder with *n* input lines can have up to 2^n output lines.

The truth table of the circuit is:

D_1	D_0	RED	YELLOW	GREEN
0	0	0	0	0
0	1	1	0	0
1	0	0	1	0
1	1	0	0	1

No traffic light is on when $D_1D_0 = 00$.

Interactive Exercise:

Open the Multisim example for the traffic light decoders of Figure 6.5. Run it as a simulation and work through all combinations of inputs. Verify that the output corresponds to the functions written in the solution for this example.

Your Turn

6.1 Draw a single-gate decoder that detects the input state $D_3D_2D_1D_0 = 1100$
   a. with active-HIGH indication.
   b. with active-LOW indication.

6.2 Create a Multisim circuit for the two decoders in Your Turn 6.1. Run the designs as a simulation to test their operation.

TABLE 6.1 Truth Table of a 2-to-4 Decoder

D_1	D_0	Y_0	Y_1	Y_2	Y_3
0	0	**1**	0	0	0
0	1	0	**1**	0	0
1	0	0	0	**1**	0
1	1	0	0	0	**1**

FIGURE 6.6 2-Line-to-4-Line Decoder © CENGAGE LEARNING 2012.

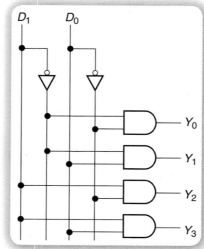

Multiple-Output Decoders

Decoder circuits often are constructed with multiple outputs. In effect, such a device is a collection of decoding gates controlled by the same inputs. A decoder circuit with *n* inputs can activate up to $m = 2^n$ load circuits. Such a decoder is usually described as an *n*-line-to-*m*-line decoder.

Figure 6.6 shows the logic circuit of a 2-line-to-4-line decoder. The circuit detects the presence of a particular state of the 2-bit input D_1D_0, as shown by the truth table in *Table 6.1*. One and only one output is HIGH for any input combination. The active input of each line is shown in **boldface** type. The subscript of the active output is the same as the value of the 2-bit input. For example, if $D_1D_0 = 10$, output Y_2 is active because 10 (binary) = 2 (decimal).

If we are using the decoder to activate one of four output loads, there might be situations where we don't want any output to be active. This is not possible in the circuit shown in Figure 6.6, as there is always one output active. A variation of the decoder that allows this type of control is shown in *Figure 6.7*.

The 2-line-to-4-line decoder shown in Figure 6.7 has an additional input called $\overline{G}$ (for "gating") that controls whether or not any input is active. If we make $\overline{G}$ LOW (active), the decoder acts the same as the one in Figure 6.6. If $\overline{G}$ is HIGH, then all outputs are deactivated (made LOW). The truth table for this decoder is shown in *Table 6.2*.

FIGURE 6.7 2-Line-to-4-Line Decoder with Active-LOW Enable

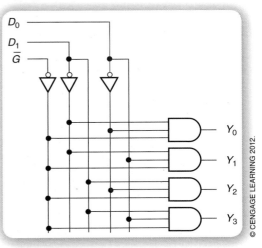

TABLE 6.2 Truth Table of a 2-Line-to-4-Line Decoder with Enable

$\overline{G}$	D_1	D_0	Y_0	Y_1	Y_2	Y_3
0	0	0	1	0	0	0
0	0	1	0	1	0	0
0	1	0	0	0	1	0
0	1	1	0	0	0	1
1	X	X	0	0	0	0

Figure 6.8 shows the circuit for a 3-line-to-8-line decoder, again with an active-LOW enable, $\overline{G}$. In this case, the decoder outputs are active LOW. One and only one output is active for any given combination of $D_2D_1D_0$. *Table 6.3* shows the truth table for this decoder. Again if the enable line is HIGH, no output is active.

FIGURE 6.8 3-Line-to-8-Line Decoder with Active-LOW Enable © CENGAGE LEARNING 2012.

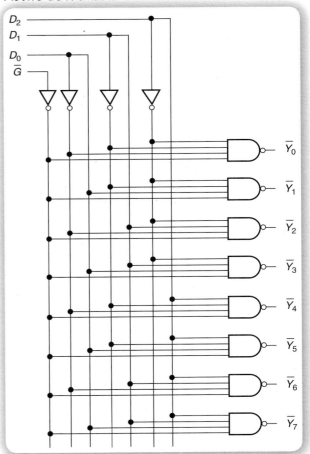

TABLE 6.3 Truth Table of a 3-to-8 Decoder with Enable

$\overline{G}$	D_2	D_1	D_0	$\overline{Y}_0$	$\overline{Y}_1$	$\overline{Y}_2$	$\overline{Y}_3$	$\overline{Y}_4$	$\overline{Y}_5$	$\overline{Y}_6$	$\overline{Y}_7$
0	0	0	0	0	1	1	1	1	1	1	1
0	0	0	1	1	0	1	1	1	1	1	1
0	0	1	0	1	1	0	1	1	1	1	1
0	0	1	1	1	1	1	0	1	1	1	1
0	1	0	0	1	1	1	1	0	1	1	1
0	1	0	1	1	1	1	1	1	0	1	1
0	1	1	0	1	1	1	1	1	1	0	1
0	1	1	1	1	1	1	1	1	1	1	0
1	X	X	X	1	1	1	1	1	1	1	1

Example 6.5

Multisim Example

Multisim File: 06.04 3to8 Decoder active HIGH.ms11

Figure 6.9 shows a 3-line-to-8-line decoder created with Multisim.

FIGURE 6.9 Example 6.5: 3-Line-to-8-line Active-HIGH Decoder

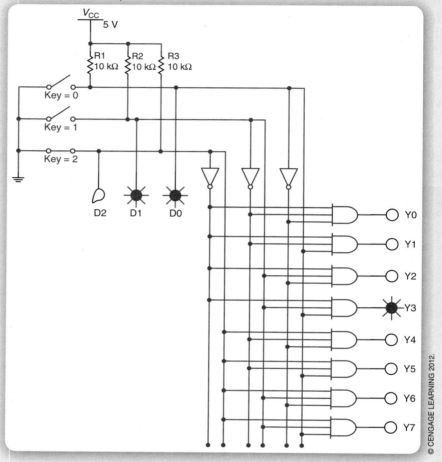

© CENGAGE LEARNING 2012.

Write the Boolean expression for each decoder output.

Which output is active when the input code is $D_2D_1D_0 = 110$? Is this output HIGH or LOW when active?

Solution

The Boolean equations for the decoder in Figure 6.9 are as follows:

$$Y_0 = \bar{D}_2\bar{D}_1\bar{D}_0$$

$$Y_1 = \bar{D}_2\bar{D}_1D_0$$

$$Y_2 = \bar{D}_2D_1\bar{D}_0$$

$$Y_3 = \bar{D}_2D_1D_0$$

$$Y_4 = D_2\bar{D}_1\bar{D}_0$$

$$Y_5 = D_2\bar{D}_1D_0$$

$$Y_6 = D_2D_1\bar{D}_0$$

$$Y_7 = D_2D_1D_0$$

continued...

Output Y_6 is active when the input code is $D_2D_1D_0 = 110$. When active, the decoder outputs are HIGH.

Interactive Exercise:

Open the Multisim file for this example and run it as a simulation. Operate the three logic switches in such a way as to activate the outputs one after the other, starting at Y_0. What sequence do you observe at the inputs when the outputs are activated in this order?

FIGURE 6.10 **74138** Decoder

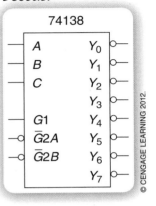

A 3-line-to-8-line decoder that is commercially available in a single chip is the 74138 decoder, shown in *Figure 6.10*. Depending on the logic family, this device will be designated 74LS138, 74HC138, or some other such number.

This decoder has active-LOW outputs and three enable inputs, one active-HIGH and two active-LOW, all of which must be active to enable any of the outputs. The truth table for this decoder is shown in *Table 6.4*.

Notice that the Y outputs are active-LOW and that they are selected by a combination of inputs CBA, where C is the most significant bit.

TABLE 6.4 Truth Table for a 74138 Decoder © CENGAGE LEARNING 2012.

G1	$\overline{G}2A$	$\overline{G}2B$	C	B	A	$\overline{Y}_0$	$\overline{Y}_1$	$\overline{Y}_2$	$\overline{Y}_3$	$\overline{Y}_4$	$\overline{Y}_5$	$\overline{Y}_6$	$\overline{Y}_7$
0	X	X	X	X	X	1	1	1	1	1	1	1	1
1	1	X	X	X	X	1	1	1	1	1	1	1	1
1	X	1	X	X	X	1	1	1	1	1	1	1	1
1	0	0	0	0	0	0	1	1	1	1	1	1	1
1	0	0	0	0	1	1	0	1	1	1	1	1	1
1	0	0	0	1	0	1	1	0	1	1	1	1	1
1	0	0	0	1	1	1	1	1	0	1	1	1	1
1	0	0	1	0	0	1	1	1	1	0	1	1	1
1	0	0	1	0	1	1	1	1	1	1	0	1	1
1	0	0	1	1	0	1	1	1	1	1	1	0	1
1	0	0	1	1	1	1	1	1	1	1	1	1	0

Example 6.6

Multisim Example

Multisim File: 06.05 74138 Decoder.ms11

Figure 6.11 shows a Multisim circuit for demonstrating the operation of a 74138 decoder.

Assume all the enable inputs are set to the logic levels shown in Figure 6.11. Which output is active when the input switches are changed to $D_2D_1D_0 = 101$? What is the logic level of the active output in this condition?

Assume the D inputs are set to the logic levels shown in Figure 6.11. What are the logic levels of the outputs when input G1 is changed to 0?

continued...

FIGURE 6.11 Example 6.6: 74138 Decoder © CENGAGE LEARNING 2012.

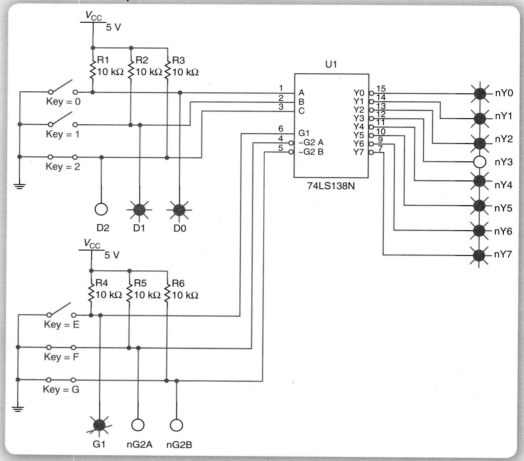

Solution

Output $nY5$ is active when $D_2D_1D_0 = 101$. The output is LOW in this case. All outputs are HIGH (inactive) when input G1 is set to 0.

Interactive Exercise:

a. Open the Multisim file for this example and run it as a simulation. Set the enable switches to the positions shown in Figure 6.11 and the data switches to all LOW. What are the logic levels of the outputs?

b. Set the switches in turn so that they advance the inputs in a binary sequence (000, 001, 010, 011, ...). What happens to the decoder outputs?

Your Turn

6.3 How many inputs are required for a binary decoder with 16 outputs?

6.4 How many inputs are required for a decoder with 32 outputs?

Programmable Logic Devices and Software

In modern digital designs, a circuit is often created using a **Programmable Logic Device** or **PLD**, a digital device whose logic function is created by the user, not the device manufacturer. A specific type of PLD that is often used is the **Field Programmable Gate Array (FPGA)**. As the name suggests, this type of device is an array of gates that is programmed "in the field", that is, by the user, to make the required logic function.

A digital designer must create the logic function by using special software and then program the design information into the FPGA. The simplest way for a beginner to learn to program a FPGA is to use a software drawing tool, similar to Multisim, and draw the components of the design on the software's workspace. We will not learn how to use FPGA design software in this chapter, but we will look at some examples of circuits created using design software called Xilinx ISE to give you a flavor of what the FPGA design process is like.

Simulation Criteria for *n*-Line-to-*m*-Line Decoders

An important part of the FPGA design process is **simulation** of the design. A simulation tool allows us to see whether the output responses to a set of circuit inputs are what we expected in our initial design idea. The simulator works by creating a **timing diagram**. We specify a set of input (**stimulus**) **waveforms**. The simulator looks at the relationship between inputs and outputs, as defined by the design file, and generates a set of **response waveforms**.

Figure 6.12 shows a set of simulation waveforms created in the Xilinx ISE Simulator (ISim) for the LOW-enabled 2-line-to-4-line decoder in Figure 6.6. The inputs D_1 and D_0 are combined as a single 2-bit value, to which an increasing binary count is applied as a stimulus. The decoder output waveforms are observed individually to determine the decoder's response. Once we have entered the design in Xilinx ISE and compiled it, we can create the waveforms of Figure 6.12.

The effectiveness of a simulation depends on our choosing input waveforms that will tell us everything we need to know about the operation of our design. For a decoder with an enable input, this means applying input values that will show how the outputs activate in the correct order when all possible input combinations are applied. However, we must also test that no outputs activate when the enable input is not active. In other words, to use a simulation effectively, we must be sure

FIGURE 6.12 Simulation Waveforms for a 2-Line-to-4-Line Decoder with Active-HIGH Outputs and Active-LOW Enable © CENGAGE LEARNING 2012.

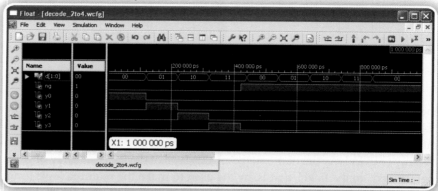

to understand the correct operation of a design and test it for all possible working and failure modes.

Let us list some reasonable criteria for the simulation of a 2-line-to-4-line decoder with an active-LOW enable input and active-HIGH decoded outputs, such as the decoder in Figure 6.7.

To test these criteria, we apply an ascending 2-bit count to inputs $D1$ and $D0$, with $\bar{G} = 0$, then apply the same count when $\bar{G} = 1$. As shown in Figure 6.12, the outputs activate in the correct order when the decoder is enabled and no outputs activate when the decoder is disabled.

Example 6.7

The 2-line-to-4-line decoder in Figure 6.7 is tested with input waveforms meeting the simulation criteria previously discussed. The simulation result, shown in *Figure 6.13*, indicates an error condition in the decoder. What is likely to be the circuit fault?

FIGURE 6.13 Example 6.7: Simulation Waveforms for a 2-Line-to-4-Line Decoder with Error in Enable Function © CENGAGE LEARNING 2012.

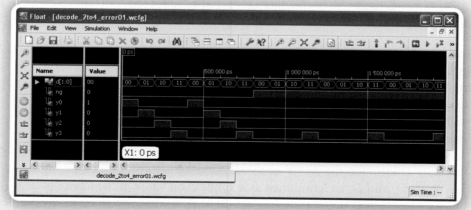

Solution

The simulation waveforms show the outputs behaving correctly when the $\bar{G}$ input is LOW, that is, when the decoder is enabled. When the decoder should be disabled (when $\bar{G}$ is HIGH), the $Y3$ output activates whenever $D1 = 1$ and $D0 = 1$. This is the correct decoding for this output, but it

continued...

should not be decoding at all when the decoder is disabled. This could result from an improper connection from $\overline{G}$ to the gate that decodes $Y3$, where the connection from $\overline{G}$ to $Y3$ is either open-circuited or stuck HIGH. The error can be corrected by examining the gate for the $Y3$ output and fixing the connection to the gate inputs.

Example 6.8

The 2-line-to-4-line decoder of Figure 6.7 is tested with input waveforms meeting appropriate simulation criteria. The simulation result, shown in *Figure 6.14*, indicates an error condition in the decoder. What is likely to be the circuit fault?

FIGURE 6.14 Example 6.8: Simulation Waveforms for a 2-Line-to-4-Line Decoder with Improperly Assigned D Inputs © CENGAGE LEARNING 2012.

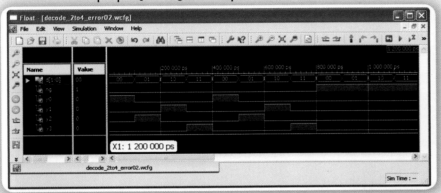

Solution

The outputs are enabled and disabled when they are supposed to be, but, when enabled, the outputs activate in the wrong order. We can sort this out by comparing the actual outputs to the expected outputs, as shown in *Table 6.5*.

FIGURE 6.15 Example 6.8: Incorrectly Labeled

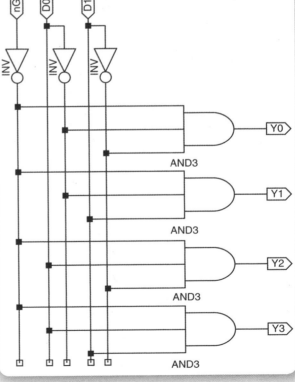

© CENGAGE LEARNING 2012.

TABLE 6.5 Actual and Expected Decoder Outputs

Output	Should activate when D =	Actually activates when D =
Y0	00	00
Y1	01	10
Y2	10	01
Y3	11	11

© CENGAGE LEARNING 2012.

Notice that the values in the actual and expected columns are mirror images of one another. This implies that when we entered the design in the Xilinx ISE Schematic Editor, we mixed up the order of $D1$ and $D0$, as shown in *Figure 6.15*. Reconnecting (or renaming) the inputs will fix the problem.

Example 6.9

A circuit equivalent to the 74138 3-line-to-8-line decoder shown in Figure 6.10 is created using Xilinx ISE software. Make a list of simulation criteria that will fully test the decoder and create a Xilinx ISE simulation to verify the operation of the device.

Solution

The simulation criteria for the decoder are shown in the box at the left. *Figure 6.16* shows a simulation that meets these criteria. An increasing binary count on *CBA*, shown grouped as *decode_inputs*[2..0], activates the outputs in sequence, as long as all three enable inputs are active. The enable inputs are grouped as *enable_inputs*[2..0], *G1* assigned to enable_inputs[2], $\bar{G}\,2A$ to enable_inputs[1], and $\bar{G}\,2B$ to enable_inputs[0]. Any one of the enable inputs inactive prevents any output from activating.

FIGURE 6.16 **Example 6.9: Simulation Waveforms for a 74138 3-Line-to-8-Line Decoder** © CENGAGE LEARNING 2012.

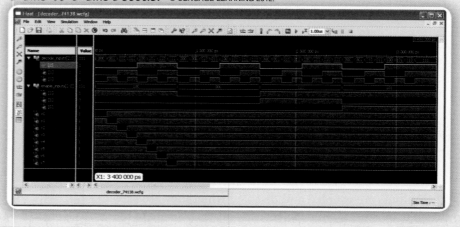

Seven-Segment Decoders

KEY TERMS

Seven-segment display An array of seven independently controlled light-emitting diode (LED) or liquid crystal display (LCD) elements, shaped like a figure-8, which can be used to display decimal digits and other characters by turning on the appropriate elements.

Common cathode display A seven-segment display in which the cathodes of all LEDs are connected together and grounded. A logic HIGH illuminates a segment when applied to its anode.

Common anode display A seven-segment LED display where the anodes of all the LEDs are connected to the circuit supply voltage. Each segment is illuminated by a logic LOW at its cathode.

Display

The **seven-segment display**, shown in *Figure 6.17*, is a numerical display device used to show digital circuit outputs as decimal digits (and sometimes hexadecimal digits or other alphabetic characters). It is called a seven-segment display because it consists of seven luminous segments, usually LEDs or liquid crystals, arranged in a figure-8. We can display any decimal digit by turning on the appropriate elements, designated by lowercase letters, *a* through *g*. It is conventional to designate the top segment as *a* and progress clockwise around the display, ending with *g* as the center element.

Figure 6.18 shows the usual convention for decimal digit display. Some variation from this convention is possible. For example, we could have drawn the digits 6 and 9 with "tails" (i.e., with segment *a* illuminated for 6 or segment *d* for 9). We typically display digit 1 by illuminating segments *b* and *c*, although segments *e* and *f* would also work.

The electrical requirements for an LED circuit are simple. Because an LED is a diode, it conducts when its anode is positive with respect to its cathode, as shown in *Figure 6.19a*. A decoder/driver for an LED display will illuminate an element by completing this circuit, either by supplying V_{CC} or ground. A series resistor must be used to limit the current to prevent the diode from burning out and to regulate its brightness. If the anode is +5 volts with respect to cathode, the resistor value should be in the range of 150 Ω to 470 Ω. High-efficiency LEDs are also available. These can use higher-valued series resistors (e.g., 1 kΩ) to achieve the same brightness with less current.

Seven-segment displays are configured as **common cathode** or **common anode**, as shown in *Figure 6.19b* and *Figure 6.19c*. In a **common cathode display**, the cathodes of all LEDs are connected together and brought out to one or more pin connections on the display package. The cathode pins are wired externally to the circuit ground. We illuminate individual segments by applying logic HIGHs to each segment's anode.

Similarly, the **common anode display** has the anodes of the segments brought out to one or more common pins. These pins must be tied to the circuit power supply (V_{CC}). The segments illuminate when a decoder/driver makes their individual

FIGURE 6.17
Seven-Segment Numerical Display

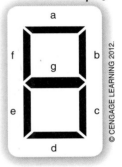

© CENGAGE LEARNING 2012.

FIGURE 6.19 **Electrical Requirements for LED Displays**

FIGURE 6.18 *Convention for Displaying Decimal Digits*

© CENGAGE LEARNING 2012.

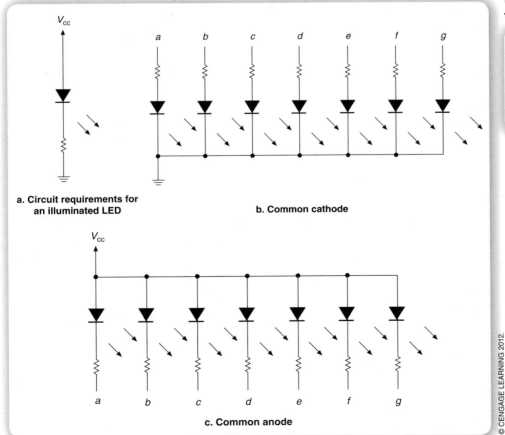

a. Circuit requirements for an illuminated LED

b. Common cathode

c. Common anode

© CENGAGE LEARNING 2012.

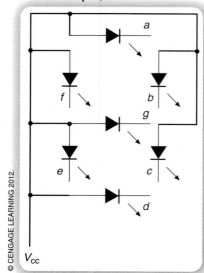

FIGURE 6.20 Physical Placement of LEDs in a Common Anode Display

V_{CC}

cathodes LOW. *Figure 6.20* shows how the diodes could be physically laid out in a common anode display.

The two types of displays allow the use of either active-HIGH or active-LOW circuits to drive the LEDs, thus giving the designer some flexibility. However, it should be noted that the majority of seven-segment decoders are for common anode displays, which require active-LOW drivers.

Example 6.10

Sketch the segment patterns required to display all 16 hexadecimal digits on a seven-segment display. What changes from the patterns in Figure 6.18 need to be made?

Solution

The segment patterns are shown in *Figure 6.21*.

Hex digits B and D must be displayed as lowercase letters, b and d, to avoid confusion between B and 8 and between D and 0. To make 6 distinct from b, 6 must be given a tail (segment *a*) and to make 6 and 9 symmetrical, 9 should also have a tail (segment *d*).

FIGURE 6.21 Example 6.10: Hexadecimal Display Format

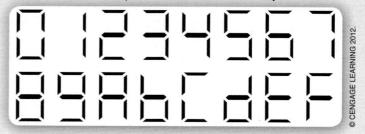

Decoder

A BCD-to-seven-segment decoder is a circuit with a 4-bit input for a **binary coded decimal (BCD)** digit and seven outputs for segment selection. To display a number, the decoder must translate the input bits to a combination of active outputs. For example, the input digit $D_3D_2D_1D_0 = 0000$ must illuminate segments *a*, *b*, *c*, *d*, *e*, and *f* to display the digit 0. We can make a truth table for each of the outputs, showing which must be active for every digit we wish to display. The truth table for a common anode decoder (active-LOW outputs) is given in *Table 6.6*.

The illumination of each segment is determined by a Boolean function of the input variables, $D_3D_2D_1D_0$. From the truth table, the function for segment *a* is

$$a = \bar{D}_3\bar{D}_2\bar{D}_1D_0 + \bar{D}_3D_2\bar{D}_1\bar{D}_0 + \bar{D}_3D_2D_1\bar{D}_0$$

(The display is active-LOW, so this means segment *a* is OFF for digits 1, 4, and 6.)

If we assume that inputs 1010 to 1111 are never going to be used ("don't care states," symbolized by X), we can make any of these states produce HIGH or LOW

TABLE 6.6   Truth Table for Common Anode BCD-to-Seven-Segment Decoder

Digit	D_3	D_2	D_1	D_0	a	b	c	d	e	f	g
0	0	0	0	0	0	0	0	0	0	0	1
1	0	0	0	1	1	0	0	1	1	1	1
2	0	0	1	0	0	0	1	0	0	1	0
3	0	0	1	1	0	0	0	0	1	1	0
4	0	1	0	0	1	0	0	1	1	0	0
5	0	1	0	1	0	1	0	0	1	0	0
6	0	1	1	0	1	1	0	0	0	0	0
7	0	1	1	1	0	0	0	1	1	1	1
8	1	0	0	0	0	0	0	0	0	0	0
9	1	0	0	1	0	0	0	1	1	0	0
	1	0	1	0	X	X	X	X	X	X	X
	1	0	1	1	X	X	X	X	X	X	X
	1	1	0	0	X	X	X	X	X	X	X
Invalid Range	1	1	0	1	X	X	X	X	X	X	X
	1	1	1	0	X	X	X	X	X	X	X
	1	1	1	1	X	X	X	X	X	X	X

© CENGAGE LEARNING 2012.

outputs, depending on which is most convenient for simplifying the segment functions. *Figure 6.22a* shows a Karnaugh map simplification for segment *a*. The resultant function is

$$a = \bar{D}_3\bar{D}_2\bar{D}_1D_0 + D_2\bar{D}_0$$

FIGURE 6.22   Decoding Segments a, b, and c of a Seven-Segment Display

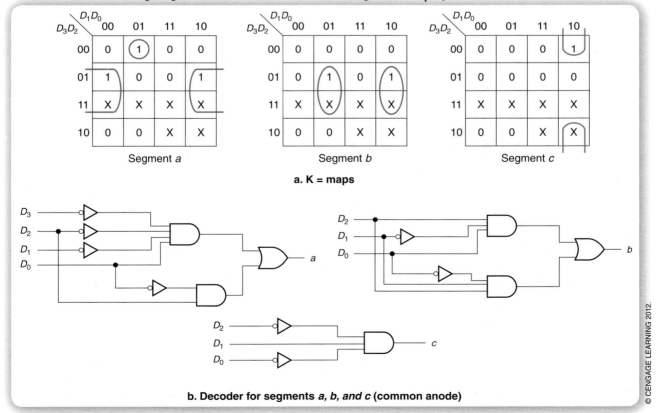

a. K = maps

b. Decoder for segments *a, b, and c* (common anode)

© CENGAGE LEARNING 2012.

The corresponding partial decoder is shown in *Figure 6.22b,* along with segments *b* and *c.*

We could do a similar analysis for each of the other segments.

Example 6.11

Multisim Example

Multisim File: 06.07 BCD-to-7-Segment Decoder (CA).ms11

Figure 6.23 shows a Multisim file that contains a common anode seven-segment decoder. Keys 3 through 0 can be combined to make a binary input. Two special-function inputs are also shown: an active-LOW Lamp Test function (nLT) and an active-LOW Blanking Input (nBI).

For the input switches set as shown, state the logic levels of the decoder inputs DCBA and outputs [OA..OG].

Solution

The input switches indicate a code of 0111, which represents the digit 7. For an active-LOW display, segments *a*, *b*, and *c* are ON and all other segments are OFF to display the digit 7.

D	C	B	A	OA	OB	OC	OD	OE	OF	OG
0	1	1	1	0	0	0	1	1	1	1

© CENGAGE LEARNING 2012.

FIGURE 6.23 *Example 6.11: 74LS47 Common Anode Seven-Segment Decoder*

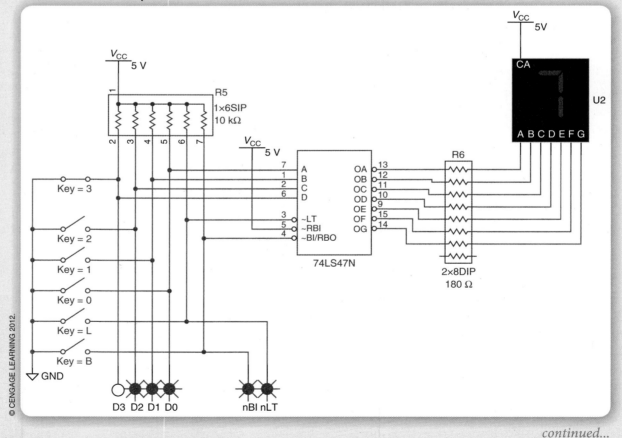

continued...

> **Note . . .**
>
> In some datasheets, decoder inputs are shown as *DCBA.* In other documents, they are given as $D_3D_2D_1D_0$.

Ripple Blanking

KEY TERMS

Ripple blanking   A technique used in a multiple-digit numerical display that suppresses or "blanks" leading or trailing zeros in the display, but allows internal zeros to be displayed.

RBI   Ripple blanking input.

RBO   Ripple blanking output.

A feature often included in seven-segment decoders is **ripple blanking**. The ripple blanking feature allows for suppression of leading or trailing zeros in a multiple digit display, while allowing zeros to be displayed in the middle of a number.

Each display decoder has a ripple blanking input ($\overline{RBI}$) and a ripple blanking output ($\overline{RBO}$), which are connected in cascade, as shown in *Figure 6.24*. If the decoder input $D_3D_2D_1D_0$ is 0000, it displays digit 0 if $\overline{RBI} = 1$ and shows a blank if $\overline{RBI} = 0$.

If $\overline{RBI} = 1$ or $D_3D_2D_1D_0$ is anything other than 0000, then $\overline{RBO} = 1$. When we cascade two or more displays, these conditions suppress leading or trailing zeros (but not both) and still display internal zeros.

To suppress leading zeros in a display, ground the $\overline{RBI}$ of the most significant digit decoder and connect the $\overline{RBO}$ of each decoder to the $\overline{RBI}$ of the next least significant digit. Any zeros preceding the first nonzero digit (9 in this case) will be blanked, as $\overline{RBI} = 0$ and $D_3D_2D_1D_0 = 0000$ for each of these decoders. The 0 inside the number 904 is displayed because its $\overline{RBI} = 1$.

Trailing zeros are suppressed by reversing the order of $\overline{RBI}$ and $\overline{RBO}$ from the previous example. $\overline{RBI}$ is grounded for the least significant digit and the $\overline{RBO}$ for each decoder cascades to the $\overline{RBI}$ of the next most significant digit.

Your Turn

6.5   When would it be logical to suppress trailing zeros in a multiple-digit display and when should trailing zeros be displayed?

FIGURE 6.24 Zero Suppression in Seven-Segment Displays

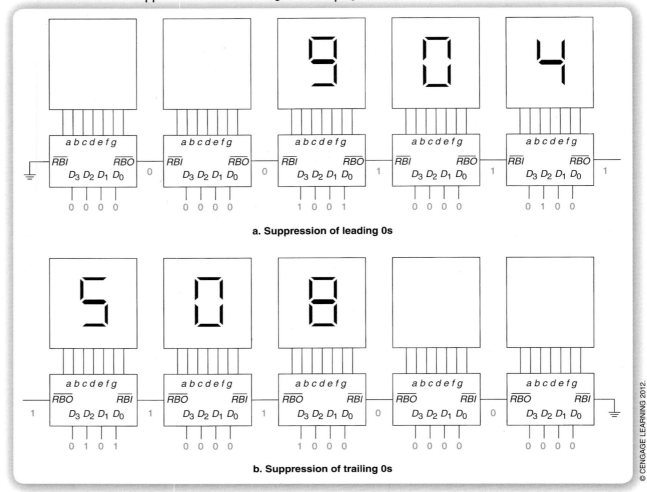

a. Suppression of leading 0s

b. Suppression of trailing 0s

6.2 ENCODERS

KEY TERMS

Encoder   A circuit that generates a binary code at its outputs in response to one or more active input lines.

Priority encoder   An encoder that generates a binary or BCD output corresponding to the subscript of the active input having the highest priority. This is usually defined as the active input with the largest subscript value.

FIGURE 6.25 3-Bit Encoder (No Input Priority)

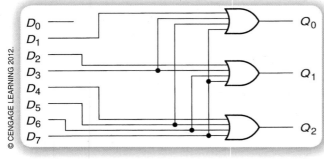

The function of a digital **encoder** is complementary to that of a digital decoder. A decoder activates a specified output for a unique digital input code. An encoder operates in the reverse direction, producing a particular digital code (e.g., a binary or BCD number) at its outputs when a specific input is activated.

Figure 6.25 shows a 3-bit binary encoder. As long as *only one input* is active at a time, the circuit generates a unique 3-bit binary output for every active input.

The encoder has only eight permitted input states out of a possible 256. *Table 6.7* shows the allowable input states, which yield the Boolean equations used to design the encoder. These Boolean equations are:

$$Q_2 = D_7 + D_6 + D_5 + D_4$$
$$Q_1 = D_7 + D_6 + D_3 + D_2$$
$$Q_0 = D_7 + D_5 + D_3 + D_1$$

The D_0 input is not connected to any of the encoding gates, as all outputs are in their LOW (inactive) state when the 000 code is selected.

TABLE 6.7 Partial Truth Table for a 3-Bit Encoder

D_7	D_6	D_5	D_4	D_3	D_2	D_1	Q_2	Q_1	Q_0
0	0	0	0	0	0	0	0	0	0
0	0	0	0	0	0	1	0	0	1
0	0	0	0	0	1	0	0	1	0
0	0	0	0	1	0	0	0	1	1
0	0	0	1	0	0	0	1	0	0
0	0	1	0	0	0	0	1	0	1
0	1	0	0	0	0	0	1	1	0
1	0	0	0	0	0	0	1	1	1

Priority Encoder

The shortcoming of the encoder circuit shown in Figure 6.25 is that it can generate incorrect codes if more than one input is active at the same time. For example, if we make D_3 and D_5 HIGH at the same time, the output is neither 011 or 101, but 111; the output code does not correspond to either active input.

One solution to this problem is to assign a priority level to each input and, if two or more are active, make the output code correspond to the highest-priority input. This is called a **priority encoder**. Highest priority is assigned to the input whose subscript has the largest numerical value.

Example 6.12

Figure 6.26 shows a priority encoder with three combinations of inputs. Determine the resultant output code for each figure. Inputs and outputs are active-HIGH.

FIGURE 6.26 Example 6.12: Priority Encoder Inputs

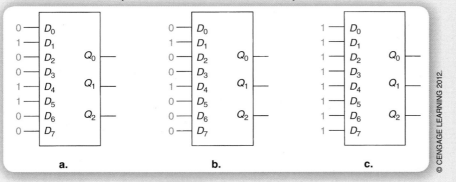

Solution

Figure 6.26a: The highest-priority active input is D_5. Inputs D_4 and D_1 are ignored. $Q_2Q_1Q_0 = 101$.

Figure 6.26b: The highest-priority active input is D_4. Input D_1 is ignored. $Q_2Q_1Q_0 = 100$.

Figure 6.26c: The highest-priority active input is D_7. All other inputs are ignored. $Q_2Q_1Q_0 = 111$.

TABLE 6.8 Truth Table for a 3-Bit Priority Encoder

D_7	D_6	D_5	D_4	D_3	D_2	D_1	Q_2	Q_1	Q_0
0	0	0	0	0	0	0	0	0	0
0	0	0	0	0	0	1	0	0	1
0	0	0	0	0	1	X	0	1	0
0	0	0	0	1	X	X	0	1	1
0	0	0	1	X	X	X	1	0	0
0	0	1	X	X	X	X	1	0	1
0	1	X	X	X	X	X	1	1	0
1	X	X	X	X	X	X	1	1	1

© CENGAGE LEARNING 2012.

The encoding principle of a priority encoder is that a low-priority input must not change the code resulting from a higher-priority input. As an example of the priority encoding principle, if inputs D_3 and D_5 are both active, the correct output code is $Q_2Q_1Q_0 = 101$. The code for D_3 would be $Q_2Q_1Q_0 = 011$. Thus, D_3 must not make $Q_1 = 1$. The Boolean expressions for Q_2, Q_1, and Q_0 covering only these two codes are:

$Q_2 = D_5$ (HIGH if D_5 is active.)

$Q_1 = D_3\bar{D}_5$ (HIGH if D_3 is active AND D_5 is NOT active.)

$Q_0 = D_3 + D_5$ (HIGH if D_3 OR D_5 is active.)

The truth table of a 3-bit priority encoder is shown in *Table 6.8.*

The equations from the 3-bit encoder of Figure 6.25 are modified by the priority encoding principle as follows:

$$Q_2 = D_7 + D_6 + D_5 + D_4$$
$$Q_1 = D_7 + D_6 + \bar{D}_5\bar{D}_4D_3 + \bar{D}_5\bar{D}_4D_2$$
$$Q_0 = D_7 + \bar{D}_6D_5 + \bar{D}_6\bar{D}_4D_3 + \bar{D}_6\bar{D}_4\bar{D}_2D_1$$

Figure 6.27 shows a 3-bit priority encoder circuit drawn in the Schematic Editor of Xilinx ISE.

FIGURE 6.27 3-Bit Priority Encoder (Xilinx ISE)

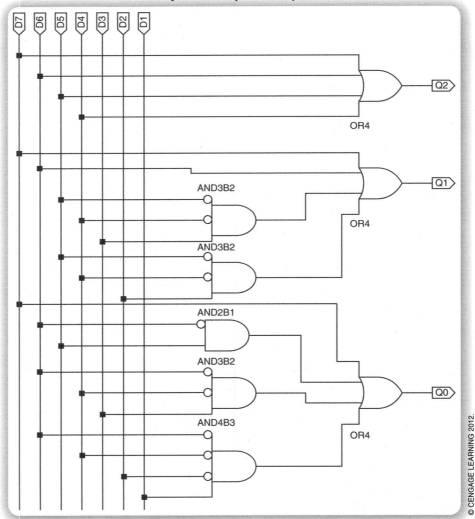

© CENGAGE LEARNING 2012.

Example 6.13

Write a set of simulation criteria for the 3-bit priority encoder shown in Figure 6.27. Use these criteria to create a simulation in Xilinx ISE.

Solution

The simulation criteria for the 3-bit priority encoder are shown in the box on the right. *Figure 6.28* shows the Xilinx ISim simulation of the 3-bit priority encoder, based on the previous criteria. The **D** inputs are shown separately, so that we can easily determine which inputs are active. The **Q** outputs are grouped so as to show the output code.

FIGURE 6.28 **Example 6.13: Simulation Waveforms for a 3-Bit Priority Encoder (Active-HIGH Inputs and Outputs)**

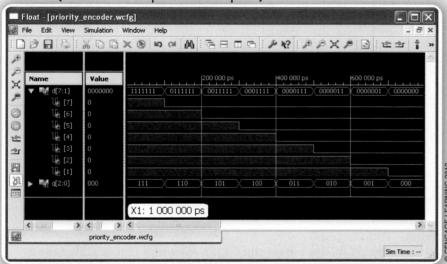

© CENGAGE LEARNING 2012.

Simulation Criteria:

► With all input bits HIGH, D7 is highest-priority of the active inputs. Expected output is 111_2 ($=7_{10}$).

► With D7 LOW and all other inputs HIGH, D6 is now the highest-priority active input. Expected output is 110_2 ($=6_{10}$).

► With D7 and D6 LOW and remaining bits HIGH, highest priority is D5. Expected output is 101_2 ($=5_{10}$).

► As each bit goes LOW in order, the output code should count down by one with each step.

Example 6.14

Multisim Example

Multisim File: 06.09 Priority Encoder.ms11

Figure 6.29 shows a Multisim design that tests the operation of a 3-bit priority encoder.

By examining the diagram, determine the following:

a. Active level of inputs and outputs
b. Highest-priority active input
b. Output code of the encoder

Solution

a. The inputs and outputs of the encoder all have bubbles, indicating that all inputs and outputs are active-LOW.
b. An active input is LOW. The highest-priority LOW input is nD5.
c. The binary code on outputs **nA[2..0]** is 010. Since the code is active-LOW, it can be inverted to show the active-HIGH version, which is shown by **A[2..0]**=101, which is the binary equivalent of 5 in decimal.

Note . . .

In the Multisim symbol for the encoder, the outputs are designated as A[2..0], rather than Q[2..0]. It is common for different manufacturers to have different terminal designations for their versions of a standard component.

continued...

FIGURE 6.29 Example 6.14: 74LS148 Priority Encoder

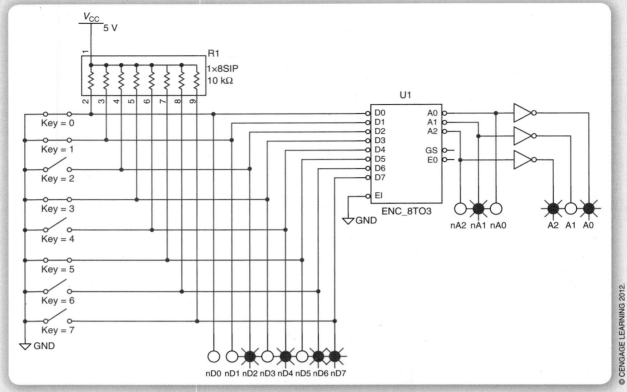

Interactive Exercise:

a. Open the Multisim file for the design shown in Figure 6.29. Set all switches LOW. What codes are displayed on **nA[2..0]** and **A[2..0]**?

b. Set the switches HIGH in the following sequence: nD7, nD6, nD5, nD4, nD3, nD2, nD1, nD0. Fill in *Table 6.9* with the binary values (0 or 1) of nA2 nA1 nA0 and A2 A1 A0.

c. Explain the results of the actions in part **b**.

TABLE 6.9 Priority Encoder Inputs © CENGAGE LEARNING 2012.

Last Switch Set HIGH	Highest Priority	nA2	nA1	nA0	A2	A1	A0
None							
7							
6							
5							
4							
3							
2							
1							

BCD Priority Encoder

A BCD priority encoder, illustrated in *Figure 6.30*, accepts 10 inputs and generates a BCD code (0000 to 1001) corresponding to the highest-priority active input. The truth table for this circuit is shown in *Table 6.10*.

TABLE 6.10   Truth Table of a BCD Priority Encoder © CENGAGE LEARNING 2012.

D_9	D_8	D_7	D_6	D_5	D_4	D_3	D_2	D_1	Q_3	Q_2	Q_1	Q_0
0	0	0	0	0	0	0	0	0	0	0	0	0
0	0	0	0	0	0	0	0	1	0	0	0	1
0	0	0	0	0	0	0	1	X	0	0	1	0
0	0	0	0	0	0	1	X	X	0	0	1	1
0	0	0	0	0	1	X	X	X	0	1	0	0
0	0	0	0	1	X	X	X	X	0	1	0	1
0	0	0	1	X	X	X	X	X	0	1	1	0
0	0	1	X	X	X	X	X	X	0	1	1	1
0	1	X	X	X	X	X	X	X	1	0	0	0
1	X	X	X	X	X	X	X	X	1	0	0	1

FIGURE 6.30   BCD Priority Encoder © CENGAGE LEARNING 2012.

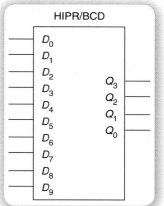

Your Turn

6.6   State the main limitation of the 3-bit binary encoder shown in Figure 6.25. How can the encoder be modified to overcome this limitation?

6.3   MULTIPLEXERS

KEY TERMS

Multiplexer   A circuit that directs one of several digital signals to a single output, depending on the states of several select inputs.

Data inputs   The multiplexer inputs that feed a digital signal to the output when selected.

Select inputs   The multiplexer inputs that select a digital input channel.

Double-subscript notation   A naming convention where two or more numerically related groups of signals are named using two subscript numerals. Generally, the first digit refers to a group of signals and the second to an element of a group (e.g., X_{03} represents element 3 of group 0 for a set of signal groups, X).

A **multiplexer** (abbreviated MUX) is a device for switching one of several digital signals to an output, under the control of another set of binary inputs. The inputs to be switched are called the **data inputs**; those that determine which signal is directed to the output are called the **select inputs**.

Figure 6.31 shows the logic circuit for a 4-to-1 multiplexer drawn in Xilinx ISE, with data inputs labeled D_0 to D_3 and the select inputs labeled S_0 and S_1. By examining the circuit, we can see that the 4-to-1 MUX is described by the following Boolean equation:

$$Y = D_0 \bar{S}_1 \bar{S}_0 + D_1 \bar{S}_1 S_0 + D_2 S_1 \bar{S}_0 + D_3 S_1 S_0$$

FIGURE 6.31   4-to-1 Multiplexer

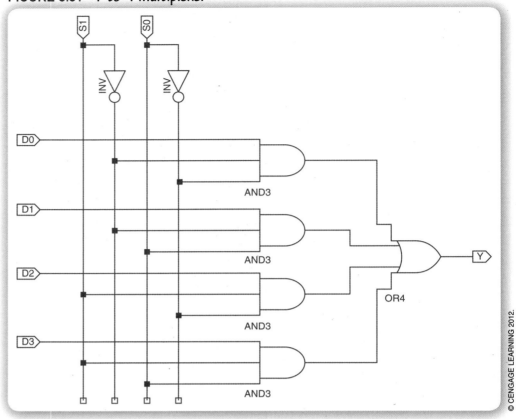

TABLE 6.11   4-to-1 MUX Truth Table

S_1	S_0	Y
0	0	D_0
0	1	D_1
1	0	D_2
1	1	D_3

© CENGAGE LEARNING 2012.

For any given combination of $S_1 S_0$, only one of the four product terms will be enabled. For example, when $S_1 S_0 = 10$, the equation evaluates to:

$$Y = (D_0 \cdot 0) + (D_1 \cdot 0) + (D_2 \cdot 1) + (D_3 \cdot 0) = D_2$$

The MUX equation can be described by a truth table as in *Table 6.11*. The subscript of the selected data input is the decimal equivalent of the binary combination $S_1 S_0$.

Figure 6.32 shows a symbol used for a 4-to-1 multiplexer.

In general, a multiplexer with *n* select inputs will have $m = 2^n$ data inputs. Thus, other common multiplexer sizes are 8-to-1 (for 3 select inputs) and 16-to-1 (for 4 select inputs). Data inputs can also be multiple-bit buses, as in *Figure 6.33*. The slash through a thick data line and the number 4 above the line indicate that it represents

> *Note . . .*
>
> A multiplexer with *n* select inputs will usually have 2^n data inputs.

FIGURE 6.32   4-to-1 Multiplexer Showing Individual Lines

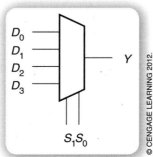

FIGURE 6.33   4-to-1 4-Bit Bus Multiplexer (Quad 4-to-1 MUX)

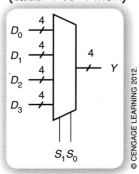

four related data signals. In this device, the select inputs switch groups of data inputs, as shown in the truth table in *Table 6.12*.

The naming convention shown in Table 6.12, known as **double-subscript notation**, is used frequently for identifying variables that are bundled in numerically related groups, the elements of which are themselves numbered. The first subscript identifies the group that a variable belongs to; the second subscript indicates which element of the group a variable represents.

TABLE 6.12 Truth Table for a 4-to-1 4-Bit Bus MUX

S_1	S_0	Y_3	Y_2	Y_1	Y_0
0	0	D_{03}	D_{02}	D_{01}	D_{00}
0	1	D_{13}	D_{12}	D_{11}	D_{10}
1	0	D_{23}	D_{22}	D_{21}	D_{20}
1	1	D_{33}	D_{32}	D_{31}	D_{30}

Multiplexing of Time-Varying Signals

We can observe the function of a multiplexer by using time-varying waveforms, such as a series of digital pulses. If we apply a different digital signal to each data input, and step the select inputs through an increasing binary sequence, we can see the different input waveforms appear at the output in a predictable sequence. It appears as though we are "changing the channel" and allowing only one of the inputs to pass through the multiplexer. This can be used as a basis for creating a simulation of a multiplexer.

Simulation Criteria:

► Each data input channel of the multiplexer will be selected in an ascending sequence by applying a binary count to the combined select inputs.

► Each data input should be easily recognizable by having a "signature" waveform applied to it. Each channel should be selected for a period no less than about three or four cycles of the signature waveform to allow us to clearly see the input signature waveform as the output waveform.

► The output waveform should display a series of unique signature waveforms, indicating the selection of the data channels in the correct sequence.

Example 6.15

Derive a set of simulation criteria that will fully test the operation of the 4-to-1 multiplexer shown in Figure 6.31. Create a simulation using Xilinx ISim, based on the criteria you derive.

Solution

The simulation criteria for the 4-to-1 multiplexer is shown in the box at the right. The simulation in Figure 6.34 is created in ISim by assigning a count value to the inputs as follows:

S[1..0]	12800 ns
D_3:	400 ns
D_2:	800 ns
D_1:	200 ns
D_0:	1600 ns

In *Figure 6.34*, we initially see the D_0 waveform appearing at the Y output when $S_1 S_0 = 00$, followed in sequence by the D_1, D_2, and D_3 waveforms when $S_1 S_0 = 01$, 10, and 11, respectively. The frequencies shown in the simulation were chosen to make as great a contrast as possible between adjacent inputs so that the different selected inputs could easily be seen. The select input waveforms are set to allow four cycles of the longest waveform (D_0) to appear at Y when selected.

FIGURE 6.34 Example 6.15: Simulation of a 4-to-1 Multiplexer

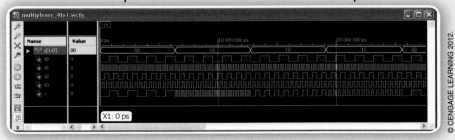

Multiplexer Applications

Multiplexers are used for a variety of applications, including selection of one data stream out of several choices, switching multiple-bit data from several channels to one multiple-bit output, sharing data on one output over time, and generating bit patterns or waveforms. We will look at the first two applications in this section.

Single-Channel Data Selection

The simplest way to use a multiplexer is to switch the select inputs manually, to direct one data source to the MUX output. Example 6.16 shows a pair of single-pole single-throw (SPST) switches supplying the select input logic for this type of application.

Example 6.16

Multisim Example

Multisim File: 06.11 4-to-1 MUX.ms11

Figure 6.35 shows a Multisim design that tests the operation of a 4-to-1 MUX. Four TTL-level pulse sources, each with a different frequency, are applied to the data inputs of the active multiplexer. The MUX output is monitored by a virtual oscilloscope. Note that the MUX select inputs are labeled differently than in Figure 6.31. Input S_1 is called B and input S_0 is called A.

Make a table listing which digital signal source in Figure 6.35 is routed to the oscilloscope for each combination of the multiplexer select inputs, S_1 and S_0.

FIGURE 6.35 Example 6.16: Multisim Circuit to Test a 4-to-1 Multiplexer

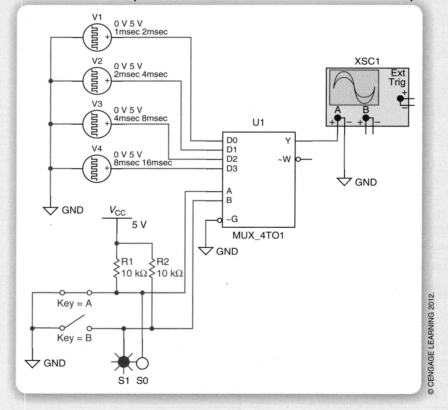

© CENGAGE LEARNING 2012.

Solution

"*Table 6.13* shows which data inputs are selected by each combination select inputs."

continued...

TABLE 6.13 *Sources Selected by a 4-to-1 MUX in Figure 6.35* © CENGAGE LEARNING 2012.

S_1 (B)	S_0 (A)	Selected Data Input	Selected Source
0	0	D_0	V1
0	1	D_1	V2
1	0	D_2	V3
1	1	D_3	V4

Interactive Exercise:

Open the Multisim file for this example and run it as a simulation. Switches A and B control the selection of the data inputs. To see the oscilloscope trace for the MUX output, open the oscilloscope window by double-clicking on the component labeled XSC1. Clicking the **Reverse** button in the oscilloscope window will change the background of the oscilloscope screen from black to white. Tapping the space bar again will change it back.

a. *Figure 6.36* shows an example of the oscilloscope window. The first half shows the MUX output when input D_0 is selected. The second half shows the output when D_2 is selected. What positions of switches A and B do these represent?

b. With the oscilloscope window open, click on the Multisim schematic window to select it. Operate switches A and B to go through all possible combinations. What do you see on the oscilloscope? How can you identify each signal you see?

FIGURE 6.36 **Example 6.16: Oscilloscope Trace Showing Multiplexer Output** © CENGAGE LEARNING 2012.

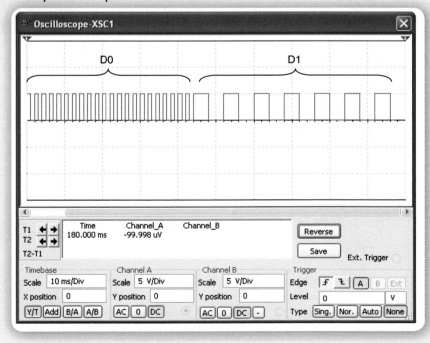

FIGURE 6.37 Quadruple 2-to-1 MUX Selecting 4-Bit Groups of Data

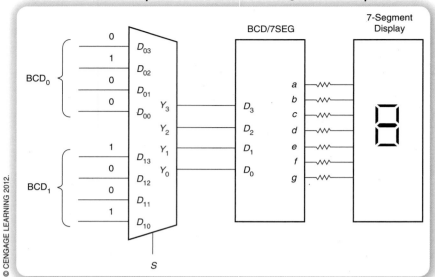

Multichannel Data Selection

Example 6.16 assumes that the output of a multiplexer is a single bit or stream of bits. Some applications require several bits to be selected in parallel, such as when data would be represented on a numerical display.

Figure 6.37 shows a circuit, based on a quadruple (4-channel) 2-to-1 multiplexer, that will direct one of two BCD digits to a seven-segment display. The MUX inputs are labeled in double-subscript notation. The bits $D_{03}D_{02}D_{01}D_{00}$ act as a 4-bit group input because the first digit of all four subscripts is 0. When the MUX select input (S) is 0, these inputs are all connected to the outputs $Y_3Y_2Y_1Y_0$. Similarly, when the select input is 1, inputs $D_{13}D_{12}D_{11}D_{10}$ are connected to the Y outputs. The seven-segment display in Figure 6.37 will display "4" if $S = 0$ (D_0 inputs selected) and "9" if $S = 1$ (D_1 inputs selected).

Figure 6.38 shows this function as a Multisim circuit. The bits 4A 3A 2A 1A act as a 4-bit group input, equivalent to the inputs $D_{03}D_{02}D_{01}D_{00}$ in Figure 6.36.

FIGURE 6.38A 74LS157 Quad 4-to-1 MUX (Selecting A Inputs)

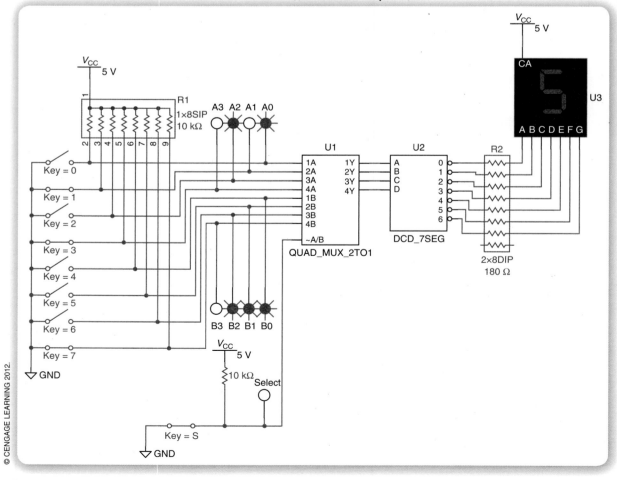

FIGURE 6.38B 74LS157 Quad 4-to-1 MUX (Selecting B Inputs)

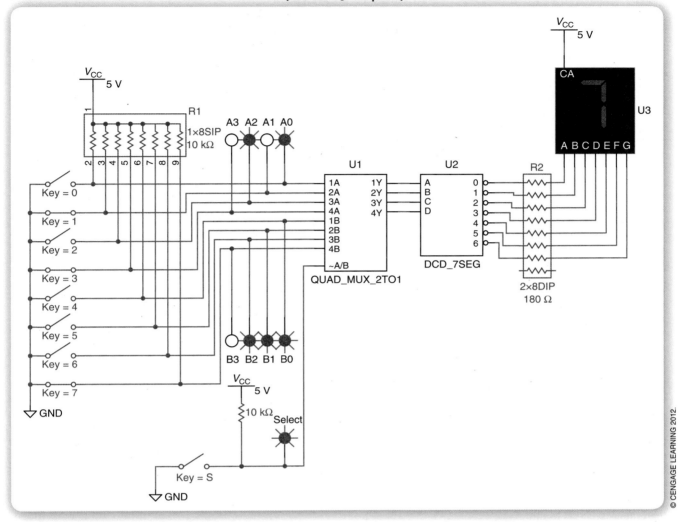

When the MUX select input ($\sim A/B$) is 0, these inputs are all connected to the outputs *4Y 3Y 2Y 1Y*. Similarly, when the select input is 1, inputs *4B 3B 2B 1B* (the equivalent of $D_{13}D_{12}D_{11}D_{10}$ in Figure 6.37) are connected to the *Y* outputs.

Example 6.17

Multisim Example

Multisim File: 06.13 Quad 2-to-1 MUX

Interactive Exercise:

Open the Multisim file for this example and set the switches as shown in Figure 6.38A. The output should display the digit 5. Change the position of the Select switch. You should see a digit 7, as shown in Figure 6.38B.

Change the switches so that the display shows the digit "3" when S = 0 and the digit "9" when S = 1. What are the switch logic levels for these cases?

6.4 DEMULTIPLEXERS

FIGURE 6.39 *4-Bit Demultiplexer/Decoder*

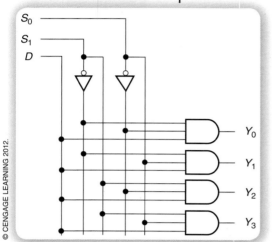

A **demultiplexer** performs the reverse function of a multiplexer. A multiplexer (MUX) directs one of several input signals to a single output; a demultiplexer (DMUX) directs a single input signal to one of several outputs. In both cases, the selected input or output is chosen by the state of an internal decoder.

Figure 6.39 shows the logic circuit for a 1-to-4 demultiplexer. Compare this to Figure 6.7, a 4-output decoder. These circuits are the same except that the active-LOW enable input has been changed to an active-HIGH data input. The circuit in Figure 6.39 could still be used as a decoder, except that its enable input would be active-HIGH.

Each AND gate in the demultiplexer enables or inhibits the signal output according to the state of the select inputs, thus directing the data to one of the output lines. For instance, $S_1 S_0 = 10$ directs incoming digital data to output Y_2, because 10 (binary) is 2 in decimal.

Figure 6.40 illustrates the use of a single device as either a decoder or a demultiplexer. In Figure 6.40a, input D is tied HIGH. When an output is selected by S_1 and S_0, it goes HIGH, acting as a decoder with active-HIGH outputs. In Figure 6.40b, D acts as a demultiplexer data input. The data are directed to the output selected by S_1 and S_0.

Multiplexer/Demultiplexer Systems

A multiplexer and demultiplexer can be used together in a system that transmits data from several digital channels along a single transmission line. The MUX selects data from several sources and directs it to one output. The DMUX accepts a single data source and distributes it to several destinations. This is used to save connection hardware between the transmitter and receiver parts of a digital communication system.

Figure 6.41 shows how a multiplexer, such as the one in Figure 6.31 can be used with a demultiplexer, such as the circuit shown in Figure 6.39. In order to keep the digital channels straight from source to destination, note that the select inputs of both the MUX and DMUX have the same source.

Note . . .

Because a single device can be used either way, this implies that all binary decoder designs used in this chapter can also be used as demultiplexers, provided they include an enable input.

A decoder/demultiplexer can have active-LOW outputs, but only if the D input is also active-LOW. This is important because the demultiplexer data must be inverted twice to retain its original logic values.

FIGURE 6.40 *Same Device Used as a Decoder or Demultiplexer*

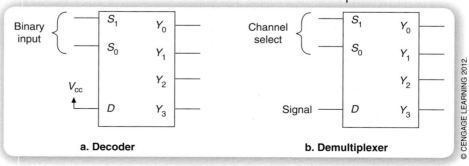

a. Decoder b. Demultiplexer

FIGURE 6.41   *Demultiplexing a Multiplexed Signal*

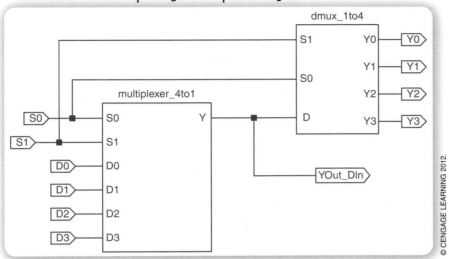

© CENGAGE LEARNING 2012.

Example 6.18

Write a set of simulation criteria for the MUX/DMUX system shown in Figure 6.40. Use these criteria to create a simulation using Xilinx ISim.

Solution

The simulation criteria for the MUX/DMUX circuit are shown at the box on the right. Figure 6.42 shows the simulation of the MUX/DMUX system as described by the simulation criteria we derived.

FIGURE 6.42   **Example 6.18: Simulation of a Demultiplexed Signal**

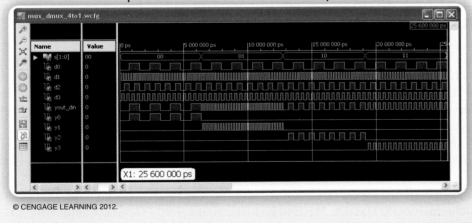

© CENGAGE LEARNING 2012.

Simulation Criteria:

► Each MUX input should have a recognizable signature waveform, so that we know which input waveform we are observing.

► Generate an increasing binary count on the MUX/DMUX select inputs. The length of the count period should be about 3 or 4 cycles of the slowest MUX input waveform.

► The MUX/DMUX connection, labeled YOut_DIn, should show the MUX input waveforms in sequence as they are selected, beginning with input D0 and progressing to D3.

► The DMUX outputs should show the selected portions of the YOut_DIn waveforms on the DMUX outputs in sequence, beginning with D0 on Y0 and progressing to D3 on Y3. When the DMUX outputs are not selected, they should remain in the LOW state.

CMOS Analog Multiplexer/Demultiplexer

KEY TERM

CMOS analog switch   A CMOS device that will pass an analog or digital signal in either direction, when enabled. Also called a transmission gate or bilateral switch. There is no TTL equivalent.

An interesting device used in some CMOS medium-scale integration multiplexers and demultiplexers, and in other applications, is the **CMOS analog switch**, also called a *transmission gate* or *bilateral switch*. This device has the property of

FIGURE 6.43 Line Drivers

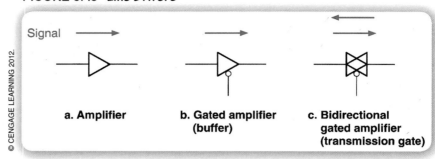

a. Amplifier

b. Gated amplifier (buffer)

c. Bidirectional gated amplifier (transmission gate)

© CENGAGE LEARNING 2012.

FIGURE 6.44 4-Channel Analog MUX/DMUX

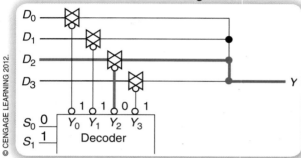

© CENGAGE LEARNING 2012.

allowing signals to pass in two directions, instead of only one, thus allowing both positive and negative voltages and currents to pass. It also has no requirement that the voltages be of a specific value such as $+5$ volts, as long as they fall within specified maximum and minimum values. These properties make the device suitable for passing analog signals.

Figure 6.43 shows several symbols, indicating the development of the transmission gate concept. Figure 6.43a and Figure 6.43b show amplifiers whose output and input are clearly defined by the direction of the triangular amplifier symbol. A signal has one possible direction of flow. Figure 6.43b includes an active-LOW gating input, which can turn the signal on and off.

Figure 6.43c shows two opposite-direction overlapping amplifier symbols, with a gating input to enable or inhibit the bidirectional signal flow. The signal through the transmission gate may be either analog or digital.

Analog switches are available in packages of four switches with part numbers such as 4066B (standard CMOS) or 74HC4066 (high-speed CMOS).

Several available CMOS MUX/DMUX chips use analog switches to send signals in either direction. *Figure 6.44* illustrates the design principle as applied to a 4-channel MUX/DMUX.

If four signals are to be multiplexed, they are connected to inputs D_0 to D_3. The decoder, activated by S_1 and S_0, selects which one of the four switches is enabled. Figure 6.44 shows channel 2 active ($S_1S_0 = 10$).

Because all analog switch outputs are connected together, any selected channel connects to Y, resulting in a multiplexed output. To use the circuit in Figure 6.44 as a demultiplexer, the inputs and outputs are merely reversed.

Some analog MUX/DMUX devices in high-speed CMOS include: 74HC4051 8-channel MUX/DMUX, 74HC4052 dual 4-channel MUX/DMUX, and 74HC4053 triple 2-channel MUX/DMUX.

Your Turn

6.7 Refer to the symbol for the 74138 3-line-to-8-line decoder in Figure 6.10. Show how to connect the inputs of this device so that it can be used as a 1-to-8 demultiplexer.

6.5 MAGNITUDE COMPARATORS

KEY TERM

Magnitude comparator A circuit that compares two *n*-bit binary numbers, indicates whether or not the numbers are equal, and, if not, which one is larger.

If we are interested in finding out whether or not two binary numbers are the same, we can use a **magnitude comparator**. The simplest comparison circuit is the Exclusive NOR gate, whose circuit symbol is shown in *Figure 6.45* and whose truth table is given in *Table 6.14*.

The output of the XNOR gate is 1 if its inputs are the same and 0 if they are different. For this reason, the XNOR gate is sometimes called a coincidence gate. The output of the XNOR gate can be symbolized by *AEQB*, indicating that the output is active when *A* equals *B*.

We can use several XNORs to compare each bit of two multibit binary numbers. *Figure 6.46* shows a 2-bit comparator with one output that goes HIGH if all bits of *A* and *B* are identical.

If the most significant bit (MSB) of *A* equals the MSB of *B*, the output of the upper XNOR is HIGH. If the least significant bits (LSBs) are the same, the output of the lower XNOR is HIGH. If both these conditions are satisfied, then *A* = *B*, which is indicated by a HIGH at the AND output. This general principle applies to any number of bits:

$$AEQB = (\overline{A_{n-1} \oplus B_{n-1}}) \cdot (\overline{A_{n-2} \oplus B_{n-2}}) \ldots (\overline{A_1 \oplus B_1}) \cdot (\overline{A_0 \oplus B_0})$$

for two *n*-bit numbers, *A* and *B*.

Some magnitude comparators also include an output that activates if *A* is greater than *B* (symbolized $A > B$ or *AGTB*) and another that is active when *A* is less than *B* (symbolized $A < B$ or *ALTB*). *Figure 6.47* shows the comparator of Figure 6.46 expanded to include the "greater than" and "less than" functions.

Let us analyze the *AGTB* circuit. The *AGTB* function has two AND-shaped gates that compare *A* and *B* bit-by-bit to see which is larger.

TABLE 6.14 XNOR Truth Table

A	B	AEQB
0	0	1
0	1	0
1	0	0
1	1	1

© CENGAGE LEARNING 2012.

FIGURE 6.45 Exclusive NOR Gate © CENGAGE LEARNING 2012.

FIGURE 6.46 2-Bit Magnitude Comparator © CENGAGE LEARNING 2012.

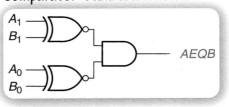

FIGURE 6.47 2-Bit Magnitude Comparator with AEQB, AGTB, and ALTB Outputs

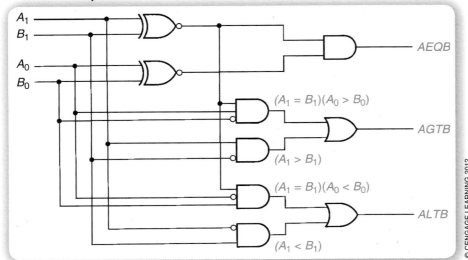

© CENGAGE LEARNING 2012.

1. The 2-input AND-shaped gate examines the MSBs of A and B. If $A_1 = 1$ AND $B_1 = 0$, then this gate output goes HIGH and we know that $A > B$. (This implies one of the following inequalities: $10 > 00$; $10 > 01$; $11 > 00$; or $11 > 01$.)

2. If $A_1 = B_1$, then we don't know whether or not $A > B$ until we compare the next most significant bits, A_0 and B_0. The 3-input gate makes this comparison. Because this gate is enabled by the XNOR, which compares the two MSBs, it is active only when $A_1 = B_1$. This yields the term $(\overline{A_1 \oplus B_1}) A_0 \overline{B_0}$ in the Boolean expression for the $AGTB$ function.

3. If $A_1 = B_1$ AND $A_0 = 1$ AND $B_0 = 0$, then the 3-input gate has a HIGH output, telling us, via the OR gate, that $A > B$. The only possibilities are ($01 > 00$) and ($11 > 10$).

Similar logic works in the $ALTB$ circuit, except that inversion is on the A, rather than the B bits. Alternatively, we can simplify either the $AGTB$ or the $ALTB$ function by using a NOR function. For instance, if we have developed a circuit to indicate $AEQB$ and $ALTB$, we can make the $AGTB$ function from the other two, as follows:

$$AGTB = \overline{AEQB + ALTB}$$

This Boolean expression implies that if A is not equal to or less than B, then it must be greater than B.

Expansion to more bits would use the same principle of comparing bits one at a time, beginning with the MSBs.

Example 6.19

TABLE 6.15   Truth Table for a 2-Bit Magnitude Comparator

A_1	A_0	B_1	B_0	ALTB	AEQB	AGTB
0	0	0	0	0	1	0
0	0	0	1	1	0	0
0	0	1	0	1	0	0
0	0	1	1	1	0	0
0	1	0	0	0	0	1
0	1	0	1	0	1	0
0	1	1	0	1	0	0
0	1	1	1	1	0	0
1	0	0	0	0	0	1
1	0	0	1	0	0	1
1	0	1	0	0	1	0
1	0	1	1	1	0	0
1	1	0	0	0	0	1
1	1	0	1	0	0	1
1	1	1	0	0	0	1
1	1	1	1	0	1	0

© CENGAGE LEARNING 2012.

a. Based on the operational description of a 2-bit magnitude comparator with active-HIGH outputs, make a truth table for the comparator. Boolean algebra is not required. Remember that you are comparing two 2-bit numbers in each case.

b. Use the truth table to create a simulation of the comparator.

Solution

a. *Table 6.15* shows the truth table of the magnitude comparator. $A_1 A_0$ is compared to $B_1 B_0$. Only one of the comparator outputs can be HIGH at one time, because, for example, if A equals B, it cannot be greater or less than B. Also, one output is always HIGH, because A must always be less than, equal to, or greater than B.

b. *Figure 6.48* shows a simulation of the comparator created with Xilinx ISim, based on Table 6.15 as a set of simulation criteria.

continued...

FIGURE 6.48 Example 6.19: Simulation of a 2-Bit Magnitude Comparator

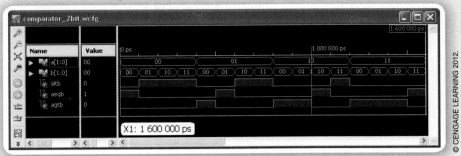

© CENGAGE LEARNING 2012.

Example 6.20

Multisim Example

Multisim File: 06.15 4-Bit Comparator

Interactive Exercise:

Open the Multisim file for this exercise, shown in *Figure 6.49*. The design uses a 4-bit comparator, labeled *MAG_CMP4*. The comparator has three outputs,

FIGURE 6.49 Example 6.20: 4-Bit Magnitude Comparator (A Is Greater Than B)

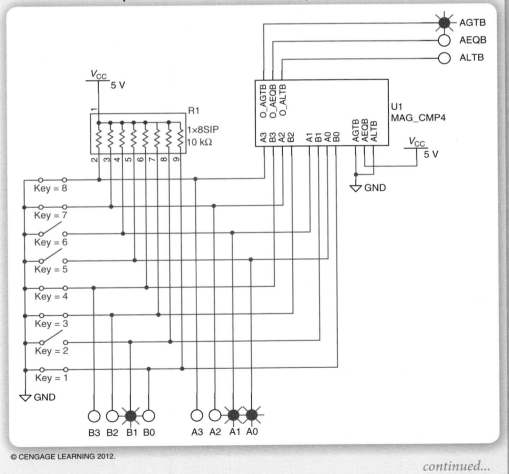

© CENGAGE LEARNING 2012.

continued...

OALTB, OAEQB, and OAGTB, as well as three "cascading inputs," ALTB, AEQB, and AGTB. The cascading inputs will be further described in the next example. For a single comparator, they must be connected so that AEQB is HIGH and the other two inputs are LOW.

a. Set all binary inputs LOW. Go through all binary combinations of $A_3A_2B_3B_2$. What do you observe?

b. Try the following binary combinations. What do you observe?

   i. $A_3A_2A_1A_0 = 1010$, $B_3B_2B_1B_0 = 1010$
  ii. $A_3A_2A_1A_0 = 1010$, $B_3B_2B_1B_0 = 0101$
 iii. $A_3A_2A_1A_0 = 0101$, $B_3B_2B_1B_0 = 1010$

Some magnitude comparators can be expanded to compare larger numbers of inputs by using the three cascading inputs, which pass information from low-order bits to high-order bits.

Example 6.21

Show how to create an 8-bit priority encoder from two 4-bit priority encoders by using their cascading inputs and outputs.

Solution

Figure 6.50 shows the connection between two comparators. The outputs of the low-bit comparator connect to the inputs of the high-bit comparator.

FIGURE 6.50 Example 6.21: 8-Bit Magnitude Comparator

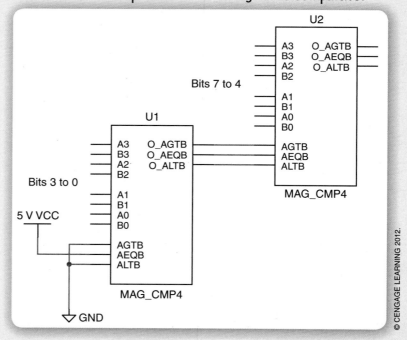

6.8 Describe the decision-making process that the circuit of a 3-bit magnitude comparator uses to determine whether input A is less than, equal to, or greater than B.

6.6 PARITY GENERATORS AND CHECKERS

KEY TERMS

Parity A system that checks for errors in a multibit binary number by counting the number of 1's.

Parity bit A bit appended to a binary number to make the number of 1's even or odd, depending on the type of parity.

Even parity An error-checking system that requires a binary number to have an even number of 1's.

Odd parity An error-checking system that requires a binary number to have an odd number of 1's.

When data are transmitted from one device to another, it is necessary to have a system of checking for transmission errors. These errors, which appear as incorrect bits, usually occur as a result of electrical limitations such as line capacitance or induced noise.

Parity error checking is a way of encoding information about the correctness of data before they are transmitted. The data can then be verified at the system's receiving end. *Figure 6.51* shows a block diagram of a parity error-checking system.

The parity generator in Figure 6.51 examines the outgoing data and adds a bit called the **parity bit** that makes the number of 1's in the transmitted data odd or even, depending on the type of parity. Data with **even parity** have an even number of 1's, including the parity bit, and data with **odd parity** have an odd number of 1's.

FIGURE 6.51 *Parity Error Checking*

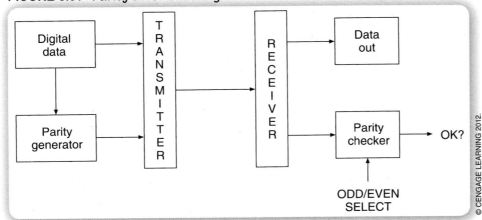

The data receiver "knows" whether to expect even or odd parity. If the incoming number of 1's matches the expected parity, the parity checker responds by indicating that correct data have been received. Otherwise, the parity checker indicates an error.

FIGURE 6.52 Exclusive OR Gate © CENGAGE LEARNING 2012.

An Exclusive OR gate can be used as a parity generator or a parity checker. *Figure 6.52* shows the gate, and *Table 6.16* is the XOR truth table. Notice that each line of the XOR truth table has an even number of 1's if we include the output column.

Figure 6.53 shows the block diagram of a circuit that will generate an even parity bit from 2 data bits, A and B, and transmit the three bits one after the other (that is, serially), to a data receiver.

Figure 6.54 shows a parity checker for the parity generator in Figure 6.53. Data are received serially, but read in parallel. The parity bit is recreated from the received values of A and B, and then compared to the received value of P to give an error indication, P'. If P and $A \oplus B$ are the same, then $P' = 0$ and the transmission is correct. If P and $A \oplus B$ are different, then $P' = 1$ and there has been an error in transmission.

TABLE 6.16 Exclusive OR Truth Table

A	B	$A \oplus B$
0	0	0
0	1	1
1	0	1
1	1	0

© CENGAGE LEARNING 2012.

FIGURE 6.53 Even Parity Generation

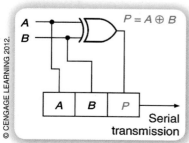

FIGURE 6.54 Even Parity Checking © CENGAGE LEARNING 2012.

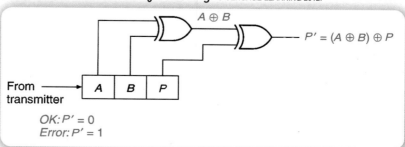

Example 6.23

The following data and parity bits are transmitted four times: $ABP = 101$. Data and parity are checked by the circuit in Figure 6.54.

1. State the type of parity used.

2. The transmission line over which the data are transmitted is particularly noisy and the data arrive differently each time as follows:

   a. $ABP = 101$
   b. $ABP = 100$
   c. $ABP = 111$
   d. $ABP = 110$

Indicate the output P' of the parity checker in Figure 6.54 for each case and state what the output means.

Solution

1. The system is using even parity. We know this because the transmitted data ABP has an even number of 1's.
2. The parity checker produces the following responses:

   a. $ABP = 101$
   $A \oplus B = 1 \oplus 0 = 1$
   $P' = (A \oplus B) \oplus P = 1 \oplus 1 = 0$ Data received correctly.

   b. $ABP = 100$
   $A \oplus B = 1 \oplus 0 = 1$
   $P' = (A \oplus B) \oplus P = 1 \oplus 0 = 1$ Transmission error. (Parity bit incorrect.)

   c. $ABP = 111$
   $A \oplus B = 1 \oplus 1 = 0$
   $P' = (A \oplus B) \oplus P = 0 \oplus 1 = 1$ Transmission error. (Data bit B incorrect.)

   d. $ABP = 110$
   $A \oplus B = 1 \oplus 1 = 0$
   $P' = (A \oplus B) \oplus P = 0 \oplus 0 = 0$ Transmission error undetected. (B and P incorrectly received.)

The second and third cases in Example 6.23 show that parity error detection cannot tell which bit is incorrect. Both cases yield the same output result, but a different bit was in error in each case.

The fourth case points out the major flaw of parity error detection: An even number of errors cannot be detected. This is true whether the parity is even or odd. If a group of bits has an even number of 1's, a single error will change that to an odd number of 1's, but a double error will change it back to even. (Try a few examples to convince yourself that this is true.)

An odd parity generator and checker can be made using an Exclusive NOR, rather than an Exclusive OR, gate. If a set of transmitted data bits requires a 1 for even parity, it follows that they require a 0 for odd parity. This implies that even and odd parity generators must have opposite-sense outputs.

Example 6.24

Modify the circuits in Figure 6.53 and Figure 6.54 to operate with odd parity. Verify their operation with the data bits $AB = 11$ transmitted twice and received once as $AB = 11$ and once as $AB = 01$.

Solution

Figure 6.55a shows an odd parity generator and *Figure 6.55b* shows an odd parity checker. The checker circuit still has an Exclusive OR output because it presents the same error codes as an even parity checker. The parity bit is recreated at the receive end of the transmission path and compared with the received parity bit. If they are the same, $P' = 0$ (correct transmission). If they are different, $P' = 1$ (transmission error).

Verification

Generator:
$$\text{Data: } AB = 11 \quad \text{Parity: } P = \overline{A \oplus B} = \overline{1 \oplus 1} = 1$$

Checker:
Received data: $AB = 11$
$$P' = (\overline{A \oplus B}) \oplus P = (\overline{1 \oplus 1}) \oplus 1 = 1 \oplus 1 = 0 \quad \text{(Correct transmission.)}$$

Generator:
$$\text{Data: } AB = 11 \quad \text{Parity: } P = \overline{A \oplus B} = \overline{1 \oplus 1} = 1$$

Checker:
Received data: $AB = 01$
$$P' = (\overline{A \oplus B}) \oplus P = (\overline{0 \oplus 1}) \oplus 1 = 0 \oplus 1 = 1 \quad \text{(Incorrect transmission.)}$$

FIGURE 6.55 Example 6.24: Odd Parity Generator and Checker

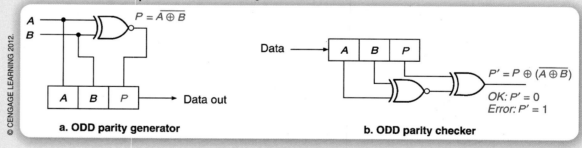

© CENGAGE LEARNING 2012.

a. ODD parity generator

b. ODD parity checker

Parity generators and checkers can be expanded to any number of bits by using an XOR gate for each pair of bits and combining the gate outputs in further stages of 2-input XOR gates. The true form of the generated parity bit is P_E, the even parity bit. The complement form of the bit is P_O, the odd parity bit.

Table 6.17 shows the XOR truth table for 4 data bits and the odd and even parity bits. The even parity bit P_E is given by $(A \oplus B) \oplus (C \oplus D)$. The odd parity bit P_O is given by $\overline{P_E} \, (\overline{(A \oplus B) \oplus (C \oplus D)})$. For every line in Table 6.17, the bit combination $ABCDP_E$ has an even number of 1's and the group $ABCDP_O$ has an odd number of 1's.

TABLE 6.17 Even and Odd Parity Bits for 4-Bit Data

A	B	C	D	$A \oplus B$	$C \oplus D$	P_E	P_O
0	0	0	0	0	0	0	1
0	0	0	1	0	1	1	0
0	0	1	0	0	1	1	0
0	0	1	1	0	0	0	1
0	1	0	0	1	0	1	0
0	1	0	1	1	1	0	1
0	1	1	0	1	1	0	1
0	1	1	1	1	0	1	0
1	0	0	0	1	0	1	0
1	0	0	1	1	1	0	1
1	0	1	0	1	1	0	1
1	0	1	1	1	0	1	0
1	1	0	0	0	0	0	1
1	1	0	1	0	1	1	0
1	1	1	0	0	1	1	0
1	1	1	1	0	0	0	1

© CENGAGE LEARNING 2012.

Example 6.25

Use Table 6.17 to draw a 4-bit parity generator and a 4-bit parity checker that can generate and check either even or odd parity, depending on the state of one select input.

Solution

Figure 6.56 shows the circuit for a 4-bit parity generator. The XOR gate at the output is configured as a programmable inverter to give P_E or P_O. When EVEN/ODD = 0, the parity output is not inverted and the circuit generates P_E. When EVEN/ODD = 1, the XOR inverts the parity bit, giving P_O.

The 4-bit parity checker, shown in *Figure 6.57*, is the same circuit, with an additional XOR gate to compare the parity bit recreated from data and the previously encoded parity bit.

FIGURE 6.56 Example 6.25: 4-Bit Parity Generator © CENGAGE LEARNING 2012.

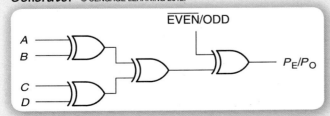

FIGURE 6.57 Example 6.25: 4-Bit Parity Checker © CENGAGE LEARNING 2012.

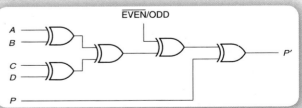

Example 6.26

Multisim Example

Multisim File: 06.17 Parity Generator and Checker with Error Generator

Figure 6.58 shows a Multisim design that demonstrates the operations of parity generation and checking. Gates U2A, U2B, and U2C generate a parity bit with

FIGURE 6.58 Example 6.26: 4-Bit Parity System (No Errors)

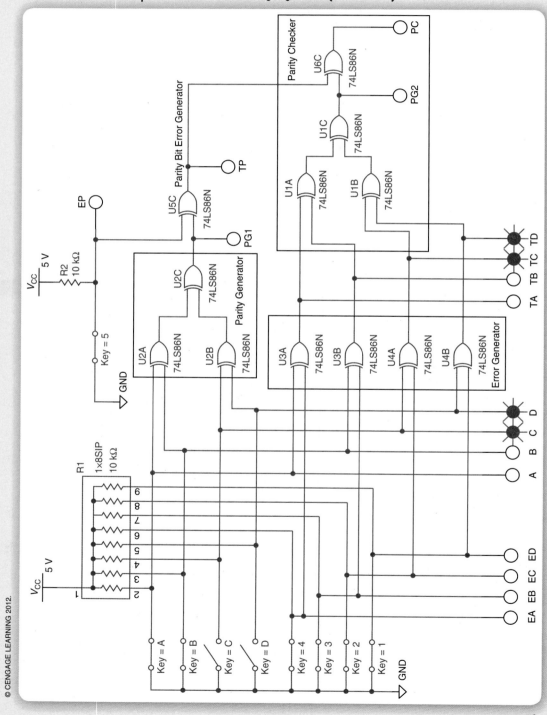

continued...

FIGURE 6.59   Example 6.26: 4-Bit Parity System (Error in Data Bit)

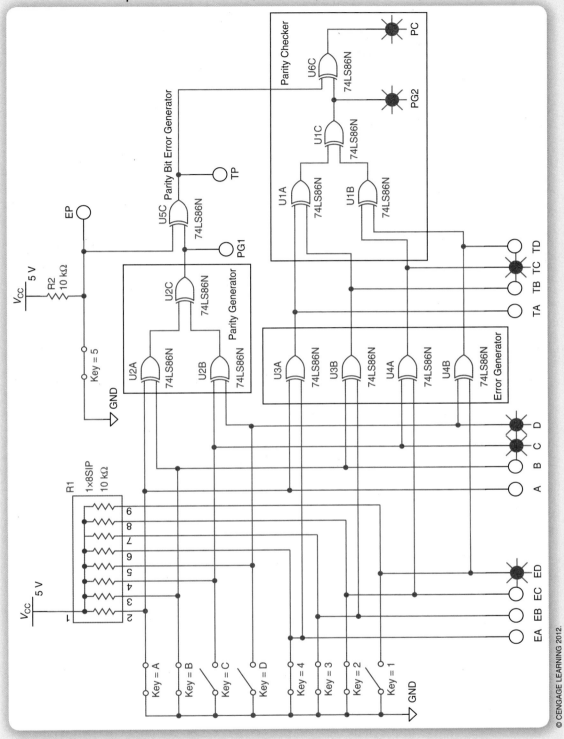

even parity from four data bits ABCD. Gates U1A, U1B, U1C, and U6C act as a parity checker. Errors can be introduced onto any transmitted data or parity bit through gates U3A, U3B, U4A, U4B, or U5C.

Digital probes monitor the various bits as shown in *Table 6.18*.

continued...

FIGURE 6.60 Example 6.26: 4-Bit Parity System (Error in Parity Bit)

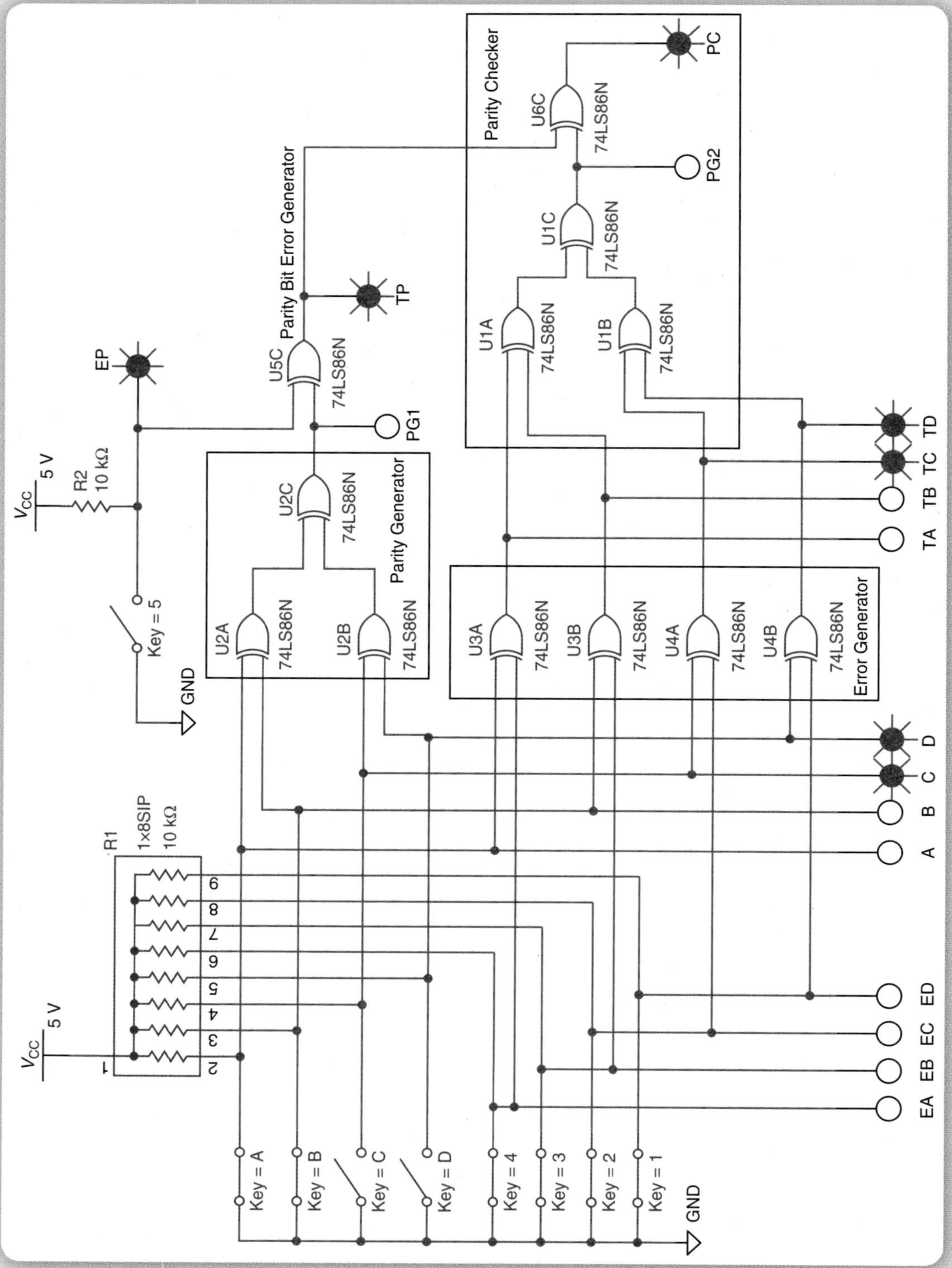

Examine Figure 6.58, *Figure 6.59*, and *Figure 6.60*.

a. Figure 6.58 shows the system with no errors introduced. How can you tell?

b. Figure 6.59 has an error on bit D only. How can you tell?

c. Figure 6.60 has an error on the parity bit only. How can you tell?

continued...

TABLE 6.18 Functions of Parity System Test Circuit

Digital Probe(s)	Function
A B C D	Original 4-bit data, determined by switches *A*, *B*, *C*, and *D*.
EA EB EC ED	Indicates an error is introduced onto a data bit. Probe ON means there is an error on that bit. Errors are determined by switches 1, 2, 3, and 4.
TA TB TC TD	Actual transmitted data after errors have been introduced or not.
PG1	Parity generated from original data. Even parity.
EP	Error introduced to parity bit when ON, determined by switch 5.
TP	Actual transmitted parity bit after error has been introduced or not.
PG2	Parity bit generated from actual received data.
PC	Parity checker output. ON if a parity error is detected.

Solution

a. We know the data in Figure 6.58 has no errors because all of the following conditions are true:
   ▶ The transmitted data *TA TB TC TD* is the same as the original data *A B C D*.
   ▶ The transmitted parity bit *TP* is the same as the parity bit generated from the data, *PG1*.
   ▶ The parity bit generated by the received data, *PG2*, is the same as the transmitted parity bit, *TP*.
   ▶ The parity checker output, *PC*, is LOW.

b. We know the data in Figure 6.59 has errors in the data because:
   ▶ The transmitted data *TA TB TC TD* is not the same as the original data *A B C D*.
   ▶ Error light *ED* is ON.
   ▶ The parity bit generated by the received data, *PG2*, is not the same as the transmitted parity bit, *TP*.
   ▶ The parity checker output, *PC*, is HIGH.

c. We know the data in Figure 6.60 has an error in the parity bit because:
   ▶ The transmitted parity bit *TP* is not the same as the parity bit generated from the data, *PG1*.
   ▶ Error light *EP* is ON.
   ▶ The parity bit generated by the received data, *PG2*, is not the same as the transmitted parity bit, *TP*.
   ▶ The parity checker output, *PC*, is HIGH.

Interactive Example:

Open the Multisim file for this example and run it as a simulation.

a. Set the switches exactly the same as in Figure 6.58. Now make the input data *A B C D* = 0111. Introduce an error onto bit *B* only. What happens?

b. Set the switches exactly the same as in Figure 6.58. Now make the input data *A B C D* = 0111. Introduce errors onto bits *B* and *D* only. What happens?

c. Set the switches exactly the same as in Figure 6.58. Now make the input data *A B C D* = 0111. Introduce errors onto bit *B* and the parity bit only. What happens?

Your Turn

6.9 Data (including a parity bit) are detected at a receiver configured for checking odd parity. Which of the following data do we know are incorrect? Could there be errors in the remaining data? Explain.

a. 010010
b. 011010
c. 1110111
d. 1010111
e. 1000101

6.7 TROUBLESHOOTING COMBINATIONAL LOGIC FUNCTIONS

When a digital circuit does not behave as expected, there may be a variety of causes for the malfunction. At the design phase, faults might result from errors in the schematic or other design information. During circuit construction, errors can be introduced by wiring mistakes. If the circuit is known to be designed and constructed properly, we may find circuit faults due to component failure.

Design and Construction Errors

Many of the functions studied here are built as a series of smaller combinational circuits, then assembled into a larger, more-complex circuit that performs some function. One example is the BCD-to-seven-segment decoder. Each segment, *a–g*, can be designed separately using the same four decoder inputs. When the circuit for each segment decoder is correctly drawn, the seven-segment display accurately shows the number specified by the BCD input. One small error in drawing the schematic can result in incorrect numbers being displayed. When this happens, many of the numbers will display correctly, further confusing the circuit designer. Rather than erasing an entire drawing, it is usually possible to troubleshoot and fix the error by identifying specifically where problems occur. Paying attention to the actual output vs. the expected output can be a valuable tool in finding the error in the drawing.

FIGURE 6.61 *BCD-to-Seven-Segment Display with a Correct and Incorrect Digit 6*

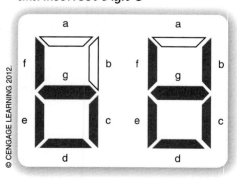

For example, consider a common anode BCD-to-seven-segment decoder with an error that occurs only when displaying the number 6. *Figure 6.61* shows the error; although we expect a 6 on the display, we see a character that resembles an upside-down letter A.

Assume that we have tried all other input combinations and they all result in the correct number displayed on the seven-segment display. A close inspection shows that only segment *b* is malfunctioning when we display the number 6. This allows us to inspect only a small portion of the entire schematic: we need only look at the equation and schematic for the driver for segment *b*, and can assume that the other segments are working correctly.

We can begin by testing each input combination and building a truth table for segment *b*. To test where the malfunction occurs, we should recreate the truth table and compare it with the original result. *Figure 6.62a* shows the original K-map. *Figure 6.62b* shows a version with the erroneous output found when the input combination is 0110. Reviewing the desired circuit found in Figure 6.22, we inspect the schematic and find the circuit in *Figure 6.63* instead. Although it looks very similar, the inverter on the bottom is connected to the wrong input. As a result, we are decoding 0101 twice, but never decoding 0110. Fixing this error results in a perfectly functional BCD-to-seven-segment decoder.

Errors like this can occur in any circuit, and as the circuits grow more complex, errors are more likely. Notice that in this case we were able to find the small wiring mistake without redrawing the entire circuit. A few minutes of carefully reviewing the expected output vs. the actual output can help identify errors quickly and efficiently.

Component Failure

If a circuit is known to have been designed and built correctly, it still might fail due to a failure of one or more of its components.

FIGURE 6.62 (a) K-Map from Original Segment *b* (b) K-Map Showing Missed Term from Segment *b* Function (Red Highlight)

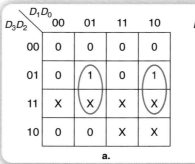

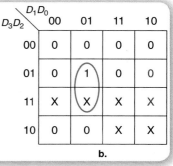

a.   b.

© CENGAGE LEARNING 2012.

FIGURE 6.63 Incorrect Decoder Circuit for Segment *b* in a BCD-to-Seven-Segment Decoder

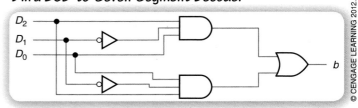

© CENGAGE LEARNING 2012.

Example 6.27

Multisim Example

Multisim File: 06.18 BCD-to-7-Segment Decoder (CA) with fault.ms11

Interactive Example:

Open the Multisim file for this example and check all input combinations for decimal digits 0–9. You should see that the digits are all correct, except for the displayed digits 2, 5, and 6, as shown in *Figure 6.64*. If we examine the correct digit patterns in Figure 6.18, we will see that the only display segments with problems are segments *b* and *c*.

A quick check with a logic probe, as shown in *Figure 6.65*, shows a LOW on *b* and a HIGH on *c*, which is correct for the code shown. However, on the opposite side of the resistors, we note that both *b* and *c* are HIGH, as shown in *Figure 6.66*. This does not tell us too much, though. For an ON segment, there should be a voltage of about 4.3 volts (equivalent to the power supply minus the voltage dropped across a diode) on the display side of the resistor and, for an OFF segment, there should be a voltage of about 5 volts. These are both enough to turn on the Multisim logic probe, which senses a value of 2.5 volts as HIGH.

We should check circuit voltages of the display inputs. Insert a voltmeter across points *b* and *c*, as shown in Figure 6.66. The meter shows a value of several picovolts. This value is so small that it is effectively zero. (Most voltmeters

FIGURE 6.64 Example 6.27: Correct and Incorrect Displays of Digits 2, 5, and 6 on a Seven-Segment Display

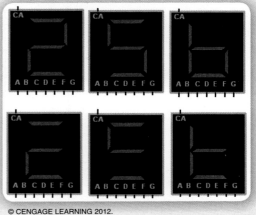

© CENGAGE LEARNING 2012.

continued...

FIGURE 6.65 Example 6.27: Troubleshooting: Decoder Output OK

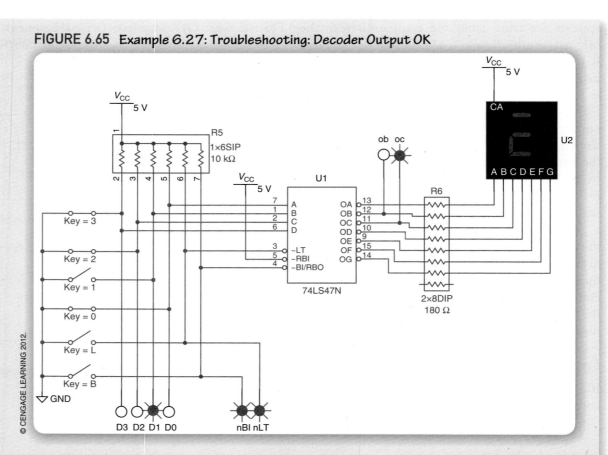

FIGURE 6.66 Example 6.27: Troubleshooting: Incorrect Level on One Display Input

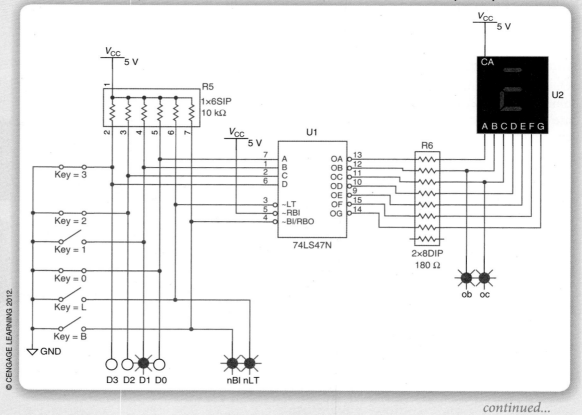

continued...

FIGURE 6.67 Example 6.27: Troubleshooting: Display Pins *b* and *c* Almost Same Voltage

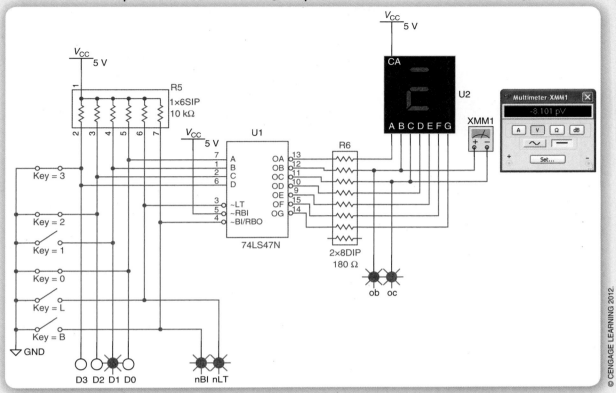

FIGURE 6.68 Example 6.27: Troubleshooting: Display Pins *a* and *c* Differ by 35 mV

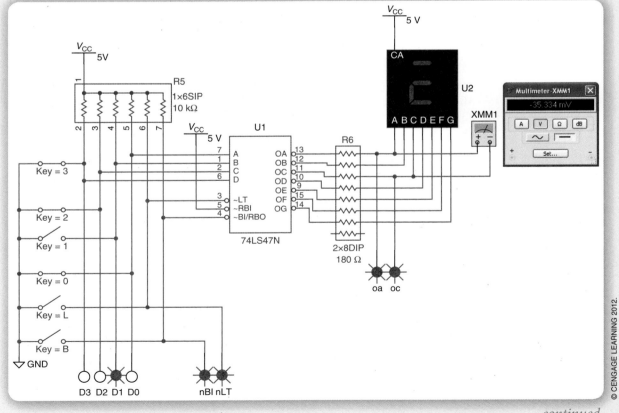

continued...

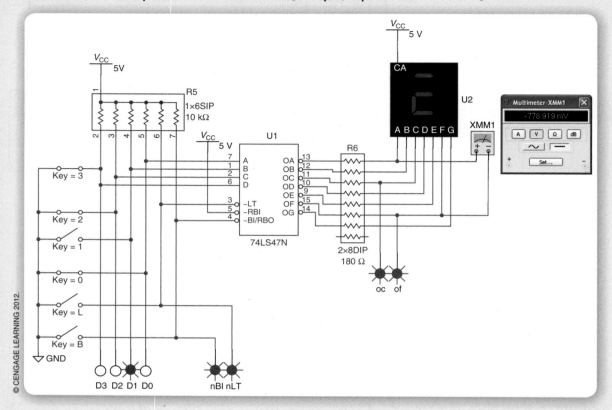

would not be able to measure a value this small.) Since *b* and *c* are supposed to be at different levels, but they are not, it seems that there is a short-circuit between the two points.

Let us check voltages from point *c* to a correctly functioning segment that is ON (segment *a*) and to a segment that is OFF (segment *f*). *Figure 6.68* shows the voltage between segments *a* and *c* (35 mV). This would be 0 volts without the short-circuit between *b* and *c*. However, the short changes the resistance value for segments *b* and *c*, resulting in a slight voltage difference. *Figure 6.69* shows the voltage between segments *c* and *f* (about 780 mV). This is a reasonable difference between an ON-state and OFF-state LED.

These tests lead us to believe that segments *b* and *c* in the seven-segment decoder are shorted together.

Delete the seven-segment display and replace it with another 74LS47N component. The circuit should work correctly with the new component.

Your Turn

6.10 If a BCD-to-seven-segment decoder displayed a 9 that appeared to be the letter *A*, the equation for which segment would be investigated first?

SUMMARY

1. A decoder detects the presence of a particular binary code. The simplest decoder is an AND or NAND gate, which can detect a binary code when combined with the right combination of input inverters.

2. Multiple-output decoders are implemented by a series of single-gate decoders, each of which responds to a different input code.

3. For an n-input decoder, there can be as many as 2^n unique outputs.

4. Some multiple-input decoders have an enable input that allows all decoder outputs to be deactivated when the enable input is inactive.

5. The effectiveness of the simulation of an FPGA design in software depends on selecting good criteria for a simulation. Ideally, good simulation criteria must test the design in all working modes and all potential failure modes.

6. A seven-segment display is an array of seven luminous segments (usually LED or LCD), arranged in a figure-8 pattern, used to display numerical digits.

7. The segments in a seven-segment display are designated by lowercase letters a through g. The sequence of labels goes clockwise, starting with segment a at the top and ending with g in the center.

8. Seven-segment displays are configured as common anode (active-LOW inputs) or common cathode (active-HIGH segments).

9. A seven-segment decoder can be described with a truth table or Boolean equation for each segment function.

10. Ripple blanking is a technique that allows leading or trailing zeros in a multiple-digit numerical display to be suppressed, while allowing internal zeros to be displayed. The technique works by cascading a ripple blanking output (*RBO*) to the ripple blanking input (*RBI*) of the next decoder.

RBO is LOW only if the corresponding decoder input is 0000 AND *RBI* on the same decoder is also LOW. A zero is suppressed if the decoder input is 0000 and $RBI = 0$. A zero is displayed if the decoder input is 0000 and $RBI = 1$.

11. An encoder is the complementary device to a decoder. It generates a binary code corresponding to the number of an active input. Without introducing priority circuitry, only one input can be active at a time, or erroneous codes can be generated.

12. A priority encoder allows more than one input to be active at a time. The output binary code corresponds to the active input with the highest priority, usually the one with the highest number.

13. A low-priority input must not change the code resulting from a higher-priority input. This is done by using OR gates to create the required output codes and AND gates to block certain code combinations.

14. A multiplexer (MUX) is a circuit that directs a signal or group of signals (called the data inputs) to an output, based on the status of a set of select inputs.

15. Generally, for n select inputs in a multiplexer, there are $m = 2^n$ data inputs. Such a multiplexer is referred to as an m-to-1 multiplexer.

16. The selected data input in a MUX is usually denoted by a subscript that is the decimal equivalent of the combined binary value of the select inputs. For example, if the select inputs in an 8-to-1 MUX are set to $S_2 S_1 S_0 = 100$, data input D_4 is selected because 100 (binary) = 4 (decimal).

17. A MUX can be designed to switch groups of signals to a multibit output.

18. A demultiplexer (DMUX) receives data from a single source and directs the data to one of several outputs, which is selected by the status of a set of select inputs.

continues...

continued...

19. A decoder with an enable input can also act as a demultiplexer if the enable input of the decoder is used as a data input for a demultiplexer.

20. A CMOS analog multiplexer or demultiplexer works by using a decoder to enable a set of analog data transmission switches. It can be used in either direction.

21. A magnitude comparator determines whether two binary numbers are equal and, if not, which one is greater.

22. The simplest equality comparator is an XNOR gate, whose output is HIGH if both inputs are the same.

23. A pair of multiple-bit numbers can be compared by a set of XNOR gates whose outputs are ANDed. The circuit compares the two numbers bit-by-bit.

24. Given two numbers A and B, the Boolean function $\overline{A}_n B_n$, if true, indicates that the nth bit of A is less than the nth bit of B.

25. Given two numbers A and B, the Boolean function $A_n \overline{B}_n$, if true, indicates that the nth bit of A is greater than the nth bit of B.

26. The less-than and greater-than functions can be combined with an equality comparator to determine, bit-by-bit, how two numbers compare in magnitude to one another.

27. Parity checking is a system of error detection that works by counting the number of 1's in a group of bits.

28. Even parity requires a group of bits to have an even number of 1's. Odd parity requires a group of bits to have an odd number of 1's. This is achieved by counting the 1's in the data bits, calculating the value of the parity bit needed for odd or even parity, and appending it to the data.

29. An XOR gate is the simplest even parity generator. Each line in its truth table has an even number of 1's, if the output column is included.

30. An XNOR gate can be used to generate an odd parity bit from two data bits.

31. A parity checker consists of a parity generator on the receive end of a transmission system and a comparator to determine if the locally generated parity bit is the same as the transmitted parity bit.

32. Parity generators and checkers can be expanded to any number of bits by using an XOR gate for each pair of bits and combining the gate outputs in further stages of 2-input XOR gates.

33. Digital circuit errors require troubleshooting to restore a circuit to working condition. Circuit faults can derive from errors of design or construction or from component failure. Design errors can be fixed by changing design information such as a schematic. Construction errors and component failure can be detected by circuit measurement and inspection.

BRING IT HOME

6.1 Decoders

6.1 When a HIGH is on the outputs of each of the decoding circuits shown in *Figure 6.70*, what is the binary code appearing at the inputs? Write the Boolean expression for each decoder output.

FIGURE 6.70 Problem 6.1: Decoding Circuits

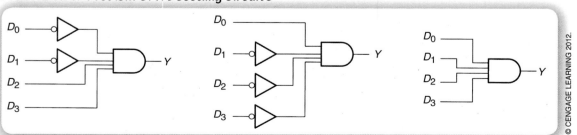

6.2 Draw the decoding circuit for each of the following Boolean expressions:

a. $\overline{Y} = \overline{D_3}\overline{D_2}\overline{D_1}D_0$

b. $\overline{Y} = \overline{D_3}\overline{D_2}D_1D_0$

c. $Y = \overline{D_3}\overline{D_2}D_1D_0$

d. $\overline{Y} = D_3D_2\overline{D_1}\overline{D_0}$

e. $Y = D_3D_2\overline{D_1}D_0$

6.3 a. Draw the schematic of a 2-line-to-4-line decoder with active-HIGH outputs and an active-LOW enable input.

b. Write a list of simulation criteria for the decoder.

6.4 a. Draw the schematic of a 3-line-to-8-line decoder with active-HIGH outputs and an active-LOW enable input.

b. Write a list of simulation criteria for the decoder.

6.5 For a generalized n-line-to-m-line decoder, state the value of m if n is:

a. 5

b. 6

c. 8

Write the equation giving the general relation between n and m.

6.6 a. Draw the schematic of a 3-line-to-8-line decoder with active-LOW outputs and no enable.

b. Write a list of simulation criteria for the decoder.

6.7 **Multisim Problem**
Multisim File: 06.04 3to8 Decoder Active HIGH.ms11

Interactive Exercise:

a. Figure 6.9 shows an active-HIGH 3-line-to-8-line decoder created with Multisim, saved in the Multisim file for this problem. Open the Multisim file and save it as **06.05 3to8 Decoder Active LOW.ms11**. Change the decoder so that it has active-LOW outputs.

b. Write the Boolean expression for each decoder output.

c. Which output is active when the input code is $D_2D_1D_0 = 110$? Is this output HIGH or LOW when active?

d. Operate the three logic switches in such a way as to activate the outputs one after the other. What sequence do you observe at the inputs when the outputs are activated in this order?

6.8 **Multisim Problem**
Multisim File: 06.06 74138 Decoder.ms11

Interactive Exercise:

Open the Multisim file for this problem and run it as a simulation.

a. Assume all the enable inputs are set to the logic levels shown in Figure 6.11. Which output is active when the input switches are changed so that $D_2D_1D_0 = 010$? What is the logic level of the active output in this condition?

b. Assume the D inputs are set to the logic levels shown in Figure 6.11. What are the logic levels of the outputs when input $\overline{G2A}$ is changed to 1?

c. Try all eight combinations of the three enable inputs. Which combination is required to enable the decoder outputs?

6.9 Write a truth table for a hexadecimal-to-seven-segment decoder for a common anode display. Use the digit patterns of Figure 6.21 as a model.

6.2 Encoders

6.10 *Figure 6.71* shows a BCD priority encoder with three sets of inputs. Determine the resulting output code for each input combination. Inputs and outputs are active-HIGH.

6.11 Which input must be active for a BCD priority encoder for its active-HIGH outputs to have the following code:

a. $Q_3Q_2Q_1Q_0 = 0111$

b. $Q_3Q_2Q_1Q_0 = 0001$

c. $Q_3Q_2Q_1Q_0 = 1001$

d. $Q_3Q_2Q_1Q_0 = 0000$

6.12 Which inputs can be determined and which are unknown for a BCD priority encoder for its active-LOW outputs to have the following code:

a. $Q_3Q_2Q_1Q_0 = 0111$

b. $Q_3Q_2Q_1Q_0 = 1110$

c. $Q_3Q_2Q_1Q_0 = 1100$

d. $Q_3Q_2Q_1Q_0 = 1111$

6.13 *Figure 6.72* shows a set of simulation waveforms for a 3-bit priority encoder. The encoder has complementary outputs (i.e., both active-LOW and active-HIGH output). Write a set of simulation criteria that could be used to create these waveforms.

continues...

continued...

FIGURE 6.71 Problem 6.10: BCD Priority Encoder

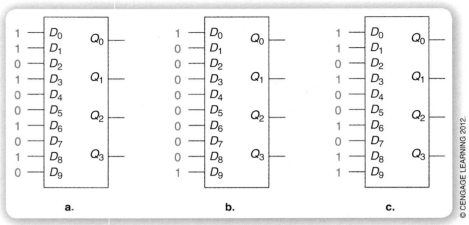

a.   b.   c.

© CENGAGE LEARNING 2012.

FIGURE 6.72 Problem 6.13: Simulation of a 3-Bit Priority Encoder with Active-LOW Inputs and Complementary Outputs © CENGAGE LEARNING 2012.

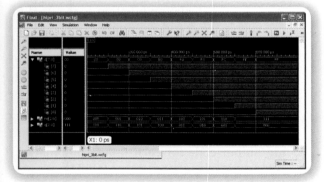

6.14 *Figure 6.73* shows a set of simulation waveforms for a 4-bit priority encoder. The encoder has complementary outputs (i.e., both active-LOW and active-HIGH output). Write a set of simulation criteria that could be used to create these waveforms.

FIGURE 6.73 Problem 6.14: Simulation of a 4-Bit Priority Encoder with Active-LOW Inputs and Complementary Outputs © CENGAGE LEARNING 2012.

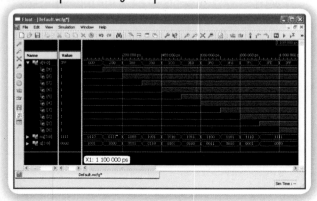

6.3 Multiplexers

6.15 *Figure 6.74* shows an 8-to-1 multiplexer used as a switcher for a digital video system, where the digital information from one of eight DVD players is switched to a digital output line. Make a table listing which digital video source in Figure 6.74 is routed to output Y for each combination of the multiplexer select inputs.

FIGURE 6.74 Problem 6.15: 8-to-1 Audio MUX

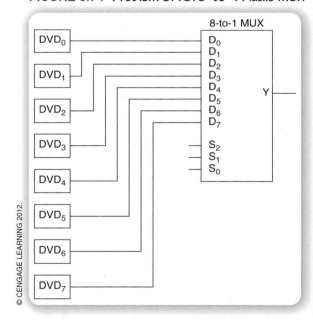

© CENGAGE LEARNING 2012.

6.16 Draw the symbol for an 8-to-1 multiplexer. Write the truth table for the multiplexer, showing which data input is selected for every binary combination of the select inputs.

6.17 Draw the symbol for a 16-to-1 multiplexer. Write the truth table for the multiplexer, showing which data input is selected for every binary combination of the select inputs.

6.18 Write the Boolean expression describing an 8-to-1 multiplexer. Evaluate the equation for the case where input D_5 is selected.

6.19 a. *Figure 6.75* shows the circuit for a 2-to-1 multiplexer. Write the Boolean expression for this circuit. Evaluate the case where input D_1 is selected.

FIGURE 6.75 Problem 6.19: Circuit for a 2-to-1 MUX

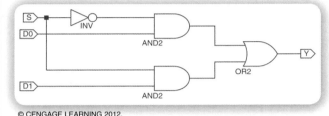

© CENGAGE LEARNING 2012.

b. *Figure 6.76* shows a set of simulation waveforms that can be used to test the design of the 2-to-1 multiplexer in Figure 6.74. Write a set of simulation criteria that could be used to create these waveforms.

FIGURE 6.76 Problem 6.19: Simulation for a 2-to-1 MUX

© CENGAGE LEARNING 2012.

6.4 Demultiplexers

6.20 a. Draw a circuit for a 1-to-2 demultiplexer with active-HIGH outputs.

b. Make a truth table that describes the function of the 1-to-2 demultiplexer drawn in part **a** of this question.

6.21 a. Draw a circuit for a 1-to-8 demultiplexer with active-HIGH outputs.

b. Make a truth table that describes the function of the 1-to-8 demultiplexer drawn in part **a** of this question.

6.22 Draw a diagram showing how eight analog switches can be connected to a decoder to form an 8-channel MUX/DMUX circuit. Briefly explain why the same circuit can be used as a multiplexer or as a demultiplexer.

6.23 Briefly state what characteristics of an analog switch make it suitable for transmitting analog signals.

6.5 Magnitude Comparators

6.24 What are the output values of a 3-bit magnitude comparator with active-HIGH output functions $AEQB$, $AGTB$, and $ALTB$ for the following input values:

   a. $A = 010$, $B = 101$
   b. $A = 000$, $B = 111$
   c. $A = 010$, $B = 010$
   d. $A = 110$, $B = 101$

6.25 For a magnitude comparator with active-HIGH output functions $AEQB$, $AGTB$, and $ALTB$, is it ever possible for more than one output to be HIGH at any time? Explain your answer.

6.6 Parity Generators and Checkers

6.26 What parity bit, P, should be added to the following data if the parity is even? If the parity is odd?

   a. 1111100
   b. 1010110
   c. 0001101

6.27 The following data are transmitted in a serial communication system (P is the parity bit). What parity is being used in each case?

   a. $ABCDEFGHP = 010000101$
   b. $ABCDEFGHP = 011000101$
   c. $ABCDP = 01101$
   d. $ABCDEP = 101011$
   e. $ABCDEP = 111011$

6.28 The data $ABCDEFGHP = 110001100$ are transmitted in a serial communication system. Give the output P' of a receiver parity checker for the following received data. State the meaning of the output P' for each case.

   a. $ABCDEFGHP = 110101100$
   b. $ABCDEFGHP = 110001101$
   c. $ABCDEFGHP = 110001100$
   d. $ABCDEFGHP = 110010100$

6.1 Decoders

6.29 a. Write the truth table for a binary-coded decimal (BCD) decoder. The decoder should decode a 4-bit input with a binary value between 0000 and 1001 by making one of its ten outputs LOW. The outputs are labeled $\overline{Y}_0$ to $\overline{Y}_9$. Unused codes should disable all decoder outputs.

b. Write a list of simulation criteria for this decoder.

6.30 Draw a diagram consisting of four seven-segment displays, each driven by a BCD-to-seven-segment decoder with ripple blanking. The circuit should be configured to suppress all leading zeros. Show the displayed digits and *RBO/RBI* logic levels for each of the following displayed values: 100, 217, 1024.

6.31 Multisim Problem
Multisim File: 06.07 BCD-to-7-Segment Decoder (CA).ms11

Interactive Exercise:

Open the Multisim file for this problem and save it as **06.08 BCD-to-7-Segment Decoder (CC).ms11.** Modify the file to use a 74LS48 common cathode BCD-to-seven-segment and a common cathode display. Run the file to show the operation of the display for all valid numeric codes. Also try the lamp test and blanking input functions.

6.2 Encoders

6.32 Multisim Problem
Multisim File: 06.09 Priority Encoder.ms11

Interactive Exercise:

a. Open the Multisim file for this problem. The circuit is the one shown in Figure 6.29. Set all switches LOW. Set the switches HIGH in the following sequence: $nD7$, $nD6$, $nD5$, $nD4$, $nD3$, $nD2$, $nD1$, $nD0$. Fill in *Table 6.19* with the binary values (0 or 1) of $nA2$ $nA1$ $nA0$ and $A2$ $A1$ $A0$. (This is the same exercise as in Example 6.14.)

b. Make sure all switches are set HIGH. Set the switches LOW in the following sequence $nD7$, $nD6$, $nD5$, $nD4$, $nD3$, $nD2$, $nD1$, $nD0$. Fill in *Table 6.20* with the binary values (0 or 1) of $nA2$ $nA1$ $nA0$ and $A2$ $A1$ $A0$.

c. Explain the difference between the two tables.

TABLE 6.19 Problem 6.32: Priority Encoder Inputs

Last Switch Set HIGH	Highest Priority	nA2	nA1	nA0	A2	A1	A0
None							
nD7							
nD6							
nD5							
nD4							
nD3							
nD2							
nD1							

© CENGAGE LEARNING 2012.

TABLE 6.20 Problem 6.32: Priority Encoder Inputs

Last Switch Set LOW	Highest Priority	nA2	nA1	nA0	A2	A1	A0
None							
nD7							
nD6							
nD5							
nD4							
nD3							
nD2							
nD1							

© CENGAGE LEARNING 2012.

6.33 Multisim Problem
Multisim File: 06.09 Priority Encoder.ms11

Interactive Exercise:

Open the Multisim file for this exercise and save it with a new name: **06.10 Priority Encoder with Seven-Segment Decoder and Rotary Sw.ms11.**

Modify the design to include an active-LOW seven-segment decoder and common anode display, as shown in *Figure 6.77.*

Notes:

▶ Input *D* of the seven-segment decoder and the center terminal of the rotary switch must be grounded by using a digital ground (DGND) component. Do not use the analog ground (GND) component or the display will not work correctly.

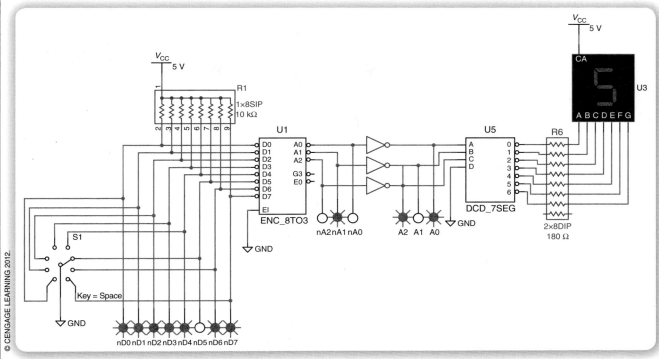

- ▶ The SPST switches from the original file are replaced by an 8-position rotary switch (8POS_ROTARY).
- ▶ Use the component 2x8DIP for the series resistors. Change the resistor value to 180 Ω.
- **a.** Run the design as a simulation. Operate the space bar to change the position of the rotary switch. Describe the action of the circuit.
- **b.** How many inputs of the priority encoder are active at any time?
- **c.** What does the seven-segment display show?
- **d.** Why is input *D* of the seven-segment display grounded?
- **e.** Why are the inverters required in the circuit?

6.3 Multiplexers

6.34 Draw the symbol for a quadruple 8-to-1 multiplexer (i.e., a MUX with eight switched groups of four bits each). Write the truth table for this device, showing which data inputs are selected for every binary combination of the select inputs. Use double-subscript notation.

6.35 Draw the symbol for an octal 4-to-1 multiplexer (i.e., a MUX with four switched groups of eight bits each). Write the truth table for this device, showing which data inputs are selected for every binary combination of the select inputs. Use double-subscript notation.

6.36 a. *Figure 6.78* shows the circuit for a dual 4-to-1 multiplexer (i.e., a MUX with two switched groups of four bits each). The circuit is created in the Xilinx ISE Schematic Editor, using two separate 4-to-1 MUX components with the select inputs of the two components tied together. Write the truth table for this circuit, showing which data inputs are selected for every binary combination of the select inputs. Use double-subscript notation.

b. *Figure 6.79* shows a set of simulation waveforms that can be used to test the design of the 2-to-1 multiplexer in Figure 6.9. Write a set of simulation criteria that could be used to create these waveforms.

continues...

continued...

FIGURE 6.78 Problem 6.36: Dual 4-to-1 Multiplexer from Components

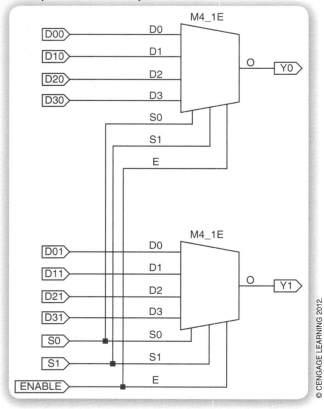

© CENGAGE LEARNING 2012.

6.37 Multisim Problem

Multisim File: 06.10 Priority Encoder with Seven-Segment Decoder and Rotary Sw.ms11

(Or **06.09 Priority Encoder.ms11** if you have not done Problem 6.33)

Interactive Exercise:

Open the Multisim file you created for Problem 6.33 and save it with a new name: **06.12 Encoder, MUX, 7seg.ms11.**

If you have not done Problem 6.33, open the file **06.07 Priority Encoder.ms11** and begin with that design. See the notes for Problem 6.33.

Modify the design to include an 8-to-1 MUX, as well as an active-LOW seven-segment decoder and common anode display, if necessary. The completed circuit is shown in *Figure 6.80*, with a more detailed view of the MUX connections shown in *Figure 6.81*. (Some of these components can be copied and pasted from other example files in this section of the chapter.)

Run the design as a simulation. Operate the space bar to change the position of the rotary switch. Observe the output of the circuit by double-clicking the oscilloscope component. (See *Figure 6.82* for a sample of the oscilloscope screen.)

a. Describe the action of the circuit when the rotary switch is rotated counter-clockwise.

b. Explain the reason for connecting the select inputs of the multiplexer to the inverter outputs.

c. Extend the function of the circuit by adding four more pulse sources to the unconnected MUX inputs.

FIGURE 6.79 Problem 6.36: Simulation for a Dual 4-to-1 Multiplexer

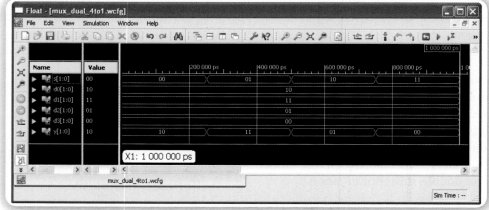

© CENGAGE LEARNING 2012.

FIGURE 6.80 Problem 6.37: Multiplexer System

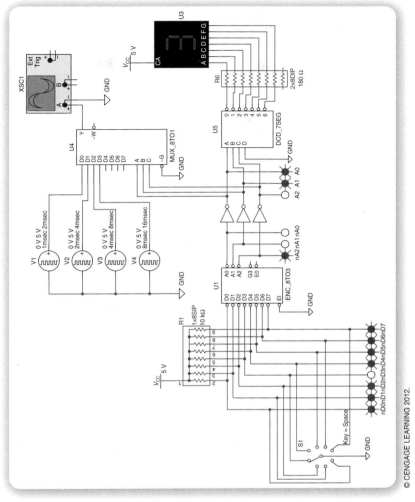

FIGURE 6.81 Problem 6.37: Detail of Multiplexer System

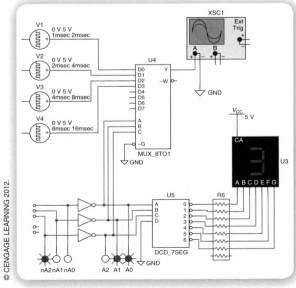

FIGURE 6.82 Problem 6.37: Oscilloscope Trace Showing Multiplexer Output

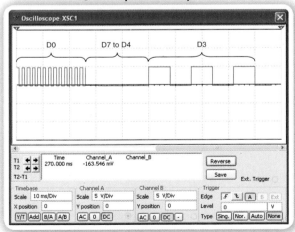

© CENGAGE LEARNING 2012.

continues...

6.4 Demultiplexers

6.38 Multisim Problem

Multisim File: 06.12 Encoder, MUX, 7seg.ms11

(Or **06.07 Priority Encoder.ms11** if you have not done Problems 6.34 or 6.38)

Interactive Exercise:

Open the Multisim file you created for Problem 6.38 and save it with a new name: **06.14 Encoder, MUX, DMUX, 7seg.ms11.**

If you have not done Problems 6.33 or 6.37, open the file **06.09 Priority Encoder.ms11** and begin with that design. See the notes for Problem 6.33.

Modify the design to include a 3-to-8 demultiplexer/decoder, as well as an 8-to-1 MUX, an active-LOW seven-segment decoder and common anode display, if necessary. The completed circuit is shown in *Figure 6.83*, with a more detailed view of

FIGURE 6.83 Problem 6.38: Demultiplexer System

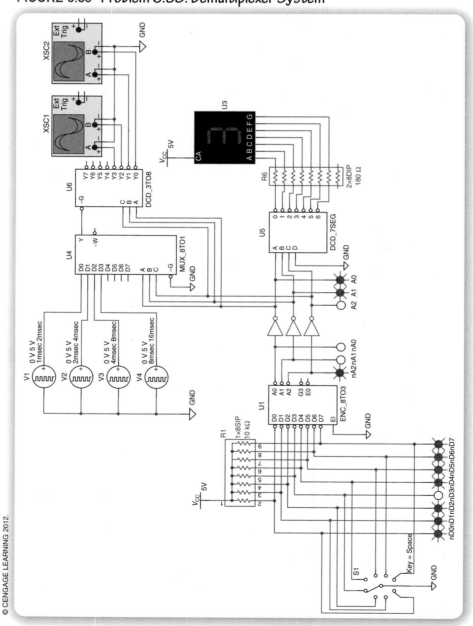

FIGURE 6.84 Problem 6.38: Detail of Demultiplexer System

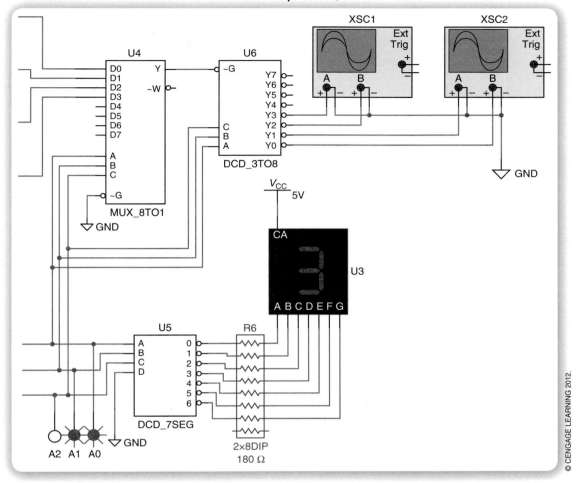

© CENGAGE LEARNING 2012.

the MUX connections shown in *Figure 6.84.* (Some of these components can be copied and pasted from other example files in this section of the chapter.)

Run the design as a simulation. Operate the space bar to change the position of the rotary switch. Observe the output of the circuit by double-clicking the oscilloscope components. Click the **Reverse** button to change the background of the oscilloscope screen from black to white, if desired. (See *Figure 6.85* for a sample of the two oscilloscope screens used in the circuit.)

a. Describe the action of the circuit when the rotary switch is rotated counterclockwise.

b. Explain the reason for connecting the active-LOW output of the MUX to the active-HIGH input of the DMUX/decoder.

c. Describe an alternative connection to the one described in part **b** of this question that will have the same effect.

6.5 Magnitude Comparators
6.39 Multisim Problem
   Multisim File: 06.15 4-Bit Comparator.ms11

Interactive Exercise:

Open the Multisim file for this exercise, shown in Figure 6.49.

a. Set all binary inputs LOW. Go through all binary combinations of $A_1A_0B_1B_0$. What do you observe?

b. Try the following binary combinations. What do you observe?
   i. $A_3A_2A_1A_0 = 0111$, $B_3B_2B_1B_0 = 1010$
   ii. $A_3A_2A_1A_0 = 0100$, $B_3B_2B_1B_0 = 0100$
   iii. $A_3A_2A_1A_0 = 1011$, $B_3B_2B_1B_0 = 0011$

continues...

continued...

FIGURE 6.85 Problem 6.38: Oscilloscope Traces Showing Demultiplexer Outputs

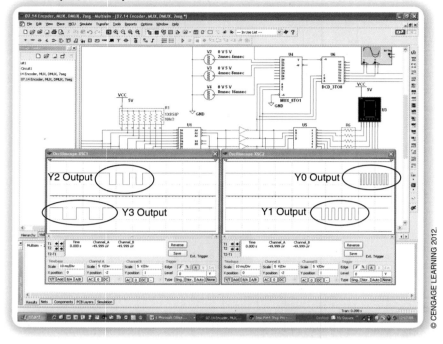

6.6 Parity Generators and Checkers

6.40 Multisim Problem

Multisim File: 06.17 Parity Generator and Checker with Error Generator.ms11

Open the Multisim file for this example and run it as a simulation. The circuit in this file is shown in Figure 6.58.

a. Set the switches exactly the same as in Figure 6.58. Now make the input data $A\ B\ C\ D = 1010$. Introduce an error onto bit A only. Explain how you can tell that an error is introduced. Refer to the states of the digital probes in your answer.

b. Set the switches exactly the same as in Figure 6.58. Now make the input data $A\ B\ C\ D = 1010$. Introduce errors onto bits A and C only. Explain how you can tell that two errors are introduced, but are not detected by the parity checker. Refer to the states of the digital probes in your answer.

c. Set the switches exactly the same as in Figure 6.58. Now make the input data $A\ B\ C\ D = 1010$. Introduce errors onto bit B and the parity bit only. Explain how you can tell that two errors are introduced, but are not detected by the parity checker. Refer to the states of the digital probes in your answer.

CHAPTER 7
Digital Arithmetic and Arithmetic Circuits

GPS DELUXE

| Menu | START LOCATION | DISTANCE | END LOCATION |

Chapter Objectives

Upon successful completion of this chapter, you will be able to:

1 Add or subtract two unsigned binary numbers.

2 Write a signed binary number in true-magnitude, 1's complement, or 2's complement form.

3 Add or subtract two signed binary numbers.

4 Explain the concept of overflow.

5 Calculate the maximum sum or difference of two signed binary numbers that will not result in an overflow.

6 Write decimal numbers in BCD code.

7 Use the ASCII table to convert alphanumeric characters to hexadecimal or binary numbers and vice versa.

8 Derive the logic gate circuits for full and half adders, given their truth tables.

9 Demonstrate the use of full and half adder circuits in arithmetic applications.

10 Add and subtract n-bit binary numbers, using parallel binary adders and logic gates.

11 Design a circuit to detect sign bit overflow in a parallel adder.

We have seen the binary number system (all 1's and 0's), and like any other number system, we can do a lot with it. We can count, add and subtract, and we can convert binary numbers to other types of numbers, which will let us do more counting and more math. Your calculator uses the techniques in this chapter to do math on the numbers you enter, all done in binary!

There are two ways to perform binary integer arithmetic: with unsigned binary numbers or with signed binary numbers. Signed binary numbers incorporate a bit defining the sign of a number; unsigned binary numbers do not. Several ways of writing signed binary numbers are true-magnitude form, which maintains the magnitude of the number in binary value, and 1's complement and 2's complement forms, which seem to change the magnitude but are more suited to digital circuitry.

In addition to positional number systems, binary numbers can be used in a variety of nonpositional number *codes*, which can represent numbers, letters, and computer control codes. Binary coded decimal (BCD) codes represent decimal digits as individually encoded groups of bits. American Standard Code for Information Interchange (ASCII) represents alphanumeric and control code characters in a 7- or 8-bit format.

There are several different digital circuits for performing digital arithmetic, most of which are based on the parallel binary adder, which in turn is based on the full adder and half adder circuits. The half adder adds two bits and produces a sum and a carry. The full adder also allows for an input carry from a previous adder stage. Parallel adders have many full adders in cascade, with carry bits connected between the stages.

7.1 DIGITAL ARITHMETIC

Digital arithmetic usually means binary arithmetic. Binary arithmetic can be performed using signed binary numbers, in which the most significant bit (MSB) of each number indicates a positive or negative sign, or unsigned binary numbers, in which the sign is presumed to be positive.

The usual arithmetic operations of addition and subtraction can be performed using signed or unsigned binary numbers. Signed binary arithmetic is often used in digital circuits for two reasons:

1. Calculations involving real-world quantities require us to use both positive and negative numbers.
2. It is easier to build circuits to perform some arithmetic operations, such as subtraction, with certain types of signed numbers than with unsigned numbers.

Unsigned Binary Arithmetic

Addition

When we add two numbers, they combine to yield a result called the sum. If the sum is larger than can be contained in one digit, the operation generates a second digit, called the carry. The two numbers being added are called the augend and the addend, or more generally, the operands.

For example, in the decimal addition $9 + 6 = 15$, 9 is the augend, 6 is the addend, and 15 is the sum. The sum cannot fit into a single digit, so a carry is generated into a second digit place.

Four binary sums give us all of the possibilities for adding two n-bit binary numbers:

$$0 + 0 = 00$$
$$1 + 0 = 01$$

$$1 + 1 = 10 \qquad (1_{10} + 1_{10} = 2_{10})$$
$$1 + 1 + 1 = 11 \qquad (1_{10} + 1_{10} + 1_{10} = 3_{10})$$

Each of these results consists of a **sum bit** and a **carry bit**. For the first two results shown, the carry bit is 0. The final sum in the table is the result of adding a carry bit from a sum in a less significant position.

When we add two 1-bit binary numbers in a logic circuit, the result *always* consists of a sum bit and a carry bit, even when the carry is 0, because each bit corresponds to a measurable voltage at a specific circuit location. Just because the value of the carry is 0 does not mean it has ceased to exist.

Your Turn

7.1 Add 11111 + 1001.
7.2 Add 10011 + 1101.

Subtraction

KEY TERMS

Subtrahend The number in a subtraction operation that is subtracted from another number.

Minuend The number in a subtraction operation from which another number is subtracted.

Difference The result of a subtraction operation.

Borrow A digit brought back from a more significant position when the subtrahend digit is larger than the minuend digit.

In unsigned binary subtraction, two operands, called the **subtrahend** and the **minuend**, are subtracted to yield a result called the **difference**. In the operation $x = a - b$, x is the difference, a is the minuend, and b is the subtrahend. To remember which comes first, think of the *min*uend as the number that is di*min*ished (i.e., something is taken away from it).

Unsigned binary subtraction is based on the following four operations:

$$0 - 0 = 0$$
$$1 - 0 = 1$$
$$1 - 1 = 0$$
$$10 - 1 = 1 \qquad (2_{10} - 1_{10} = 1_{10})$$

The last operation shows how to obtain a positive result when subtracting a 1 from a 0: **borrow** 1 from the next most significant bit.

Borrowing Rules:

1. If you are borrowing from a position that contains a 1, leave behind a 0 in the borrowed-from position.
2. If you are borrowing from a position that already contains a 0, you must borrow from a more significant digit that contains a 1. All 0's up to that point become 1's, and the last borrowed-from digit becomes a 0.

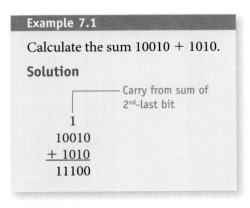

Example 7.1

Calculate the sum 10010 + 1010.

Solution

Carry from sum of 2nd-last bit

```
         1
      10010
    + 1010
     ------
      11100
```

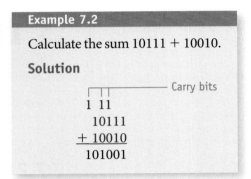

Example 7.2

Calculate the sum 10111 + 10010.

Solution

Carry bits

```
      1 11
      10111
    + 10010
     -------
     101001
```

Example 7.3

Subtract 1110 − 1001.

Solution

(New 2nd LSB)———┐ ┌———(Bit borrowed from 2nd LSB)

```
    01
  1110
− 1001
  0101
```

Example 7.4

Subtract 10000 − 101.

Solution

```
              0111
10000  (Original    10000  (After borrowing
− 101  problem)    − 101  from higher-order bits)
                    1011
```

Your Turn

7.3 Subtract 10101 − 10010.
7.4 Subtract 10000 − 1111.

7.2 REPRESENTING SIGNED BINARY NUMBERS

KEY TERMS

Sign bit A bit, usually the MSB, that indicates whether a signed binary number is positive or negative.

Magnitude bits The bits of a signed binary number that tell us how large the number is (i.e., its magnitude).

True-magnitude form A form of a signed binary number whose magnitude is represented in true binary.

1's complement A form of signed binary notation in which negative numbers are created by complementing all bits of a number, including the sign bit.

2's complement A form of signed binary notation in which negative numbers are created by adding 1 to the 1's complement form of the number.

> **Note . . .**
>
> Positive numbers are the same in all three notations.

Binary arithmetic operations are performed by digital circuits that are designed for a fixed number of bits, because each bit has a physical location within a circuit. It is useful to have a way of representing binary numbers within this framework that accounts not only for the magnitude of the number, but also for the sign.

This can be accomplished by designating one bit of a binary number, usually the most significant bit, as the **sign bit** and the rest as **magnitude bits**. When the number is negative, the sign bit is 1, and when the number is positive, the sign bit is 0.

We can write the magnitude bits in several ways, each having its particular advantages. **True-magnitude form** represents the magnitude in straight binary form, which is relatively easy for a human operator to read. Complement forms, such as **1's complement** and **2's complement**, modify the magnitude so that it is more suited to digital circuitry. Regardless of the complement form used, the number of bits to use must be specified.

True-Magnitude Form

In true-magnitude form, the magnitude of a number is translated into its true binary value. The sign is represented by the MSB, 0 for positive and 1 for negative.

Example 7.5

Write the following numbers in 6-bit true-magnitude form:

    a. $+25_{10}$ **b.** -25_{10} **c.** $+12_{10}$ **d.** -12_{10}

Solution

Translate the magnitudes of each number into 5-bit binary, padding with leading zeros as required, and set the sign bit to 0 for a positive number and 1 for a negative number.

    a. 011001   **b.** 111001   **c.** 001100   **d.** 101100

1's Complement Form

True-magnitude and 1's complement forms of binary numbers are the same for positive numbers—the magnitude is represented by the true binary value and the sign bit is 0. We can generate a negative number in one of two ways:

1. Write the positive number of the same magnitude as the desired negative number. Complement each bit, including the sign bit; or
2. Subtract the n-bit positive number from a binary number consisting of n 1's.

Either method gives us the same result.

Example 7.6

Convert the following numbers to 8-bit 1's complement form:

    a. $+57_{10}$ **b.** -57_{10} **c.** $+72_{10}$ **d.** -72_{10}

Solution

Positive numbers are the same as numbers in true-magnitude form. Negative numbers are the bitwise complements of the corresponding positive number.

    a. $+57_{10}$ = 00111001   **b.** -57_{10} = 11000110
    c. $+72_{10}$ = 01001000   **d.** -72_{10} = 10110111

We can also generate an 8-bit 1's complement negative number by subtracting its positive magnitude from 11111111 (eight 1s). For example, for part **b**:

$$
\begin{array}{r}
11111111 \\
-\ 00111001 \quad (+57_{10}) \\
\hline
11000110 \quad (-57_{10})
\end{array}
$$

2's Complement Form

Positive numbers in 2's complement form are the same as in true-magnitude and 1's complement forms. We create a negative number by adding 1 to the 1's complement form of the number.

Convert the following numbers to 8-bit 2's complement form:

a. $+57_{10}$ **b.** -57_{10} **c.** $+72_{10}$ **d.** -72_{10}

Solution

a. $+57 = 00111001$

b. $-57 = 11000110$ (1's complement)
$$\frac{1}{11000111}$$ (2's complement)

c. $+72 = 01001000$

d. $-72 = 10110111$ (1's complement)
$$\frac{1}{10111000}$$ (2's complement)

A negative number in 2's complement form can be made positive by 2's complementing it again. Try it with the negative numbers in Example 7.7.

2's Complement Form—Shortcut

There is a shortcut method to find the 2's complement of a binary number. We will get the same result using either method.

Start with the rightmost bit (or the least significant bit) of a positive number; write this bit, then work to the left, writing each bit until we write a 1. Once we write a 1, we will continue to the left but write the complement of each remaining bit until we reach the most significant bit. We can test this with some of the numbers we used in Example 7.7.

Using 8 bits, we find $+57 = 00111001$. The least significant bit is 1, which is written as the LSB of the 2's complement value. Because we have now written a 1, we will complement each remaining bit.

$$+57 = 00111001$$

1st step: least significant bit: 1
Because we have written a 1, invert each remaining bit: 11000111

Try the shortcut method with 72:

$$+72 = 01001000$$

Start with the LSB: write each bit until we write a 1: 1000
Since we have written a 1, invert each remaining bit: 10111000

Notice that the results match those found in Example 7.7.

7.3 SIGNED BINARY ARITHMETIC

KEY TERM

Signed binary arithmetic Arithmetic operations performed using signed binary numbers.

Signed Addition

Signed addition is done in the same way as unsigned addition. The only difference is that both operands *must* have the same number of magnitude bits, and each has a sign bit.

Example 7.8

Add $+30_{10}$ and $+75_{10}$. Write the operands and the sum as 8-bit signed binary numbers.

Solution

$$
\begin{array}{rl}
+30 & 00011110 \\
+75 & +01001011 \\
\hline
+105 & 01101001
\end{array}
$$

(Magnitude bits)
(Sign bit)

Subtraction

The real advantage of using complement notation for **signed binary arithmetic** becomes evident when we subtract signed binary numbers. In subtraction, we add a negative number instead of subtracting a positive number. We thus have only one kind of operation—addition—and can use the same circuitry for both addition and subtraction.

This idea does not work for true-magnitude numbers. In the complement forms, the magnitude bits change depending on the sign of the number. In true-magnitude form, the magnitude bits are the same regardless of the sign of the number.

Let us subtract $80_{10} - 65_{10} = 15_{10}$ using 8-bit 1's complement and 2's complement addition. We will also show that the method of adding a negative number to perform subtraction is not valid for true-magnitude signed numbers.

1's Complement Method

> **KEY TERM**
>
> **End-around carry** An operation in 1's complement subtraction where the carry bit resulting from a sum of two 1's complement numbers is added to that sum.

Add the 8-bit 1's complement values of 80 and -65. If the sum results in a carry beyond the sign bit, perform an **end-around carry**. That is, add the carry to the sum.

$$
\begin{array}{l}
+80_{10} = 01010000 \\
+65_{10} = 01000001 \\
-65_{10} = 10111110 \quad \text{(1's complement)}
\end{array}
$$

$$
\begin{array}{rl}
(+80) & 01010000 \\
+ (-65) & + \underline{10111110} \\
& 1\,00001110 \\
& {\longrightarrow}\ 1 \quad \text{(End-around carry)} \\
(+15) & 00001111
\end{array}
$$

2's Complement Method

Add the 8-bit 2's complement values of 80 and -65. If the sum results in a carry into the 9th bit, discard it. (We discard this bit because in a defined 8-bit size, there is no place to hold it, much as the leading digit in a car's odometer is not seen when the mileage rolls over from 99999 to 100000. The "1" is not seen because there is no place to show it.)

Note . . .

When a sum carries in to 9th bit of an 8-bit *signed* sum or difference, we discard that carry bit, because in an adding circuit, there would be no place to hold it. (The carry bit would be kept in an *unsigned* sum or difference.)

$$+80_{10} = 01010000$$
$$+65_{10} = 01000001$$
$$-65_{10} = 10111110 \quad \text{(1's complement)}$$
$$+ \quad\underline{\quad\quad 1}$$
$$10111111 \quad \text{(2's complement)}$$

$$
\begin{array}{ll}
(+80) & 01010000 \\
+(-65) & +\ \underline{10111111} \\
(+15) & 1\ 00001111
\end{array}
$$

(Discard carry)

True-Magnitude Method (not valid)

$$+80_{10} = 01010000$$
$$+65_{10} = 01000001$$
$$-65_{10} = 11000001$$

$$
\begin{array}{ll}
(+80) & 01010000 \\
+(-65) & +\ \underline{11000001} \\
(?) & 1\ 00010001
\end{array}
$$

If we perform an end-around carry, the result is $00010010 = 18_{10}$. If we discard the carry, the result is $00010001 = 17_{10}$. Neither answer is correct. Thus, adding a negative true-magnitude number is not equivalent to subtraction; in other words, it does not work.

Negative Sum or Difference

All examples to this point have given positive-valued results. When a 2's complement addition or subtraction yields a negative sum or difference, we can't just read the magnitude from the result because a 2's complement operation modifies the bits of a negative number. We must calculate the 2's complement of the sum or difference, which will give us the positive number that has the same magnitude. That is, $-(-x) = +x$.

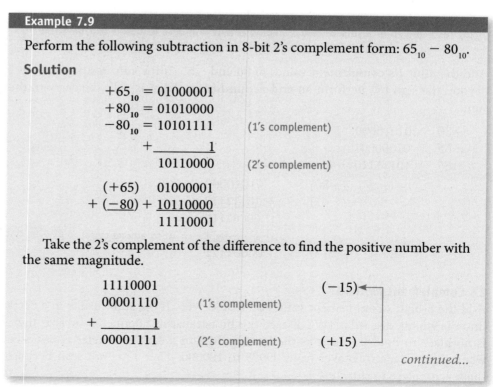

Example 7.9

Perform the following subtraction in 8-bit 2's complement form: $65_{10} - 80_{10}$.

Solution

$$+65_{10} = 01000001$$
$$+80_{10} = 01010000$$
$$-80_{10} = 10101111 \quad \text{(1's complement)}$$
$$+ \quad\underline{\quad\quad 1}$$
$$10110000 \quad \text{(2's complement)}$$

$$
\begin{array}{ll}
(+65) & 01000001 \\
+\ (-80) & +\ \underline{10110000} \\
& 11110001
\end{array}
$$

Take the 2's complement of the difference to find the positive number with the same magnitude.

$$
\begin{array}{ll}
11110001 & \quad\quad\quad\quad\quad\quad\quad\quad (-15) \\
00001110 & \text{(1's complement)} \\
+\ \underline{\quad\quad 1} \\
00001111 & \text{(2's complement)} \quad\quad (+15)
\end{array}
$$

continued...

$00001111 = +15_{10}$. We generated this number by complementing 11110001.
Thus, $11110001 = -15_{10}$.

We get the same result using the 2's complement shortcut:

$$-15 = 11110001$$

1st step: least significant bit: 1

Because we have written a 1, invert each remaining bit: $+15 = 00001111$

Range of Signed Numbers

The largest positive number in 2's complement notation is a 0 followed by n 1's for a number with n magnitude bits. For instance, the largest positive 4-bit number is $0111 = +7_{10}$. The negative number with the largest magnitude is *not* the 2's complement of the largest positive number. We can find the largest-magnitude negative number by extension of a sequence of 2's complement numbers.

The 4-bit 2's complement form of -7_{10} is $1000 + 1 = 1001$. The positive and negative numbers with the next largest magnitudes are 0110 ($= +6_{10}$) and 1010 ($= -6_{10}$). If we continue this process, we will get the list of numbers in *Table 7.1*.

We have generated the 4-bit negative numbers from -1_{10} (1111) through -7_{10} (1001) by writing the 2's complement forms of the positive numbers 1 through 7. Notice that these numbers count down in binary sequence. The next 4-bit number in the sequence (which is the only binary number we have left) is 1000. By extension, $1000 = -8_{10}$. This number is its own 2's complement. (Try it.) It exemplifies a general rule for the largest-magnitude numbers with n magnitude bits: a 2's complement number consisting of a 1 followed by n 0's is equal to -2^n.

TABLE 7.1 4-Bit 2's Complement Numbers

Decimal	2's Complement
+7	0111
+6	0110
+5	0101
+4	0100
+3	0011
+2	0010
+1	0001
0	0000
−1	1111
−2	1110
−3	1101
−4	1100
−5	1011
−6	1010
−7	1001
−8	1000

© CENGAGE LEARNING 2012.

Example 7.10

Write the largest positive and negative numbers for an 8-bit signed number in 2's complement and decimal notation.

Solution

$$01111111 = +127 \quad (\text{7 magnitude bits: } 2^7 - 1 = 127)$$
$$10000000 = -128 \quad (\text{1 followed by seven 0's: } -2^7 = -128)$$

Example 7.11

Write -16_{10}:

a. As an 8-bit 2's complement number.
b. As a 5-bit 2's complement number.

(8-bit numbers are more common than 5-bit numbers in digital systems, but it is useful to see how we must write the same number differently with different numbers of bits.)

Solution

a. An 8-bit number has 7 magnitude bits and 1 sign bit.

$$
\begin{array}{rl}
+16 = & 00010000 \\
-16 = & 11101111 \quad \text{(1's complement)} \\
+ & \underline{1} \\
& 11110000 \quad \text{(2's complement)}
\end{array}
$$

b. A 5-bit number has 4 magnitude bits and 1 sign bit. Four magnitude bits are not enough to represent $+16$. However, a 1 followed by n 0's is equal to -2^n. For a 1 and four 0's, $-2^n = -2^4 = -16$. Thus, $10000 = -16_{10}$.

The last five bits of the binary equivalent of -16 are the same in both the 5-bit and 8-bit numbers.

Note . . .

A 2's complement number consisting of a 1 followed by n 0's is equal to -2^n. Therefore, the range of a signed number, x, is $-2^n \le x \le 2^n - 1$ for a number with n magnitude bits.

Sign Extension

Example 7.11 implies that the magnitude bits of a 2's complement number remain the same, regardless of the bit size of the number, but the number may be padded out with leading 1's or 0's, depending on its sign. This is called **sign extension**. Sign extension can apply to any positive or negative number in 2's complement format. If we wish to write a given positive number with a larger number of bits, we pad the number with leading 0's (positive sign bits). If we wish to extend the number of bits in a negative number, we pad it with leading 1's (negative sign bits).

For example, the smallest bit size required to write $+25$ is six bits: a sign bit and five magnitude bits. The number is written: 011001. The 6-bit negative value in 2's complement form is 100111. In 8-bit and 12-bit form, these numbers are simply padded out with additional leading sign bits, as follows:

$$8\text{-bit: } +25 = 00011001$$
$$-25 = 11100111$$
$$12\text{-bit: } +25 = 000000011001$$
$$-25 = 111111100111$$

Your Turn

7.5 Write -32 as an 8-bit 2's complement number.
7.6 Write -32 as a 6-bit 2's complement number.

Sign Bit Overflow

Note...

A sum of positive signed binary numbers must not exceed $2^n - 1$ for numbers having n magnitude bits. Otherwise, there will be an overflow into the sign bit.

Signed addition of positive numbers is performed in the same way as unsigned addition. The only problem occurs when the number of bits in the sum of two numbers exceeds the number of magnitude bits and **overflows** into the sign bit. This causes the number to *appear* to be negative when it is not. For example, the sum $75 + 96 = 171$ causes an overflow in 8-bit signed addition. In unsigned addition the binary equivalent is:

$$1001011$$
$$+ \underline{1100000}$$
$$10101011$$

In signed addition, the sum is the same, but has a different meaning.

$$0\ 1001011$$
$$+\ \underline{0\ 1100000}$$
$$1\ 0101011$$

(Sign bit) ⌐ ⌐⎽⎽⎽⌐ (Magnitude bits)

The sign bit is 1, indicating a negative number, which cannot be true, because the sum of two positive numbers is always positive.

Overflow in Negative Sums

Overflow can also occur with large negative numbers. For example, the 8-bit addition of -80_{10} and $+65_{10}$ should produce the result:

$$-80_{10} + (-65_{10}) = -145_{10}$$

In 8-bit 2's complement notation, we get:

$$+80_{10} = 01010000$$
$$-80_{10} = 10101111 \qquad \text{(1's complement)}$$
$$+\ \underline{\qquad\ 1}$$
$$10110000 \qquad \text{(2's complement)}$$

$$+65_{10} = 01000001$$
$$-65_{10} = 10111110 \qquad \text{(1's complement)}$$
$$+\ \underline{\qquad\ 1}$$
$$10111111 \qquad \text{(2's complement)}$$

$$-80 \qquad 10110000$$
$$+(-65) \quad +\underline{10111111}$$
$$? \qquad 101101111$$

⌐⌐⌐ (Incorrect magnitude = 111_{10})
(Erroneous sign bit = 0)
(Discard carry)

This result shows a positive sum of two negative numbers—clearly incorrect. We can extend the statement we made earlier about permissible magnitudes of sums to include negative as well as positive numbers.

For an 8-bit signed number in 2's complement form, the permissible range of sums is $10000000 \le \text{sum} \le 01111111$. In decimal, this range is $-128 \le \text{sum} \le +127$.

The practical solution to an overflow problem is to use more bits when performing the math if possible. We used 8 bits in our examples. If we had used, for example, 12 bits, overflow would not have occurred. We need at least one more bit than our original values to avoid overflow: for example, we need 9 bits when using 8-bit inputs. In such a case, the 8-bit inputs would need to be sign-extended to 9 bits by padding a positive number with a leading 0 or a negative number with a leading 1.

> ### Note . . .
> A sum of signed binary numbers must be within the range of $-2^n \le \text{sum} \le 2^n - 1$ for numbers having n magnitude bits. Otherwise, there will be an overflow into the sign bit.

> ### Note . . .
> A sum of two positive numbers is always positive. A sum of two negative numbers is always negative. Any 2's complement addition or subtraction operation that appears to contradict these rules has produced an overflow into the sign bit.

Example 7.12

Which of the following sums will produce a sign bit overflow in 8-bit 2's complement notation? How can you tell?

a. $67_{10} + 33_{10}$
b. $67_{10} + 63_{10}$
c. $-96_{10} - 22_{10}$
d. $-96_{10} - 42_{10}$

Solution

A sign bit overflow is generated if the sum of two positive numbers appears to produce a negative result or the sum of two negative numbers appears to

continued...

Note . . .

The carry bit generated in 1's and 2's complement operations is not the same as an overflow bit. (See Example 7.12, parts c and d.) An overflow is a change in the sign bit, which leads us to believe that the number is opposite in sign from its true value. A carry is the result of an operation carrying beyond the physical limits of an n-bit number. It is similar to the idea of an odometer rolling over from 999999 to 1000000. There are not enough places to hold the new number, so it goes back to the beginning and starts over.

produce a positive result. In other words, overflow occurs if the operand sign bits are both 1 and the sum sign bit is 0 or vice versa. We know this will happen if an 8-bit sum is outside the range ($-128 \leq$ sum $\leq +127$).

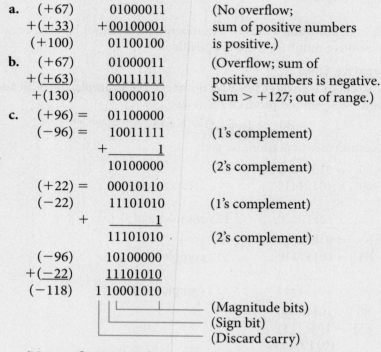

(Overflow; sum of two negative numbers is positive. Sum < -128; out of range.)

7.4 NUMERIC AND ALPHANUMERIC CODES

BCD Code

KEY TERM

Binary coded decimal (BCD) A code that represents each digit of a decimal number by a binary value.

BCD stands for **binary coded decimal**. As the name implies, BCD is a system of writing decimal numbers with binary digits. There is more than one way to do this, as BCD is a *code,* not a positional number system. That is, the various positions of the bits do not necessarily represent increasing powers of a specified number base and are used to represent a number, not (usually) to mathematically manipulate a number.

The most commonly used BCD code is 8421 code, where the bits for *each decimal digit* are weighted as they are for a 4-bit binary number. Other BCD codes include Excess-3 code and 2421 code, which have advantages for signed decimal arithmetic. We will not examine these codes.

8421 Code

> **KEY TERM**
>
> **8421 code** A BCD code that represents each digit of a decimal number by its 4-bit true binary value.

The most straightforward BCD code is the 8421 code, also called natural BCD (NBCD). Each decimal digit is represented by its 4-bit true binary value. When we talk about BCD code, this is usually what we mean.

This code is called 8421 because these are the positional weights of each binary digit. *Table 7.2* shows the decimal digits and their BCD equivalents.

8421 BCD is not a positional number system, because each decimal digit is encoded separately as a 4-bit number.

TABLE 7.2 Decimal Digits and Their 8421 BCD Equivalents

Decimal Digit	BCD (8421)
0	0000
1	0001
2	0010
3	0011
4	0100
5	0101
6	0110
7	0111
8	1000
9	1001

> **Example 7.13**
>
> Write 4987_{10} in both binary and 8421 BCD.
>
> **Solution**
>
> The binary value of 4987_{10} can be calculated by repeated division by 2:
>
> $$4987_{10} = 1\ 0011\ 0111\ 1011_2$$
>
> The BCD digits are the binary values of each decimal digit, encoded separately. We can break bits into groups of 4 for easier reading. Note that the first and last BCD digits each have a leading zero to make them 4 bits long.
>
> $$4987_{10} = 0100\ 1001\ 1000\ 0111_{BCD}$$
>
> Notice that these two representations are different.

ASCII Code

> **KEY TERMS**
>
> **Alphanumeric code** A code used to represent letters of the alphabet and numerical characters.
>
> **ASCII** American Standard Code for Information Interchange. A 7- or 8-bit code for representing alphanumeric and control characters.
>
> **Case shift** Changing letters from capitals (UPPERCASE) to small letters (lowercase) or vice versa.

Digital systems and computers could operate perfectly well using only binary numbers. However, if there is any need for a human operator to understand the input and output data of a digital system, it is necessary to have a system of communication that is understandable to both a human operator and the digital circuit.

A code that represents letters (alphabetic characters) and numbers (numeric characters) as binary numbers is called an **alphanumeric code**. The most commonly used alphanumeric code is **ASCII** ("askey"), which stands for American Standard Code for Information Interchange. ASCII code represents letters, numbers, and other printable characters in 7 bits. In addition, ASCII has a repertoire of "control characters," codes that are used to send control instructions to and from devices such as video display terminals, printers, and modems.

Table 7.3 shows the ASCII code in both binary and hexadecimal forms. The code for any character consists of the bits in the column heading, then those in the row heading. For example, the ASCII code for "A" is 1000001_2 or 41H. The code for "a" is 1100001_2 or 61H. The codes for capital (uppercase) and lowercase letters differ only by the second most significant bit, for all letters. Thus, we can make an alphabetic **case shift**, like using the Shift key on a typewriter or computer keyboard, by switching just one bit.

The codes in columns 0 and 1 are control characters. They cannot be displayed on any kind of output device, such as a printer or video monitor, although they may be used to control the device. For instance, if the codes 0AH (Line Feed) and 0DH (Carriage Return) are sent to a monitor configured as an ASCII terminal, the cursor will advance by one line and then will return to the beginning of the line.

The displayable characters begin at 20H ("space") and continue to 7EH ("tilde"). Spaces are considered ASCII characters.

Numeric characters are listed in column 3, with the least significant digit of the ASCII code being the same as the represented number value. For example, the numeric character "0" is equivalent to 30H in ASCII. The character "9" is represented as 39H.

Example 7.14

Encode the following string of characters into ASCII (hexadecimal form). Do not include quotation marks.

"Total system cost: $4,000,000. @ 10%"

Solution

Each character, including spaces, is represented by two hex digits as follows:

54	6F	74	61	6C	20	73	79	73	74	65	6D	20	63	6F	73	74	3A	20
T	o	t	a	l	SP	s	y	s	t	e	m	SP	c	o	s	t	:	SP

24	34	2C	30	30	30	2C	30	30	30	2E	20	40	20	31	30	25
$	4	,	0	0	0	,	0	0	0	.	SP	@	SP	1	0	%

Your Turn

7.7 Decode the following sequence of hexadecimal ASCII codes.

54 72 75 65 20 6F 72 20 46 61 6C 73 65 3A 20 31

2F 34 20 3C 20 31 2F 32

TABLE 7.3 *ASCII Code*

					MSBs			
	000	**001**	**010**	**011**	**100**	**101**	**110**	**111**
LSBs	**(0)**	**(1)**	**(2)**	**(3)**	**(4)**	**(5)**	**(6)**	**(7)**
0000 (0)	NUL	DLE	SP	0	@	P	'	p
0001 (1)	SOH	DC1	!	1	A	Q	a	q
0010 (2)	STX	DC2	"	2	B	R	b	r
0011 (3)	ETX	DC3	#	3	C	S	c	s
0100 (4)	EOT	DC4	$	4	D	T	d	t
0101 (5)	ENQ	NAK	%	5	E	U	e	u
0110 (6)	ACK	SYN	&	6	F	V	f	v
0111 (7)	BEL	ETB	'	7	G	W	g	w
1000 (8)	BS	CAN	(	8	H	X	h	x
1001 (9)	HT	EM	)	9	I	Y	i	y
1010 (A)	LF	SUB	*	:	J	Z	j	z
1011 (B)	VT	ESC	+	;	K	[	k	{
1100 (C)	FF	FS	,	<	L	\	I	|
1101 (D)	CR	GS	-	=	M	]	M	}
1110 (E)	SO	RS	.	>	N	^	n	~
1111 (F)	SI	US	/	?	O	_	o	DEL

Control Characters

NUL—NULL	DLE—Data Link Escape
SOH—Start of Header	DC1—Device Control 1
STX—Start Text	DC2—Device Control 2
ETX—End Text	DC3—Device Control 3
EOT—End of Transmission	DC4—Device Control 4
ENQ—Enquiry	NAK—No Acknowledgment
ACK—Acknowledge	SYN—Synchronous Idle
BEL—Bell	ETB—End of Transmission Block
BS—Backspace	CAN—Cancel
HT—Horizontal Tabulation	EM—End of Medium
LF—Line Feed	SUB—Substitute
VT—Vertical Tabulation	ESC—Escape
FF—Form Feed	FS—Form Separator
CR—Carriage Return	GS—Group Separator
SO—Shift Out	RS—Record Separator
SI—Shift In	US—Unit Separator
SP—Space	DEL—Delete

7.5 BINARY ADDERS AND SUBTRACTORS

Half and Full Adders

FIGURE 7.1 Half Adder

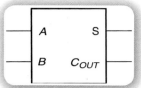

© CENGAGE LEARNING 2012.

TABLE 7.4 Half Adder
Truth Table © CENGAGE LEARNING 2012.

A	B	C_{OUT}	S
0	0	0	0
0	1	0	1
1	0	0	1
1	1	1	0

FIGURE 7.2 Half Adder Circuit

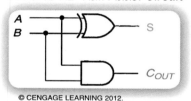

© CENGAGE LEARNING 2012.

There are only three possible sums of two 1-bit binary numbers:

$$0 + 0 = 00$$
$$0 + 1 = 01$$
$$1 + 1 = 10$$

We can build a simple combinational logic circuit to produce these sums. Let us designate the bits on the left side of these equalities as inputs to the circuit and the bits on the right side as outputs. Let us call the LSB of the output the sum bit, symbolized by S, and the MSB of the output the carry bit, designated C_{OUT}.

Figure 7.1 shows the logic symbol of the circuit, which is called a **half adder**. Its truth table is given in *Table 7.4*. Because addition is the same whichever order the inputs are added ($A + B = B + A$), the second and third lines of the truth table are the same.

The Boolean functions of the two outputs, derived from the truth table, are:

$$C_{OUT} = AB$$
$$S = \overline{A}\,B + A\,\overline{B} = A \oplus B$$

The corresponding logic circuit is shown in *Figure 7.2*.

 Example 7.15

FIGURE 7.3 Example 7.15: Half Adder
Circuit in Multisim © CENGAGE LEARNING 2012.

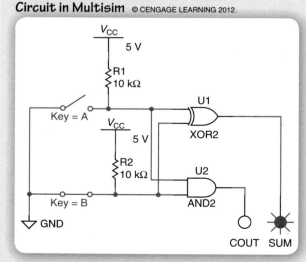

Multisim Example

Figure 7.3 shows a Multisim file for a half adder circuit. Use Multisim to create this circuit, including logic switches and digital probes for inputs A and B and digital probes for outputs COUT and SUM. Save the file as **07.01 Half Adder. ms11.**

a. Run the file and go through all combinations of the input switches. How do the input combinations and resulting outputs compare to Table 7.4?

b. What is the 2-bit output combination of COUT and SUM if $A = 0$ and $B = 1$?

c. What is the 2-bit output combination of COUT and SUM if $A = 1$ and $B = 0$?

d. What can you conclude about the order in which two bits are added from the answers to parts **b** and **c**?

e. How does the 2-bit output combination of COUT and SUM relate to A and B, in general?

continued...

Solution

a. The input combinations and output results are the same as in Table 7.4.
b. *COUT SUM* = 01
c. *COUT SUM* = 01
d. The 2-bit output always has a value of 01 if only one of *A* or *B* has a value of 1. This is equivalent to saying $0 + 1 = 1 + 0 = 01$. In other words, *A* and *B* can be added in either order without affecting the result.
e. The 2-bit output combination is the same as the sum of both inputs.

The half adder circuit works very well when we want to add two bits together, but obviously this is a very limiting condition. To add two multiple-bit numbers together, we need to add the least significant bits of each number, then the next two bits, and so on. If the addition at any stage generates a carry, then this carry must be added to the sum of the bits in the next stage. We did many examples of this process in the first three sections of this chapter.

The half adder circuit cannot account for an *input* carry, that is, a carry from a lower-order 1-bit addition. A **full adder**, shown in *Figure 7.4,* can add two 1-bit numbers *and* accept a carry bit from a previous adder stage. Operation of the full adder is based on the following sums:

$$0 + 0 + 0 = 00$$
$$0 + 0 + 1 = 01$$
$$0 + 1 + 1 = 10$$
$$1 + 1 + 1 = 11$$

Designating the left side of the above equalities as circuit inputs A, B, and C_{IN} and the right side as outputs C_{OUT} and S, we can make the truth table in *Table 7.5.* (The second and third of the above sums each account for three lines in the full adder truth table.) Notice that the combined 2-bit value of C_{OUT} and S is the binary sum of $A+B+C_{IN}$. In this case, we use the "plus" sign (+) to mean the arithmetic sum of A, B, C_{IN}, not the Boolean OR function.

The full adder can be constructed from two half adders and an OR gate, as shown in *Figure 7.5.* The first half adder adds A and B and generates a sum bit and a carry bit. The sum bit from the first half adder is added to C_{IN} in the second half adder, which generates the sum of all three inputs, consisting of a sum bit and carry bit. The carry output is HIGH whenever there is a carry from either half adder, or both. The full adder logic circuit is shown in *Figure 7.6.* In this figure, the half adders are shown as gate-level circuits, as in Figure 7.2.

FIGURE 7.4 Full Adder

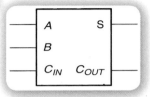

© CENGAGE LEARNING 2012.

Note . . .

The combined 2-bit value of C_{OUT} and S is the binary sum of $A+B+C_{IN}$.

TABLE 7.5 Full Adder Truth Table © CENGAGE LEARNING 2012.

A	B	C_{IN}	C_{OUT}	SUM
0	0	0	0	0
0	0	1	0	1
0	1	0	0	1
0	1	1	1	0
1	0	0	0	1
1	0	0	0	1
1	0	1	1	0
1	1	0	1	0
1	1	1	1	1

FIGURE 7.5 Full Adder from Two Half Adders

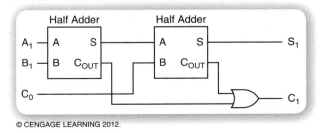

© CENGAGE LEARNING 2012.

FIGURE 7.6 Full Adder Logic Circuit

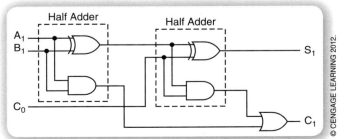

© CENGAGE LEARNING 2012.

Example 7.16

Evaluate the Boolean expression for S and C_{OUT} of the full adder in Figure 7.5 for the following input values. What is the binary value of the outputs in each case?

a. $A = 0, B = 0, C_{IN} = 1$
b. $A = 1, B = 0, C_{IN} = 0$
c. $A = 1, B = 0, C_{IN} = 1$
d. $A = 1, B = 1, C_{IN} = 0$

Solution

The output of a full adder for any set of inputs is simply given by $C_{OUT} S = A + B + C_{IN}$. For each of the stated sets of inputs:

a. $C_{OUT} S = A + B + C_{IN} = 0 + 0 + 1 = 01$
b. $C_{OUT} S = A + B + C_{IN} = 1 + 0 + 0 = 01$
c. $C_{OUT} S = A + B + C_{IN} = 1 + 0 + 1 = 10$
d. $C_{OUT} S = A + B + C_{IN} = 1 + 1 + 0 = 10$

We can verify each of these sums by plugging the specified inputs into the full adder circuit of Figure 7.6. The Boolean equations of the full adder circuit are given by:

$$C_{OUT} = (A \oplus B) C_{IN} + A B$$
$$S = (A \oplus B) \oplus C_{IN}$$

a. $C_{OUT} = (0 \oplus 0) \cdot 1 + 0 \cdot 0 = 0 \cdot 1 + 0 = 0 + 0 = 0$
   $S = (0 \oplus 0) \oplus 1 = 0 \oplus 1 = 1$ (Binary equivalent: $C_{OUT} S = 01$)
b. $C_{OUT} = (1 \oplus 0) \cdot 0 + 1 \cdot 0 = 1 \cdot 0 + 0 = 0 + 0 = 0$
   $S = (1 \oplus 0) \oplus 0 = 1 \oplus 0 = 1$ (Binary equivalent: $C_{OUT} S = 01$)
c. $C_{OUT} = (1 \oplus 0) \cdot 1 + 1 \cdot 0 = 1 \cdot 1 + 0 = 1 + 0 = 1$
   $S = (1 \oplus 0) \oplus 1 = 1 \oplus 1 = 0$ (Binary equivalent: $C_{OUT} S = 10$)
d. $C_{OUT} = (1 \oplus 1) \cdot 0 + 1 \cdot 1 = 0 \cdot 0 + 1 = 0 + 1 = 1$
   $S = (1 \oplus 1) \oplus 0 = 0 \oplus 0 = 0$ (Binary equivalent: $C_{OUT} S = 10$)

In each case, the binary equivalent is the same as the number of HIGH inputs, regardless of which inputs they are.

Creating a Full Adder from Half Adders Using Multisim

KEY TERMS

Hierarchical design A design that is ordered in layers or levels. The highest level of the design has components that are complete designs in and of themselves. These lower-level blocks might contain yet lower levels of design components.

Subcircuit A block of circuit components that can be used as a single component in a higher level of a hierarchical design.

FIGURE 7.7 Multisim
Components for a Half
Adder Subcircuit

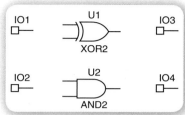

FIGURE 7.8 Renaming a
Subcircuit Connector

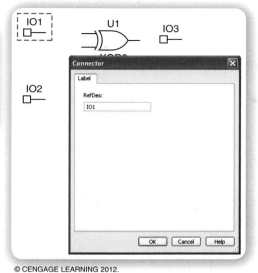

FIGURE 7.9 Renamed Subcircuit
Connector

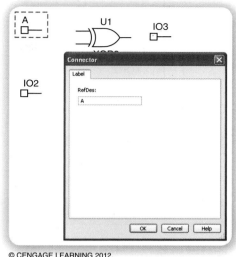

Multisim has a very useful function called **hierarchical design**, which allows us to create our own components, called **subcircuits**, from the existing list of Multisim components. A subcircuit is shown as a block with inputs and outputs when it is used as a component in a higher level of a design hierarchy.

In this section, we will learn how to use hierarchical design in Multisim, first to make a half adder subcircuit from logic gates and connectors, then to use two half adder subcircuits to make a full adder. It is recommended that you treat this section as a tutorial and follow along with the steps as they are illustrated below.

To create the half adder, begin by making a new Multisim file and adding the components shown in *Figure 7.7*. The XOR gate and the AND gate are found in **TTL** family, in the **Misc Digital** group of components. The four input and output connectors, IO1 through IO4, can be inserted by selecting the Multisim **Place** menu, then **Connectors, HB/SC Connector**.

We can rename the connectors to correspond to the input and output names for the half adder. *Figure 7.8* and *Figure 7.9* show how to change the top left connector name from IO1 to A. First select the connector by clicking it; it will be outlined by a broken line. Next, right-click the highlighted connector component. The dialog box in Figure 7.8 will open, showing the name IO1. Replace this by A, as shown in Figure 7.9 and click **OK**.

The output connectors IO3 and IO4 should be turned around so that the line points to the outputs of the XOR and AND gates. To do this, highlight the two output connectors by dragging a rectangle around them and right-click the highlighted components. The pop-up menu shown in *Figure 7.10* will appear. Select **Flip Horizontal**.

The half adder circuit can be completed by connecting and renaming the components, as shown in *Figure 7.11*. To make this into a subcircuit, highlight all the components by dragging a rectangle around them, then select **Replace by Subcircuit** ... from the **Place**

FIGURE 7.10 Select "Flip Horizontal"
from Component Pop-Up Menu

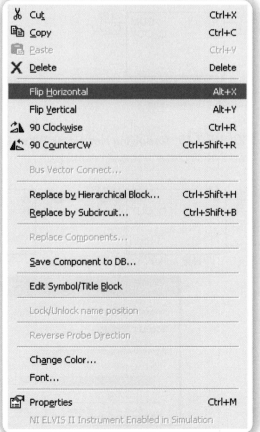

FIGURE 7.11 *Logic Gate Connections for Half Adder Subcircuit*

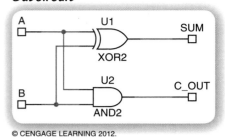

© CENGAGE LEARNING 2012.

FIGURE 7.13 *Naming a Subcircuit*

© CENGAGE LEARNING 2012.

FIGURE 7.14 *Half Adder Subcircuit Block*

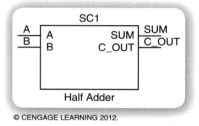

© CENGAGE LEARNING 2012.

FIGURE 7.15 *Resolving Net Name When Connecting Two Subcircuits* © CENGAGE LEARNING 2012.

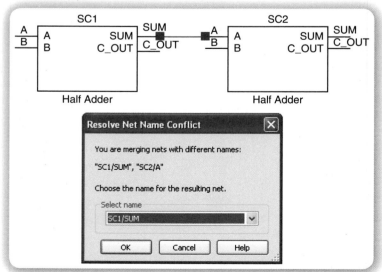

FIGURE 7.12 *"Replace by Subcircuit" from Pop-Up Menu or "Place" Menu*

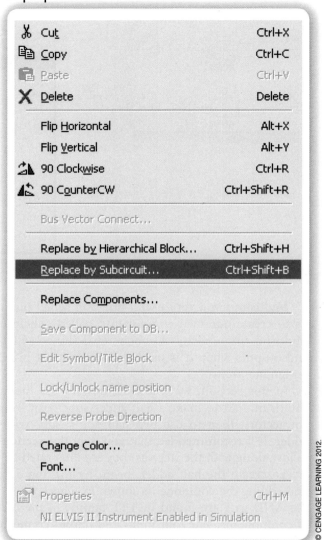

© CENGAGE LEARNING 2012.

menu, shown in *Figure 7.12.* You could also right-click the highlighted components to get the same result. When you select this menu option, you will see a dialog box, shown in *Figure 7.13,* which allows you to name the subcircuit. Call it **Half Adder** and click OK. The resulting subcircuit symbol is shown in *Figure 7.14.*

To make the full adder, copy the half adder block and start connecting the various pins together. When two pins with different names are connected, as shown in *Figure 7.15,* you will be asked to choose one of the names. After the half adder subcircuits are connected to an OR gate and connectors, the full adder circuit is as shown in *Figure 7.16.* This can now be made into a full adder component, as shown in *Figure 7.17.*

FIGURE 7.16 Connecting Two Half Adder Subcircuits to Make a Full Adder

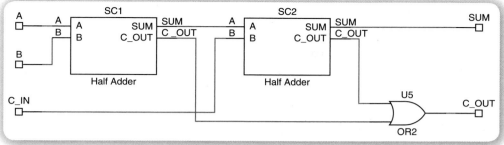

© CENGAGE LEARNING 2012.

FIGURE 7.17 Full Adder Subcircuit

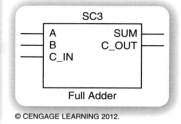

© CENGAGE LEARNING 2012.

Example 7.17

Multisim Example

Multisim File: 07.03 Full Adder Block with Output Decode.ms11

Figure 7.18 shows a full adder subcircuit connected to logic switch inputs and a seven-segment decoded output. Both input and output logic levels of the full adder are indicated with logic probes.

a. Open the Multisim file for this example and run it as a simulation. Set the input switches to the same values as shown in Figure 7.18. Briefly explain why the binary and decimal values of the outputs are as shown in the diagram.

b. Set the switches to all possible binary input combinations. For each combination, how many inputs are HIGH and what binary and decimal values are shown at the outputs?

FIGURE 7.18 Example 7.17: *Decoding the Output of a Full Adder* © CENGAGE LEARNING 2012.

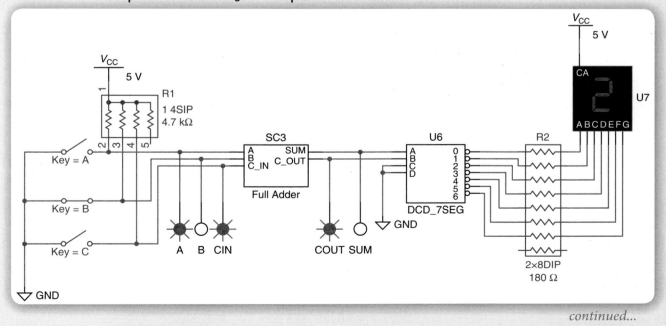

continued...

Solution

a. The full adder adds the three input bits to generate a sum and a carry bit. For the set of values shown in Figure 7.18, this is given by:

$$A + C + C_{IN} = C_{OUT} \, SUM$$
$$1 + 0 + 1 = 10$$

The binary output "10" is equivalent to the decimal digit "2," which is shown on the seven-segment display.

b. The full adder will behave as described in Table 7.5. The binary value of the full adder outputs and the decimal value on the numerical display will always be the same as the number of HIGH inputs on the full adder. For example, for inputs $0 + 1 + 0$, the binary outputs will be 01 and the decimal display will read 1, indicating that 1 input is HIGH. If the inputs are $1 + 1 + 1$, then the binary outputs will be 11 and the decimal display will be at 3, showing that all three inputs are HIGH.

Example 7.18

Multisim Example

Multisim File: 07.04 Three Full Adder Blocks.ms11

Combine three full adders to make a circuit that will add two 3-bit numbers to yield a 4-bit output consisting of three sum bits and a carry bit. This can be written as:

$$A_3 A_2 A_1 + B_3 B_2 B_1 = C_3 S_3 S_2 S_1$$

Check that the circuit will work by adding the following numbers and writing the binary equivalents of the inputs and outputs:

a. $A_3 A_2 A_1 = 011, B_3 B_2 B_1 = 010$

b. $A_3 A_2 A_1 = 101, B_3 B_2 B_1 = 011$

Solution

The 3-bit adder is shown in *Figure 7.19*. The first full adder combines A_1 and B_1, with the input carry connected to an input labeled C_0; A_2, B_2, and C_1 are added in the second full adder. The carry output, C_1, of the first full adder is connected to the carry input of the second full adder. A similar connection is made between the second and third adders. (Note that we could also use a half adder to add the LSBs of a multiple-bit addition, since there is no carry input to the least significant bits.)

Sums:

a. $011 + 010 = 0101$ (This is shown in Figure 7.19.)

$A_1 = 1, B_1 = 0$ $C_1 = 0, S_1 = 1$

$A_2 = 1, B_2 = 1, C_1 = 0$ $C_2 = 1, S_2 = 0$

$A_3 = 0, B_3 = 0, C_2 = 1$ $C_3 = 0, S_3 = 1$

(Binary equivalent: $A_3 A_2 A_1 + B_3 B_2 B_1 = C_3 S_3 S_2 S_1 = 0101$)

continued...

b. $101 + 011 = 1000$

$$A_1 = 1, B_1 = 1 \qquad C_1 = 1, S_1 = 0$$
$$A_2 = 0, B_2 = 1, C_1 = 1 \qquad C_2 = 1, S_2 = 0$$
$$A_3 = 1, B_3 = 0, C_2 = 1 \qquad C_3 = 1, S_3 = 0$$

(Binary equivalent: $A_3 A_2 A_1 + B_3 B_2 B_1 = C_3 S_3 S_2 S_1 = 1000$)

FIGURE 7.19 Example 7.18: 3-Bit Adder from Full Adder Subcircuits

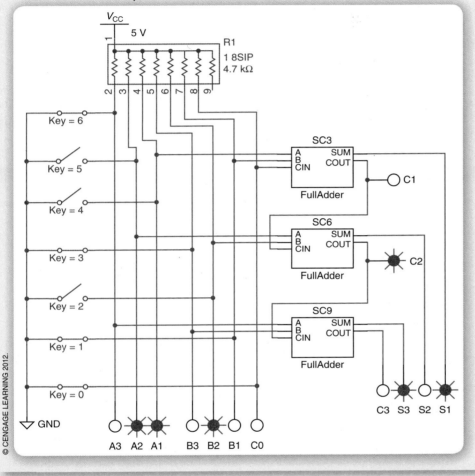

Parallel Binary Adder/Subtractor

KEY TERMS

Cascade To connect an output of one device to an input of another, often for the purpose of expanding the number of bits available for a particular function.

Parallel binary adder A circuit, consisting of *n* full adders, that will add two *n*-bit binary numbers. The output consists of *n* sum bits and a carry bit.

As Example 7.18 implies, a binary adder can be expanded to any number of bits by using a full adder for each bit addition and connecting their carry inputs and outputs in **cascade**. *Figure 7.20* shows four full adders connected as a 4-bit **parallel binary adder.**

FIGURE 7.20 4-Bit Parallel Binary Adder

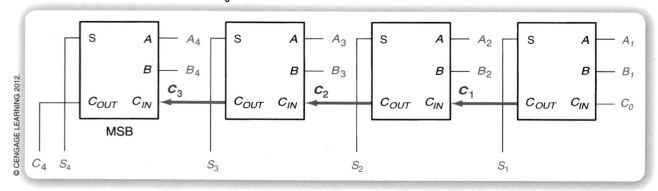

The first stage (LSB) can be either a full adder with its carry input forced to logic 0 or a half adder, as there is no previous stage to provide a carry. The addition is done one bit at a time, with the carry from each adder passed to the next stage.

Example 7.19

Verify the summing operation of the circuit in Figure 7.20 by calculating the output for the following sets of inputs:

a. $A_4 A_3 A_2 A_1 = 1111$, $B_4 B_3 B_2 B_1 = 0001$
b. $A_4 A_3 A_2 A_1 = 0101$, $B_4 B_3 B_2 B_1 = 1001$

Solution

At each stage, $A + B + C_{IN} = C_{OUT} S$.

a. $1010 + 1100 = 10110$
$(10_{10} + 12_{10} = 22_{10})$
$A_1 = 0, B_1 = 0, C_0 = 0; C_1 = 0, S_1 = 0$
$A_2 = 1, B_2 = 0, C_1 = 0; C_2 = 0, S_2 = 1$
$A_3 = 0, B_3 = 1, C_2 = 0; C_3 = 0, S_3 = 1$
$A_4 = 1, B_4 = 1, C_3 = 0; C_4 = 1, S_4 = 0$

(Binary equivalent: $C_4 S_4 S_3 S_2 S_1 = 10110$)

b. $0101 + 1001 = 1110$
$(5_{10} + 9_{10} = 14_{10})$
$A_1 = 1, B_1 = 1, C_0 = 0; C_1 = 1, S_1 = 0$
$A_2 = 0, B_2 = 0, C_1 = 1; C_2 = 0, S_2 = 1$
$A_3 = 1, B_3 = 0, C_2 = 0; C_3 = 0, S_3 = 1$
$A_4 = 0, B_4 = 1, C_3 = 0; C_4 = 0, S_4 = 1$

(Binary equivalent: $C_4 S_4 S_3 S_2 S_1 = 01110$)

Parallel Adder/Subtractor

2's Complement Subtractor

Recall the technique for subtracting binary numbers in 2's complement notation. For example, to find the difference $0101 - 0011$ by 2's complement subtraction:

1. Find the 2's complement of 0011:

$$
\begin{array}{ll}
0011 & \\
1100 & \text{(1's complement)} \\
\underline{+1} & \\
1101 & \text{(2's complement)}
\end{array}
$$

2. Add the 2's complement of the subtrahend to the minuend:

$$
\begin{array}{ll}
0101 & (+5) \\
\underline{+\ 1101} & \underline{+(-3)} \\
1\ 0010 & (+2)
\end{array}
$$

(Discard carry) ———┘

We can easily build a circuit to perform 2's complement subtraction, using a parallel binary adder and an inverter for each bit of one of the operands. The Multisim circuit shown in *Figure 7.21* performs the operation $(A - B)$. The 4-bit binary adder in the diagram is constructed of full adder subcircuits, somewhat like the 3-bit adder in Figure 7.19.

The four inverters generate the 1's complement of B. The parallel adder generates the 2's complement by adding the carry bit (held at logic 1) to the 1's complement at the B inputs. Algebraically, this is expressed as:

$$A - B = A + (-B) = A + \overline{B} - 1$$

where $\overline{B}$ is the 1's complement of B, and $(\overline{B} + 1)$ is the 2's complement of B.

FIGURE 7.21 4-Bit 2's Complement Subtractor

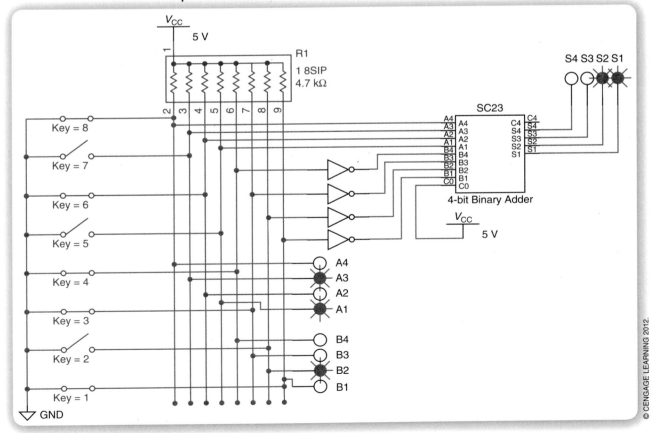

© CENGAGE LEARNING 2012.

Example 7.20

Multisim Example

Multisim File: 07.06 4-Bit Parallel Subtractor.ms11

a. Verify the operation of the 2's complement subtractor in Figure 7.21 by manually calculating the following subtraction:

$$0101 - 0011$$

b. Open the Multisim file for this example and run it as a simulation. Enter the inputs specified in part **a** of this example to verify correct operation of the circuit.

c. Does the Multisim subtractor work for unsigned subtractions? Under what conditions? Use the subtraction $1010 - 0100$ as an example.

d. Does the Multisim subtractor work for subtractions that give negative results? Use the subtraction $0100 - 0110$ as an example.

Solution

a. Let $\overline{B}$ be the 1's complement of B.

Inverter inputs (B):	0011
Inverter outputs ($\overline{B}$):	1100

Sum ($A + \overline{B} + 1$):	0101	($+5$)
	1100	$+ (-2)$
	$+\quad 1$	
Result:	0011	($+3$)

b. The Multisim circuit for the conditions in part **a** of this example is shown in Figure 7.21.

c. When the inputs of the Multisim subtractor circuit are set to calculate $1010 - 0100$, the output result is 0110. This is equivalent to the decimal subtraction $10 - 4 = 6$. Because the signed value of $+10$ (decimal) cannot fit into a 4-bit signed number, this must be an unsigned operation. The unsigned subtraction is valid at least for these values. By trying additional examples, you would find that the result is valid for any difference in the range $0 \leq x \leq 15$. Negative numbers cannot be represented by unsigned numbers.

d. When the inputs of the Multisim subtractor circuit are set to calculate $0100 - 0110$, the output result is 1110. This is equivalent to the decimal subtraction $(+4) - (+6) = (-2)$. We can find the magnitude of the result by taking its 2's complement. The valid range of 4-bit signed numbers is $-8 \leq x \leq +7$.

Parallel Binary Adder/Subtractor

Figure 7.22 shows a parallel binary adder configured as a programmable adder/subtractor. The Exclusive OR gates work as programmable inverters to pass B to the parallel adder in either true or complement form, as shown in *Figure 7.23*. When one input of an XOR gate is held LOW, the logic level at the other input is passed straight through the gate ($0 \oplus B = B$). When one input of an XOR gate

FIGURE 7.22 *2's Complement Adder/Subtractor*

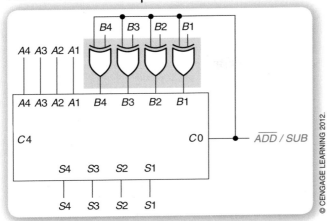

© CENGAGE LEARNING 2012.

FIGURE 7.23 *XOR Gate as a Programmable Inverter*

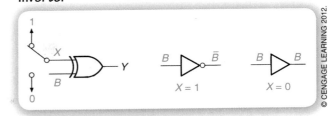

© CENGAGE LEARNING 2012.

is held HIGH, the logic level at the other input is passed through the gate and inverted ($0 \oplus B = \bar{B}$).

The $\overline{ADD}/SUB$ input is tied to the XOR inverter/buffers and to the carry input of the parallel adder. When $\overline{ADD}/SUB = 1$, B is complemented and the 1 from the carry input is added to the complement sum. The effect is to subtract ($A - B$). When $\overline{ADD}/SUB = 0$, the B inputs are presented to the adder in true form and the carry input is 0. This produces an output equivalent to ($A + B$).

This circuit can add or subtract 4-bit signed or unsigned binary numbers.

Example 7.21

Multisim Example

Multisim File: 07.07 4-Bit Parallel Adder Subtractor.ms11

a. Verify the operation of the 2's complement adder/subtractor in *Figure 7.24a* by manually calculating the following unsigned sum:

$$1010 + 0100$$

b. Open the Multisim file for this example and run it as a simulation. Enter the inputs specified in part **a** of this example to verify correct operation of the circuit.

c. Verify the operation of the 2's complement adder/subtractor in *Figure 7.25* by manually calculating the following unsigned subtraction:

$$1010 - 0100$$

d. Open the Multisim file for this example and run it as a simulation. Enter the inputs specified in part **c** of this example to verify correct operation of the circuit.

e. Try some other input combinations to further explore the operation of this circuit. What is the range of signed and unsigned addition and subtraction operations for this circuit?

continued...

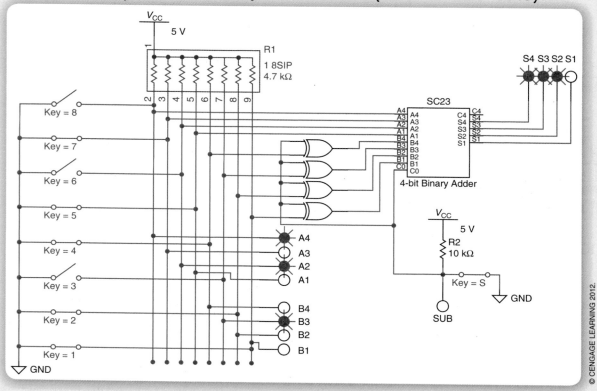

FIGURE 7.24 Example 7.21: 4-Bit Binary Adder/Subtractor (1010 + 0100 = 1110)

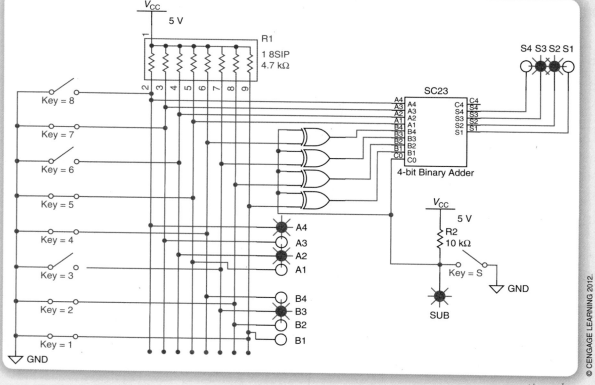

FIGURE 7.25 Example 7.21 4-Bit Binary Adder/Subtractor (1010 − 0100 = 0110)

continued...

Solution

a. **SUB $= 0$**

XOR-gate inputs (B):	0100
XOR-gate outputs (B):	0100

Sum ($A + B + 0$):

$$
\begin{array}{rr}
1010 & (10) \\
0100 & + (4) \\
+ \quad 0 & \underline{} \\
\end{array}
$$

Result: 1110   (14)

b. The Multisim circuit for the conditions in part **a** of this example is shown in Figure 7.24.

c. **SUB $= 1$. Let $\overline{B}$ be the 1's complement of B.**

Inverter inputs (B):	0100
Inverter outputs ($\overline{B}$):	1011

Sum ($A + \overline{B} + 1$):

$$
\begin{array}{rr}
1010 & (10) \\
1011 & +(-4) \\
+ \quad 1 & \underline{} \\
\end{array}
$$

Result: 0110   (6)

(Carry is discarded.)

d. The Multisim circuit for the conditions in part **a** of this example is shown in Figure 7.25.

e. The ranges of signed and unsigned results from the 4-bit adder/subtractor are determined as follows.

An unsigned result (sum or difference) has four magnitude bits: $0 \leq result \leq 15$.

A signed result (sum or difference) has a sign bit and three magnitude bits: $-8 \leq result \leq +7$.

Overflow Detection

Recall from Example 7.12 the condition for detecting a sign bit overflow in a sum of two binary numbers.

If the sign bits of both operands are the same and the sign bit of the sum is different from the operand sign bits, an overflow has occurred.

This implies that overflow is not possible if the sign bits of the operands are different from each other. This is true because the sum of two opposite-sign numbers will always be smaller in magnitude than the larger of the two operands.

Here are two examples:

1. $(+15) + (-7) = (+8)$; $+8$ has a smaller magnitude than $+15$.
2. $(-13) + (19) = (-4)$; -4 has a smaller magnitude than -13.

No carry into the sign bit will be generated in either case.

A 4-bit parallel binary adder will add two signed binary numbers as follows:

$$
\begin{array}{ll}
S_A\, A_3\, A_2\, A_1 & (S_A = \text{Sign bit of } A) \\
+ S_B\, B_3\, B_2\, B_1 & (S_B = \text{Sign bit of } B) \\
\hline
S_S\, S_3\, S_2\, S_1 & (S_S = \text{Sign bit of sum})
\end{array}
$$

TABLE 7.6 Truth Table for a Sign Bit Overflow Detector

S_A	S_B	S_S	V
0	0	0	0
0	0	1	1
0	1	0	0
0	1	1	0
1	0	0	0
1	0	1	0
1	1	0	1
1	1	1	0

© CENGAGE LEARNING 2012.

FIGURE 7.26 Overflow Detector

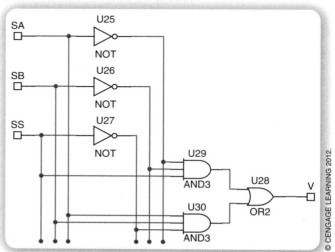

© CENGAGE LEARNING 2012.

From our condition for overflow detection, we can make a truth table for an overflow variable, V, in terms of S_A, S_B, and S_S. Let us specify that $V = 1$ when there is an overflow condition. This condition occurs when $(S_A = S_B) \neq S_S$. Table 7.6 shows the truth table for the overflow detector function.

The SOP Boolean expression for the overflow detector is:

$$V = \overline{S}_A \overline{S}_B S_S + S_A S_B \overline{S}_S$$

Figure 7.26 shows a logic circuit that will detect a sign bit overflow in a parallel binary adder. The detector is drawn as a Multisim subcircuit. The inputs S_A, S_B, and S_S are the MSBs (sign bits) of the adder A and B inputs and S outputs, respectively.

Example 7.22

Multisim Example

Multisim File: 07.08 4-Bit Parallel Adder Subtractor with Overflow Detection.ms11

a. Combine the 4-bit adder/subtractor shown in Figure 7.24 and other logic to make a 4-bit adder/subtractor that includes a circuit to detect sign bit overflow.

b. Test the adder/subtractor with overflow detection by entering the following input combinations. Which combinations result in sign bit overflow? Explain why sign bit overflow is or is not generated.

   i. 1000 + 1000
  ii. 0111 + 0001
 iii. 0111 + 1000
 iv. 0111 + 1100

Solution

a. *Figure 7.27* represents the 4-bit adder/subtractor with an overflow detector of the type shown in Figure 7.26.

b. *Table 7.7* shows the inputs specified above, with the resultant sum and overflow for each one. Sign bit overflow is detected when $V = 1$.

continued...

FIGURE 7.27 4-Bit Binary Adder/Subtractor with Overflow Detector

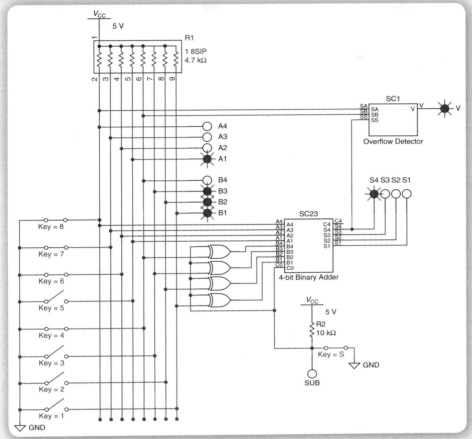

© CENGAGE LEARNING 2012.

TABLE 7.7 Sums and Overflow States © CENGAGE LEARNING 2012.

A	B	S	V	Reason
1000	1000	0000 (Discard carry bit)	1	Sum of two negative numbers appears to be positive
0111	0001	1000	1	Sum of two positive numbers appears to be negative
0111	1000	1111	0	Result is within acceptable range: $(+7) + (-8) = (-1)$
0111	1100	0011 (Discard carry bit)	0	Result is within acceptable range: $(+7) + (-4) = (+3)$

Your Turn

7.8 What is the permissible range of values of a sum or difference, x, in a 12-bit parallel binary adder if it is written as:
a. A signed binary number?
b. An unsigned binary number?

1. Addition combines an augend (x) and an addend (y) to get a sum ($z = x + y$).

2. Binary addition is based on four sums:

$$0 + 0 + 0 = 00$$
$$0 + 0 + 1 = 01$$
$$0 + 1 + 1 = 10$$
$$1 + 1 + 1 = 11$$

3. A sum of two bits generates a sum bit and a carry bit. (For the first two sums just shown, the carry bit is 0; the last two sums have a carry of 1. The last sum includes a carry from a lower-order bit.)

4. Subtraction combines a minuend (x) and a subtrahend (y) to get a difference ($z = x - y$).

5. Binary subtraction is based on the following four differences:

$$0 - 0 = 0$$
$$1 - 0 = 1$$
$$1 - 1 = 0$$
$$10 - 1 = 1$$

6. If the subtrahend bit is larger than the minuend bit, as in the fourth difference above, a 1 must be borrowed from the next higher-order bit.

7. Binary addition or subtraction can be unsigned, where the magnitudes of the operands and result are presumed to be positive, or signed, where the operands and result can be positive or negative. The sign is indicated by a sign bit.

8. The sign bit (usually MSB) of a binary number indicates that the number is positive if it is 0 and negative if it is 1.

9. Signed binary numbers can be written in true-magnitude, 1's complement, or 2's complement form. True magnitude has the same binary value for positive and negative numbers, with only the sign bit changed. A 1's complement negative number is generated by inverting all bits of the positive number of the same magnitude. A 2's complement negative number is generated by adding 1 to the equivalent 1's complement number. Positive numbers are the same in all three forms.

10. 1's complement or 2's complement binary numbers are used in signed addition or subtraction. Subtraction is performed by adding a negative number in complement form to another number in complement form [i.e., $x - y = x + (-y)$]. This technique does not work for true-magnitude form.

11. A negative sum or difference in 2's complement subtraction must be converted to a positive form to read its magnitude [i.e., $-(-x) = +x$].

12. A signed binary number, x, with n magnitude bits has a valid range of $-2^n \leq x \leq + (2^n - 1)$.

13. A negative number with a power-of-2 magnitude (i.e., -2^n) is written in 2's complement form as n 0's preceded by all 1's to fill the defined size of the number (e.g., in 8-bit 2's complement form, $-128 = 10000000$ (1 followed by seven 0's; $128 = 2^7$); in 8-bit 2's complement form, $-8 = 11111000$ (all 1's, followed by three 0's; $8 = 2^3$).

14. If the sum of two n-bit 2's complement numbers carries beyond the nth bit, discard the carry. The carry is not retained because in an n-bit number there is no $(n + 1)^{\text{th}}$ position to hold the carry (e.g., there is no 9th bit in an 8-bit number). Analogy: A car odometer does not hold the leading value of "1" when the mileage goes from 99999 to 100000. The "1" is not shown because there is no place to hold it.

15. A 2's complement number can be expanded to a larger bit size by the process of sign extension. To expand a positive number to a larger bit size, pad the number with leading 0's. To expand a negative number, pad it with leading 1's.

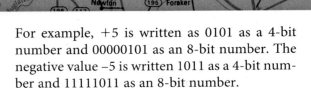

For example, +5 is written as 0101 as a 4-bit number and 00000101 as an 8-bit number. The negative value –5 is written 1011 as a 4-bit number and 11111011 as an 8-bit number.

16. If a sum or difference falls outside the permissible range of magnitudes for a 2's complement number, it generates an overflow into the sign bit of the number. The result is that the sum of two positive numbers appears to be negative (e.g., 01111111 + 00000010 = 10000001) or the sum of two negative numbers appears to be positive (e.g., 11111111 + 10000000 = 01111111, where the carry beyond the 8th place is discarded).

17. Binary numbers can be used in nonpositional codes to represent numbers or alphanumeric characters.

18. Binary coded decimal (BCD) codes represent decimal numbers as a series of 4-bit groups of numbers. Natural BCD or 8421 code does this as a positionally weighted code for each digit (e.g., 158 = 0001 0101 1000 [NBCD]).

19. ASCII code represents alphanumeric characters and computer control codes as a 7-bit group of binary numbers. Alpha characters are listed in uppercase in columns 4 and 5 of the ASCII table. Lowercase alpha characters are in columns 6 and 7. Numeric characters are in column 3.

20. A half adder combines two bits to generate a sum and a carry. It can be represented by the following truth table:

A	B	C_{OUT}	S
0	0	0	0
0	1	0	1
1	0	0	1
1	1	1	0

© CENGAGE LEARNING 2012.

21. From the half adder truth table, we can derive two equations:

$$C_{OUT} = AB$$
$$S = A \oplus B$$

22. A full adder can accept an input carry from a lower-order adder and combine the input carry with two operands to generate a sum and output carry. Its operation can be summarized in the following truth table:

A	B	C_{IN}	C_{OUT}	S
0	0	0	0	0
0	0	1	0	1
0	1	0	0	1
0	1	1	1	0
1	0	0	0	1
1	0	1	1	0
1	1	0	1	0
1	1	1	1	1

© CENGAGE LEARNING 2012.

23. Two half adders can be combined to make a full adder. Operands A and B go to the first half adder. The sum output of the first half adder and the carry input go to the inputs of the second half adder. The carry outputs of both half adders are combined in an OR gate.

24. The following Boolean equations for a full adder can be derived from the truth table and Boolean algebra:

$$C_{OUT} = (A \oplus B) C_{IN} + AB$$
$$S = (A \oplus B) \oplus C_{IN}$$

25. Multiple full adders can be cascaded to make a parallel binary adder. Operands A_1 and B_1 are applied to the first full adder. Carry bit C_0 is grounded. A_2 and B_2 go to the second adder stage, and so on. The carry output of one stage is cascaded to the carry input of the following stage. This connection is called ripple carry.

26. A parallel binary adder can be made into a 2's complement subtractor by inverting one set of inputs and tying the input carry to a logic HIGH.

27. A parallel binary adder can be made into a 2's complement adder/subtractor by using a set of XOR gates as programmable inverters and connecting the XOR control line to the carry input of the adder.

28. A sign bit overflow in a 2's complement adder/subtractor can be detected by comparing the sign bits of the operands to the sign bit of the result. If the sign bits of the operands are the same as each other, but different from the sign bit of the result, there has been an overflow. The Boolean equation for this detector is given by $V = \overline{S_A}\,\overline{S_B}S_S + S_A S_B \overline{S_S}$.

BRING IT HOME

For maximum learning benefit, do not use a calculator for the problems in this chapter, except to check your work.

7.1 Digital Arithmetic

7.1 Add the following unsigned binary numbers.
  a. $10101 + 1010$
  b. $10101 + 1011$
  c. $1111 + 1111$
  d. $11100 + 1110$
  e. $11001 + 10011$
  f. $111011 + 101001$

7.2 Subtract the following unsigned binary numbers.
  a. $1100 - 100$
  b. $10001 - 1001$
  c. $10101 - 1100$
  d. $10110 - 1010$
  e. $10110 - 1001$
  f. $10001 - 1111$
  g. $100010 - 10111$
  h. $1100011 - 100111$

7.2 Representing Signed Binary Numbers

7.3 Write the following decimal numbers as 8-bit signed binary numbers, using 2's complement form.
  a. -110
  b. 67
  c. -54
  d. -93
  e. 0
  f. -1
  g. 127
  h. -127

7.3 Signed Binary Arithmetic

7.4 Perform the following arithmetic operations in the 2's complement system. Use 8-bit numbers consisting of a sign bit and 7 magnitude bits. (The numbers shown are in the decimal system.)

  If a result is negative, convert it back to decimal to prove the correctness of the operation.
  a. $37 + 25$
  b. $85 + 40$
  c. $95 - 63$
  d. $63 - 95$
  e. $-23 - 50$
  f. $120 - 73$
  g. $73 - 120$

7.5 What are the largest positive and negative numbers, expressed in 2's complement notation, that can be represented by an 8-bit signed binary number?

7.6 a. Write $+19$ as 2's complement value with the smallest number of bits possible.
  b. Write $+19$ as an 8-bit 2's complement number.
  c. Write $+19$ as a 12-bit 2's complement number.

7.7 a. Write -19 as a 2's complement value with the smallest number of bits possible.
  b. Write -19 as an 8-bit 2's complement number.
  c. Write -19 as a 12-bit 2's complement number.

7.4 Numeric and Alphanumeric Codes

7.8 Convert the following decimal numbers to true binary and to 8421 BCD code.
  a. 709_{10}
  b. 1889_{10}
  c. 2395_{10}
  d. 1259_{10}
  e. 3972_{10}
  f. 7730_{10}

7.9 Write your name in ASCII code.

7.5 Binary Adders and Subtractors

7.10 Write the truth table for a half adder, and from the table derive the Boolean expressions for both C_{OUT} (carry output) and S (sum output) in terms of inputs A and B. Draw the half adder circuit.

7.11 Write the truth table for a full adder.

7.12 Draw a circuit showing a full adder constructed from two half adders.

7.13 Evaluate the Boolean expression for S and C_{OUT} of the full adder in Figure 7.6 for the following input values. What is the binary value of the outputs in each case?
  a. $A = 0, B = 0, C_{IN} = 0$
  b. $A = 0, B = 1, C_{IN} = 0$
  c. $A = 0, B = 1, C_{IN} = 1$
  d. $A = 1, B = 1, C_{IN} = 1$

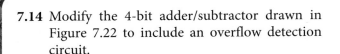

7.14 Modify the 4-bit adder/subtractor drawn in Figure 7.22 to include an overflow detection circuit.

7.15 What is the permissible range of values that a sum or difference, x, can have in a 16-bit parallel binary adder if it is written as:
  a. A signed binary number
  b. An unsigned binary number

EXTRA MILE

7.3 Signed Binary Arithmetic

7.16 Perform the following *signed* binary operations, using 2's complement notation where required. State whether or not sign bit overflow occurs. Give the signed decimal equivalent values of the sums in which overflow does *not* occur.
  a. 01101 + 00110
  b. 01101 + 10110
  c. 01110 − 01001
  d. 11110 + 00010
  e. 11110 − 00010

7.17 Without doing any binary complement arithmetic, indicate which of the following operations will result in 2's complement overflow. (Assume an 8-bit representation consisting of a sign bit and 7 magnitude bits.) Explain the reasons for each choice. (All numbers are shown in the decimal system.)
  a. −109 + 36
  b. 109 + 36
  c. 65 + 72
  d. −110 − 29
  e. 117 + 11
  f. 117 − 11

7.18 Explain how you can know, by examining sign or magnitude bits of the numbers involved, when overflow has occurred in 2's complement addition or subtraction.

7.4 Numeric and Alphanumeric Codes

7.19 Give an example of a consumer product where designing in binary coded decimal might be better than designing in true binary. Explain the reason for your answer.

7.20 Encode the following text into ASCII code: "15% off purchases over $50 (Monday only)."

7.21 Decode the following sequence of ASCII characters.
43 41 55 54 49 4F 4E 21 20 45 72 61
73 69 6E 67 20 61 6C 6C 20 64 61 74 61
21 20 41 72 65 20 79 6F 75 20 73 75 72
65 3F

7.5 Binary Adders and Subtractors

7.22 Multisim Problem
  a. Use Multisim to make a 4-bit adder from four full adder subcircuits, as shown in *Figure 7.28*.

FIGURE 7.28 **Problem 7.22: 4-Bit Adder from Full Adder Subcircuits** © CENGAGE LEARNING 2012.

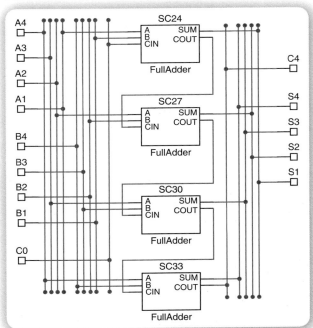

continues...

continued...

b. Make the 4-bit adder from part **a** of this problem into a subcircuit and make a Multisim test circuit, as shown in *Figure 7.29*.

7.23 Multisim Problem

Verify the summing operation of the circuit in Figure 7.29, as follows. Determine the output of each full adder based on the inputs shown below. Calculate each sum manually and compare it to the 5-bit Multisim-simulated output $(C_4 S_4 S_3 S_2 S_1)$ of the parallel adder circuit.

a. $A_4 A_3 A_2 A_1 = 0100$, $B_4 B_3 B_2 B_1 = 1001$
b. $A_4 A_3 A_2 A_1 = 1010$, $B_4 B_3 B_2 B_1 = 0110$
c. $A_4 A_3 A_2 A_1 = 0101$, $B_4 B_3 B_2 B_1 = 1101$
d. $A_4 A_3 A_2 A_1 = 1111$, $B_4 B_3 B_2 B_1 = 0111$

FIGURE 7.29 Problem 7.22: Testing a 4-Bit Unsigned Binary Adder in Multisim

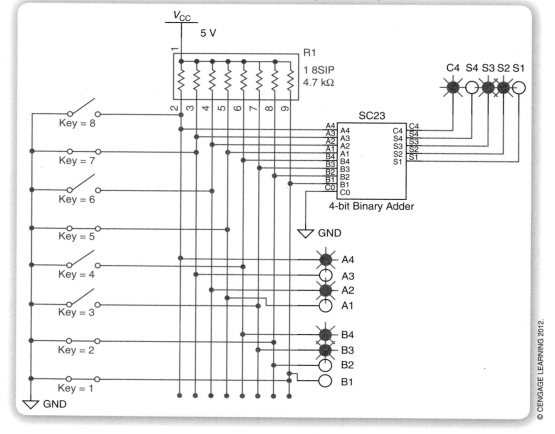

© CENGAGE LEARNING 2012.

CHAPTER 8
Digital System Application

START LOCATION · DISTANCE · END LOCATION

Menu

CHAPTER OBJECTIVES

Upon successful completion of this chapter, you will be able to:

1 Design a complex system-level circuit using smaller subcircuits.

2 Use glue logic to join subcircuits to one another.

3 Use problem-solving techniques to develop an effective solution to a circuit design problem.

4 Draw a block diagram to represent a circuit at a system level.

5 Design an adder/subtractor.

6 Design a 4-bit multiplier.

7 Effectively troubleshoot errors within a large circuit by carefully reviewing subcircuits.

PHOTO: © ISTOCKPHOTO.COM/CATENARYMEDIA.

KEY TERM

Glue logic Gates or simple circuitry whose purpose is to connect more complex blocks of a larger circuit.

Now that we know how to design some basic digital circuits, let's try to design a simple calculator. Adders, multiplexers, decoders, and other combinational logic functions are useful for performing the tasks for which they are designed. The functionality of these devices can be extended when we combine them to form larger, more complicated circuits. They are often interfaced to each other by using logic gates, which are sometimes referred to as **glue logic**.

When a problem requiring a more complex circuit is specified, the specification of the requirements are usually given verbally, that is, as a word problem, similar to those that you may have encountered in areas such as math or physics. If the person designing the solution uses sound problem-solving techniques, the solution he or she develops will more likely meet the needs as specified.

The application in this chapter demonstrates a solution to a digital problem, showing how individual components can be integrated at a system level. A complete solution is presented, including some alternative circuits and trade-offs. The solution may be downloaded to a development board when complete, to better understand the solution through demonstration.

8.1 PROBLEM-SOLVING TECHNIQUES

KEY TERM

Block diagram A high-level diagram showing a complex circuit as a combination of smaller subcircuits.

Problems have been and will continue to be defined as verbal descriptions of the final specifications the solutions must meet. "Word problems" encountered in math or science courses are good examples.

One of the most important aspects of any word problem, and one most often overlooked, is the need to *understand* the problem in its entirety. Many students (and many circuit designers) are guilty of diving right in to solve a problem, only to discover that they have overlooked a crucial aspect of the requirements.

We are ultimately in search of a digital circuit that can meet given specifications. Any combinational circuit so far can be built using AND, OR, and NOT gates. However, we have seen that some digital circuits are used so often, we have developed special circuits to perform certain *functions*: for example, adders or multiplexers. We can use these circuits with little attention to how they are designed, or "what is inside" these devices. Once we understand the requirements, or once we understand the question, we can use the tools we have learned about to complete the circuit successfully.

The first step in the design process is to find the *accurate description of the problem*. We have to determine exactly what *inputs* and *outputs* are required, and understand exactly what the performance specifications are. We have to understand what effect the inputs have on the system, and discover any constraints on the inputs; for example, whether the inputs are active-HIGH or active-LOW. We have to understand each output of the system, including its function and any constraints. The outputs may be described in a truth table form unless there are a large number of outputs, in which case we may describe them verbally.

Once the inputs and outputs of the system have been adequately described, a **block diagram** or flowchart showing the solution from a high level could be drawn. At this level, we can understand the overall solution, breaking a complex circuit into a series of smaller modules; each module can be designed and connected to the others.

Consider the development of a decoder: a description of what the circuit does is given, inputs and outputs are listed, a truth table or verbal description of the function relating the outputs to the inputs is developed, and the circuit is built. This device is used often, so there is no need to redesign it each time we need a decoder on its own or as part of a larger design. A specific example of a decoder, the 74138 Decoder, is shown in *Figure 8.1*. If we need to use a decoder in a larger design, we can use this symbol with a full understanding of the effect the inputs have on the outputs. We do not need to "look inside" to see the logic circuit or gates, and we don't need to be concerned with exactly how this device works. If we need to use a decoder, we can use the symbol without the full schematic. We can use symbols we specify to draw the circuit at a high level, then "fill in" the circuitry for each block.

When a complete circuit schematic is drawn, which may be in blocks or modules, we can test it using Multisim to simulate the design, download it to a laboratory board, or both. If possible, each input combination should be tested to verify the correct output.

FIGURE 8.1 74138 Decoder

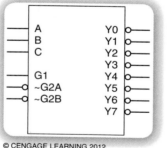

© CENGAGE LEARNING 2012.

8.2 SAMPLE APPLICATION: A SMALL CALCULATOR

One example of a complex combinational circuit could be a calculator. The performance specifications will be given and the circuit will be built step by step.

This gives the opportunity to look at trade-offs and decisions we may make as we build a circuit.

Problem Statement

Design a small calculator that takes two 4-bit binary numbers as inputs, and outputs results as shown in *Table 8.1* according to the "choice" inputs. The final design should have two 4-bit inputs (A and B), two "choice" inputs (C), and eight outputs (Z).

TABLE 8.1   *Description of C Inputs to Calculator Problem*

Choice	Function	Output
00	Add	$Z = A + B$
01	Subtract	$Z = A - B$ (negative numbers in 2's complement form)
10	Multiply	$Z = A \times B$
11	Min/Max	Upper 4 bits of Z = maximum of A or B Lower 4 bits of Z = minimum of A or B

First Step: Understand the Question

Because the first step in the process is to find an accurate description of the problem, read the problem statement one more time.

This circuit has two inputs that are each 4-bit binary numbers. The output will be one of four results: first, the output could be $A + B$. Because these are both 4-bit numbers, the highest value we could expect would be $1111 + 1111 = 11110$ binary, which easily fits into the 8 bits allowed for the output. The output could be $A - B$; in this case, we may need to deal with complementing one of the variables. Although overflow could occur if there are not enough output bits to store our subtraction result, overflow should not occur because we are using 4-bit input values and 8 bits for the output. A third possibility is multiplication. We have two 4-bit inputs, so the highest value we can expect is $1111 \cdot 1111 = 11100001$ binary, which fits into our 8-bit output. Finally, we have a requirement to compare A and B, sending the maximum of the two 4-bit inputs to the upper four bits of the output and the minimum of A and B to the lower four bits. What do we do if $A = B$? It doesn't matter in this case if we send A or B to the upper or lower four bits: the result would be identical.

We also have inputs that will allow users to choose their desired function. We may want to consider these inputs as one 2-bit value or two individual inputs. For now, we will consider this a 2-bit input value.

Your Turn

8.1 The function of one component in our calculator circuit is to send the larger value of A and B to the upper four bits, the lesser of the two to the lower four bits. What is true if $A = B$?

a. This is a different case and will require additional circuitry.

b. Either input can be directed to either 4-bit component of the output.

c. This situation cannot occur.

Second Step: Inputs and Outputs

The following inputs are described:

Inputs A and B will form the operands for our calculator.

$$A_3 \, A_2 \, A_1 \, A_0 \text{ form the 4-bit value A.}$$
$$B_3 \, B_2 \, B_1 \, B_0 \text{ form the 4-bit value B.}$$

We also have a 2-bit input variable called C that allows us to select which operation will be done:

$$C_1 \text{ and } C_0\text{: two bits used to select the operation.}$$

Notice that A and B will usually be treated as a 4-bit binary number instead of four individual bits. There is no reason to consider the value of C as a single value rather than two individual bits. A and B both represent numbers; for example, if A = 1001, we understand this to mean "A = 9." If C = 01, we understand this to mean *subtract* because C_1 = 0 and C_1 = 1, but saying "C = 1" really has no meaning.

The following outputs are given:

$$Z_7 \, Z_6 \, Z_5 \, Z_4 \, Z_3 \, Z_2 \, Z_1 \, Z_0$$

These form the output value from the inputs specified by the user. Note that for addition, subtraction, and multiplication, the output value Z is treated like one 8-bit binary number. For the Max/Min function, Z can be thought of as two 4-bit numbers: $Z_7 \, Z_6 \, Z_5 \, Z_4$ and $Z_3 \, Z_2 \, Z_1 \, Z_0$.

Third Step: Block Diagram

The next step is to develop a block diagram of the system. This will allow smaller, more manageable pieces of the project to be built rather than thinking of the whole design at once.

There may be more than one way to break the design into smaller parts. In other words, two designers drawing a block diagram may come up with different solutions—as long as they both meet all of the design specifications, they both may be equally valid.

Consider the different sections of this design. We will need to design some input structure. The input block may be as simple as listing the inputs in correct order, but we won't be sure until we begin designing. We'll call this an input block for now. Similar to the input block, an output block may not be necessary when our schematic is complete; it may simply be a set of outputs in the correct order. The final block diagram is shown in *Figure 8.2*.

Each function can be designated as its own block. It is possible that some blocks may be combined and others may turn out to be unnecessary. For now, we'll specify individual blocks for each function.

As each separate block is completed, we must ensure that all signals from one block to the next are specified exactly. For example, when our design is complete, all signals that the multiplication block requires from the input block must be available. Some of these signals may be inputs and outputs for the entire system and some may be new signals that exist only inside of the design.

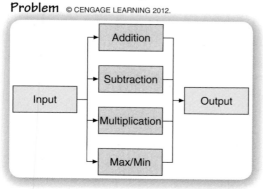

FIGURE 8.2 Block Diagram of Calculator Problem © CENGAGE LEARNING 2012.

8.3 COMPONENTS OF THE CALCULATOR

Individual components of the larger calculator will be drawn and then assembled at the end to form the complete calculator design.

Starting the Design in Multisim Software

Start the Multisim software and select File > New > Design. It is recommended that you save your work often, and store your design in a new folder or directory. If you are working on a network, check with your instructor to see how to store your work on a network drive.

We will use the Multisim feature of Hierarchical Blocks to allow us to design in modules. The advantage of this method is, once we design a piece of the circuit, we no longer have to be concerned with the internal circuitry—we can just insert it into our overall calculator circuit, giving us a neat, clean schematic.

Input Block

The input block, *Figure 8.3*, seems like a natural starting point. It is the first block encountered and controls inputs to all other blocks. Because all of the inputs to the system are specified in the problem, it should be straightforward to design. This component may need revisiting if we find that another block requires additional signals or additional information.

We know that the inputs to our system consist of:

$A_3 A_2 A_1 A_0$: the 4-bit value A
$B_3 B_2 B_1 B_0$: the 4-bit value B
$C_1 C_0$: two bits used to select the operation

Start a new design in Multisim and save it with the name **calculator**. This will be our overall design when the calculator is finished.

For now, there is no indication that the input block is anything more than a list of all inputs to the system. Because this is the most straightforward solution, this is what we will choose. If we need to make modifications later, we will be able to.

Choose Place > New Hierarchical Block from the menu. Call this **calc_input** by typing this in for the file name of the hierarchical block, and assign 10 inputs and 10 outputs (see *Figure 8.4*). Once this block is created, a component will appear on the calculator design, and you can click to place it.

We can customize the component at this point. Double-click the component and you have the opportunity to change the name by typing a new name into RefDes:. Change the name to **Input**. Click EditHB/SC to configure the inside of the component—to change the function of the component. Double-clicking on the input name allows you to change the names of the inputs to A3, A2, A1, A0, B3, etc. Connect the input pin to the output pin on the other side of the component and accept the new name (see *Figure 8.5*). When all 10 inputs are connected to 10 outputs, click save.

FIGURE 8.3 Input Block from Block Diagram

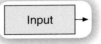

© CENGAGE LEARNING 2012.

FIGURE 8.4 Hierarchical Block for *calc_input*

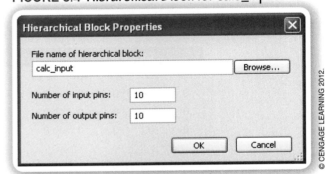

© CENGAGE LEARNING 2012.

FIGURE 8.5 Multisim Renaming Connected Signals

© CENGAGE LEARNING 2012.

Addition Block

You might recall that addition circuits have been studied from different viewpoints: a gate-by-gate level, a level using half adders and full adders, and as more complex multibit adders. We can use any of these solutions to meet the specifications of this block.

Our first choice could be to look at this block as a typical combinational circuit with known inputs and outputs. This solution requires a truth table followed by K-maps to simplify equations.

How many inputs enter this block? Obviously, there are (at least) eight inputs—four bits for A and four bits for B. The C inputs don't figure into the addition, but instead are used to determine if this block is selected. Such inputs are referred to as **control lines** because their purpose is to control something in the circuit rather than directly affecting the output.

We are allowing eight bits for our output. However, we expect the three highest bits to always be LOW because the highest addition value possible is $1111 + 1111 = 11110$, requiring five bits. If the outputs are thought of as functions of the eight input bits, a truth table whose first few rows are shown in *Table 8.2* would be needed. Table 8.2 is not the complete truth table because our full truth table would have

$$2^{\text{number of inputs}} = 2^8 = 256 \text{ rows}$$

It is apparent that there must be a more efficient way than a truth table to design this function. You may have noticed that most previous examples of combinational logic have three or four inputs; a system with even five inputs becomes difficult to work with using truth tables or K-maps.

Another option is to design this circuit using full adders to add each individual pair of bits of A and B. *Figure 8.6* shows a 4-bit parallel adder using

TABLE 8.2 Partial Truth Table for 4-Bit Adder Function

Input A				Input B				Output Z							
A3	A2	A1	A0	B3	B2	B1	B0	Z7	Z6	Z5	Z4	Z3	Z2	Z1	Z0
0	0	0	0	0	0	0	0	0	0	0	0	0	0	0	0
0	0	0	0	0	0	0	1	0	0	0	0	0	0	0	1
0	0	0	0	0	0	1	0	0	0	0	0	0	0	1	0

...

© CENGAGE LEARNING 2012.

FIGURE 8.6 4-Bit Parallel Adder

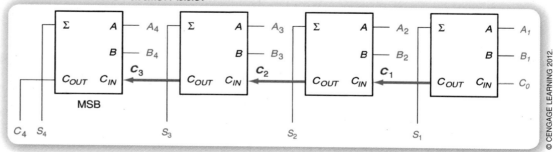

© CENGAGE LEARNING 2012.

full adders. Each full adder actually consists of the gate level circuit shown in *Figure 8.7*.

FIGURE 8.7 **Full Adder from Logic Gates**

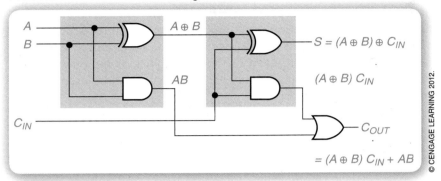

© CENGAGE LEARNING 2012.

Example 8.1

Draw a schematic of the adder function using the gate level implementation of a full adder.

Solution

The solution is shown in *Figure 8.8*.

FIGURE 8.8 **Example 8.1: Gate Level Full Adder**

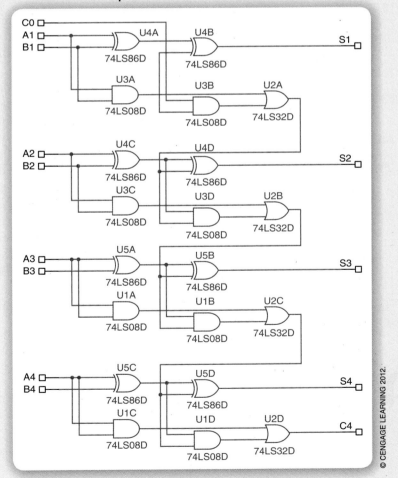

© CENGAGE LEARNING 2012.

FIGURE 8.9 7483 4-Bit Adder

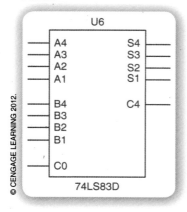

74LS83D

FIGURE 8.10 7483 4-Bit Adder with Inputs and Outputs

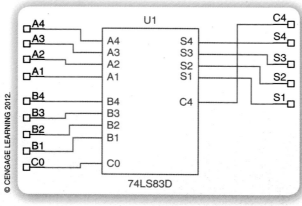

74LS83D

FIGURE 8.11 Calculator with Input and Adder Modules

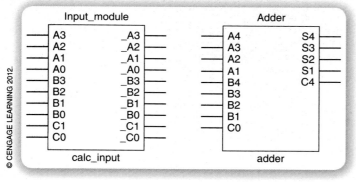

We have already seen a circuit to add two 4-bit numbers called a 4-bit adder. The 4-bit adder is implemented in a specific device labeled 7483, after the TTL part number for a device with this function. This is shown in *Figure 8.9*. Using this device allows the adder block to be drawn with only one device; this is an effective use of a device that has previously been developed. This solution still does not take the calculator function select inputs (C_1 and C_0) into account; these signals do not affect the sum, but there must be additional circuitry within our final calculator to choose the output of the addition block as the output of the calculator. Note that the 7483 symbol has a C_0 input that will be used later for this purpose.

Open a new hierarchical block and add a 7483 component, draw inputs to all of the inputs of the 7483, and draw outputs from the outputs. This is shown in *Figure 8.10*. Save this component as **adder**.

Notice the numbering scheme used in Figure 8.10. At first glance, some of the component inputs and outputs appear to be connected to the wrong input or output pins. Recall that we named our inputs A3, A2, A1, and A0. The 7483 symbol names its inputs A4, A3, A2, and A1. Our most significant bit (MSB) is A3; the MSB of the device is A4. We can either

▶ Connect our MSB to the MSB of the device and continue down to the least significant bit (LSB) (shown), or
▶ Change our numbering scheme to match that of the device.

The typical solution, and the solution we will implement here, is to match the LSB and MSB of our inputs and outputs with the LSBs and MSBs of the devices selected. Although it may seem like a good idea to change the numbering scheme of our inputs and outputs, the next component selected may use a naming convention that is the same as our original system. There is no standard naming convention for inputs, outputs, and control lines on devices such as the 7483; you may see different names for input, output, and control lines for the same device depending on where you see the symbol. We will see that the adder block uses our numbering scheme once we create the symbol.

Once the hierarchical block is created, you can close the window with the 7483 inside. We can now add the adder component any time by selecting **Hierarchical Block from File** from the Place menu. The symbol and inputs are shown in *Figure 8.11* before any connections have been drawn.

Your Turn

8.2 How many gates would be required to build the full adder implementation of the 4-bit parallel adder?

Subtraction Block

We have seen a 4-bit subtractor designed using the 4-bit and four inverters to perform 2's complement subtraction, as shown in *Figure 8.12*. We can create a new hierarchical block for the subtractor circuit shown in *Figure 8.13*. Once this block is created, we can add it to our calculator; see *Figure 8.14*.

Connecting the input lines to our adder and subtractor would begin to complete the design of the calculator. There are remaining issues: first, the select lines (C_1 and C_0) have not been taken into account. Second, after we developed an adder circuit and subtractor circuit, we followed those designs by the design of a 4-bit adder/subtractor. Is it possible to replace two blocks with one schematic, requiring only one 7483? Yes.

FIGURE 8.12 *2's Complement Subtractor*

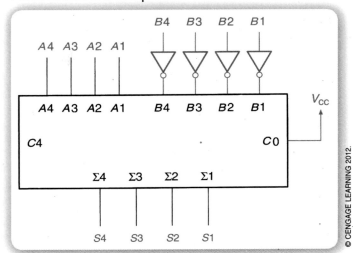

FIGURE 8.13 Subtractor Circuit

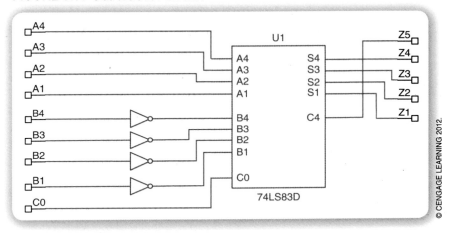

FIGURE 8.14 Calculator with Adder and Subtractor Modules

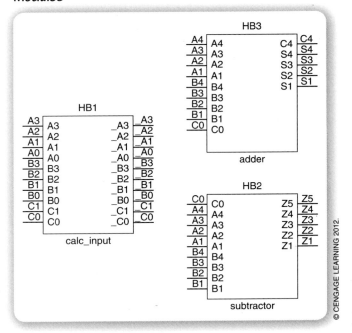

FIGURE 8.15 Block Diagram with Adder and Subtractor Modules Combined

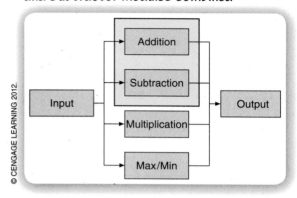

FIGURE 8.16 2's Complement Adder/Subtractor

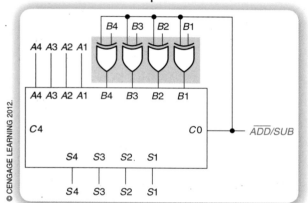

This would give us a more efficient design with one circuit meeting the requirements for two blocks. Of course, we may want to denote this in some way on our original block diagram as shown in *Figure 8.15*. Recall our adder/subtractor design from earlier, shown in *Figure 8.16*. This block will add A + B when *ADD/SUB* = 0 and subtract A − B in 2's complement form when *ADD/SUB* = 1.

Addition/Subtraction Block

The schematic for a 4-bit adder/subtractor is shown in *Figure 8.17*. Draw this schematic as a new hierarchical block and call it **AddSub**. The next step is to add this component to the overall file, **calculator**. Now, what signal from the input block should be connected to our signal named *notA_S*? Look at Figure 8.16; which line (C_1 or C_0) should be connected?

Reviewing Table 8.1 shows that input C_1 has no effect on whether this component should add or subtract. Input C_0 does have an effect: if C_0 is LOW, we want to add. If input C_0 is HIGH, we want to subtract. Compare this to input *notA_S*: if this input is LOW, we want to add. If this input is HIGH, we want to subtract. This matches input C_0 exactly! Therefore, C_0 should be connected to *notA_S*. Although this does not specifically select this function, we can see that if this component is selected, the output result is A + B if C_0 is LOW or A − B if C_0 is HIGH. The diagram (showing C_0 connected) is shown in *Figure 8.18*.

FIGURE 8.17 4-Bit 2's Complement Adder/Subtractor

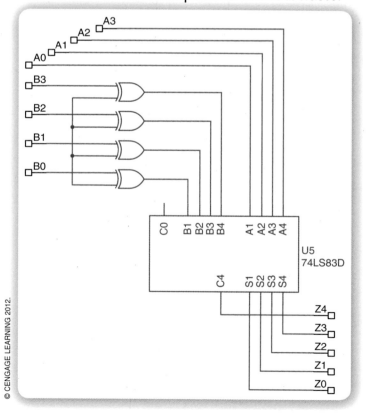

FIGURE 8.18 4-Bit 2's Complement Adder/Subtractor, with C_o Connected

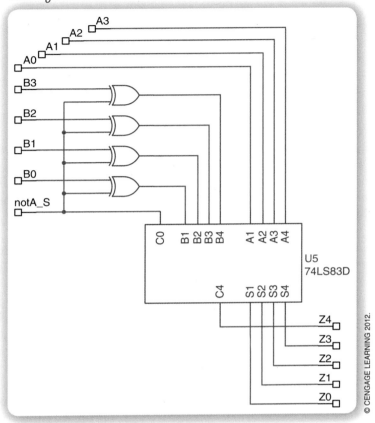

A few unresolved issues in our calculator design still need to be solved. Input C_1 has not been used, and only five of the eight outputs have been accounted for.

Multiplication Block

KEY TERMS

Multiplicand The number in a multiplication problem that is multiplied by each digit in the multiplier.

Multiplier The number which the multiplicand is multiplied by to get the result of a multiplication problem.

Start by examining a multiplication problem using decimal numbers; for example, 15×43:

$$\begin{array}{r} 15 \\ \times\, 43 \\ \hline 45 \\ 60 \\ \hline 645 \end{array}$$

In our example, the **multiplicand** is 15 and the **multiplier** is 43. Multiplication involves multiplying the multiplicand by each digit in the multiplier. For each digit of the multiplier, we shift the result by one digit to the left. When we have gone through each digit of the multiplier, we finish by *adding* each of these terms to get the final product.

Follow through an example of binary multiplication, involving the same rules as multiplication with decimal numbers. We'll use 1010 × 1011 for an example:

$$
\begin{array}{r}
1010 \\
\times\ 1101 \\
\hline
1010 \\
0000 \\
1010 \\
1010 \\
\hline
10000010
\end{array}
$$

A quick check: if we convert these binary numbers to decimal, we would get 10 × 13 = 130, showing that our binary multiplication is correct.

Notice that one advantage of using binary numbers is that each number we add is either the multiplicand itself or 0; we add the multiplicand only if the corresponding bit in the multiplier is 1. In other words, we multiply by only 0 or 1.

One possible choice to build a multiplier might be to design this circuit at a gate level, but with eight inputs and eight outputs, we would have eight truth tables with 256 rows each—not a practical solution.

We will design a solution using our observation that we *multiply* using *repeated addition*. If we review the previous example, we start with the multiplicand (because the LSB of the multiplier is 1). Next, we shift to the left and add zero (because the next bit in the multiplier is 0). We continue with this pattern: if the next bit of the multiplier is zero, shift one bit to the left and add zero. If the next multiplier bit is 1, shift one bit to the left and add the multiplicand.

Because we are adding numbers repeatedly, adders can be used instead of a gate level solution designed from scratch. Our solution will quickly exceed five bits because the multiplier and multiplicand are both four bits, so the solution requires an adder capable of adding more than two 4-bit numbers. Fortunately, it is straightforward to use multiple 4-bit adders to build adders for 8 bits, 12 bits, 16 bits, or more. We will use two 4-bit adders together to make 8-bit adders.

Multiplication through repeated addition requires the addition of the multiplicand (shifted left one bit) or zero; we can use AND gates as shown in *Figure 8.19* to select between these two values. One input to each AND gate is the appropriate bit in the multiplier: if this bit is 0, we will get LOW from this output. If this bit is 1, we will get the multiplicand as an output. Figure 8.19 shows the circuit that selects the multiplicand bits for the first multiplication in the sequence. Circuits for the second, third, and fourth multiplication will have a similar form; but with the multiplier bits shifted one, two, or three bits to the left, respectively.

FIGURE 8.19 AND Gates Used to Direct or Block Multiplier Bits

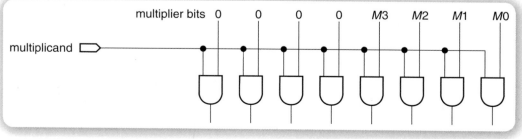

Recall our earlier example, now shown padded with zeros where necessary:

$$1010 = A$$
$$\times\ 1101 = B$$
$$0001010 = W$$
$$0000000 = X$$
$$0101000 = Y$$
$$1010000 = Z$$
$$10000010$$

Note that the final answer is $W + X + Y + Z$, the sum of four 7-bit numbers. One way to structure this repeated addition is as follows:

Add: $W + X$
Add: $Y + Z$
Add: these two sums

The overall sum would be the final answer of $A \times B$.

Example 8.2

Draw a schematic of a 2-bit multiplier using a 7483 4-bit parallel adder and other required logic. How many outputs are required?

Solution

The solution is shown in *Figure 8.20*. The 2-bit number A is multiplied by the 2-bit number B with the result Z. The Z output must be four bits wide.

(largest value $= 11 \times 11 = 1001$)

FIGURE 8.20 Example 8.2: 2-Bit Multiplier Using a 7483

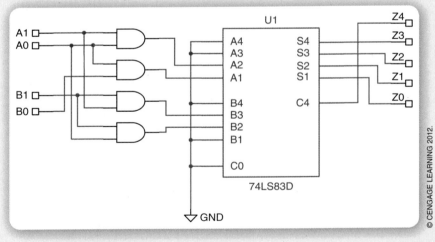

© CENGAGE LEARNING 2012.

We will design the multiplier by adding two 4-bit numbers together three times. Create a new hierarchical block called **multiplier** and enter the schematic shown in *Figure 8.21*. While this schematic may appear very full, we are able to use On-Page Connectors (see Place > Connectors > On-Page Connector) to connect two signals without drawing the line, which can clear up our drawing. Figure 8.21 shows the ground signal in green and some components are labeled to help you recreate the drawing. Back in the calculator design, we can add the multiplier block. The result is shown in *Figure 8.22*. Note that the property "Show footprint

FIGURE 8.21 4-Bit Multiplier Circuit

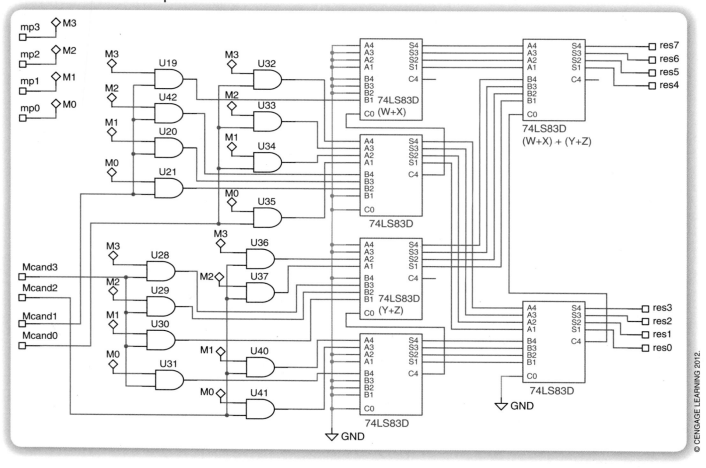

FIGURE 8.22 Calculator with AddSub and Multiplier Modules

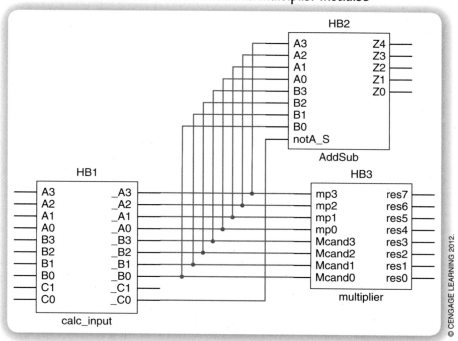

pin names" was deselected (unchecked) in the component properties to keep Figure 8.22 uncluttered.

Your Turn

8.3 What size adders would be required to multiply two 8-bit numbers with the multiplication using addition method?

Max/Min Block

The max/min block is based on a comparison of the two inputs A and B. It directs the greater value to the upper four bits of the output and the lesser to the lower four bits. *Figure 8.23* shows this pictorially for the case where B > A; our task is to design the circuitry in the center block.

Recall the 4-bit comparator. *Figure 8.24* shows the circuit for a 4-bit magnitude comparator. Magnitude comparators can serve as a basic building block within a circuit, and like other functions, there is a standard device, labeled 7485, to compare two 4-bit numbers. Select a 7485

FIGURE 8.23 Max/Min Function: B > A Is Shown

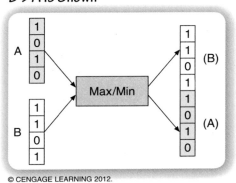

FIGURE 8.24 4-Bit Magnitude Comparator

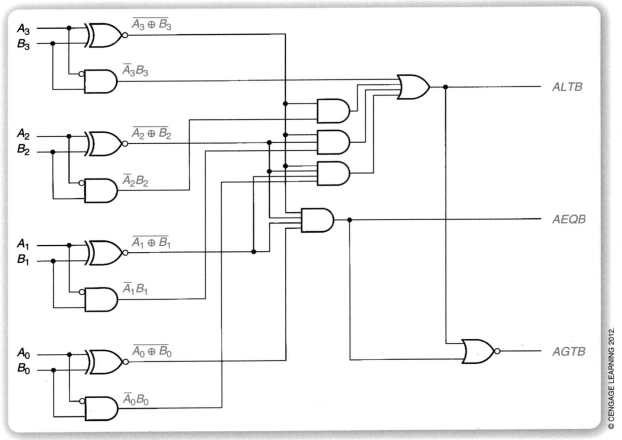

FIGURE 8.25 *7485 4-Bit Magnitude Comparator*

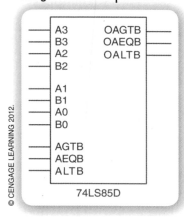

Note . . .

The 7485 has three control inputs: AGTB, AEQB, and ALTB. These are used as "tiebreakers" for lower-order bits if two or more 7485s are cascaded to make 8- or 12-bit comparators. For example, if A = B, we would expect the output OAEQB to be HIGH. However, this device looks at these "tiebreaker" inputs from another 7485 first and outputs the condition indicated by these control inputs. If A > B or A < B, the tiebreaker inputs are not considered because the relative size of A and B can be decided without them. We won't need to use this feature on our design, so we'll tie these inputs LOW, effectively disabling them.

device in Multisim to see the component shown in *Figure 8.25*. This device has two 4-bit inputs, A and B, outputs to indicate whether A > B (AGTB), A = B (AEQB), or A < B (ALTB).

Start a new hierarchical block named **maxmin** with a 7485 comparator and eight inputs for the 4-bit values A and B and eight outputs for the 4-bit larger and 4-bit smaller numbers as shown in *Figure 8.26*. The outputs of the comparator will indicate which input is larger, indicating the input we want to direct to the highest four bits of the output. However, it is not enough to know which is larger; the larger input must be selected and sent to the output. Examining one output bit, for example, the highest output bit (call it Z7) will need to be either A3 or B3. Z6 will need to be either A2 or B2, depending on whether the 4-bit value A or the 4-bit value B is larger. We have seen such a circuit before: *Figure 8.27* shows a quadruple 2-to-1 multiplexer (or MUX),

FIGURE 8.26 *7485 Comparator with Inputs*

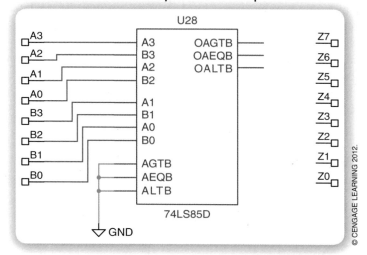

FIGURE 8.27 *Quadruple 2-to-1 MUX as a Digital Output Selector*

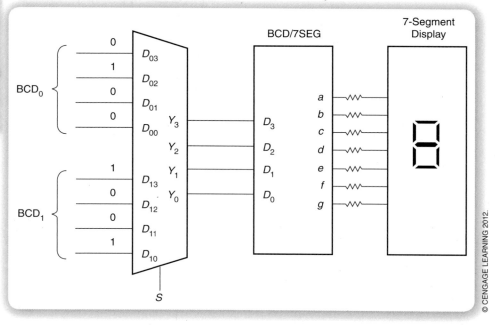

which allows one of two 4-bit inputs to pass through to the 4-bit output. Such a device is available; the 74157 (*Figure 8.28*) is a quadruple 2-to-1 MUX. This device has a control line named ~G, which is an Active-LOW enable. Therefore, to allow this device to be active, we need to tie this input to LOW. Add this to the schematic as shown in *Figure 8.29*.

The ~A/B input on the 74157 controls which 4-bit input is passed to the output. The A inputs are selected and run straight through to the outputs if SEL is LOW; the B inputs are selected and run straight to the outputs if this line is HIGH. If we run OALTB from the 7485 to the ~A/B input of the 74157, the A inputs would be selected if they are greater; if A > B, OALTB will be LOW, therefore ~A/B will be LOW, sending inputs A out of the 74157 MUX. If B is larger than A, OALTB will be HIGH, sending B to the output of the 74157. If they are equal, it doesn't matter which input is sent to the output. Do you see why this is the case?

This circuit gives us the larger of A or B at the outputs of the 74157 MUX. We also need the smaller value of A or B as the lower four bits of the output. We can add another 74157 to choose the lower four bits simply by running OAGTB to its ~A/B input as shown. The final design is shown

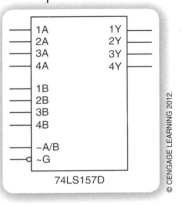

FIGURE 8.28 74157
Quadruple 2-to-1 MUX

© CENGAGE LEARNING 2012.

FIGURE 8.29 *7485 Combined with 74157*

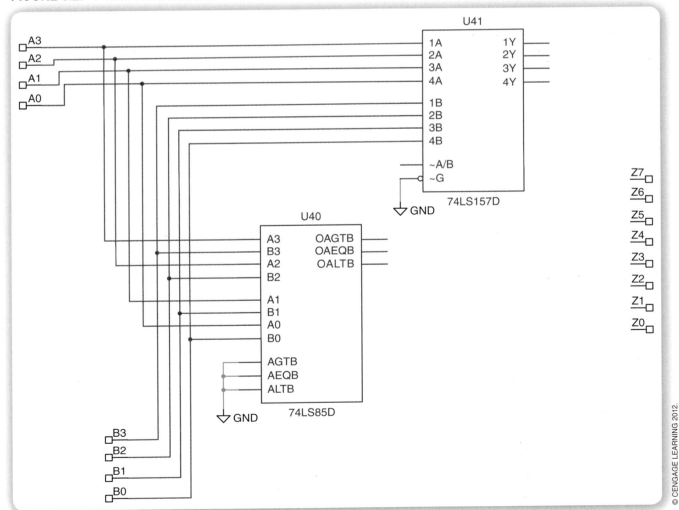

© CENGAGE LEARNING 2012.

FIGURE 8.30 Final Design of MaxMin Module

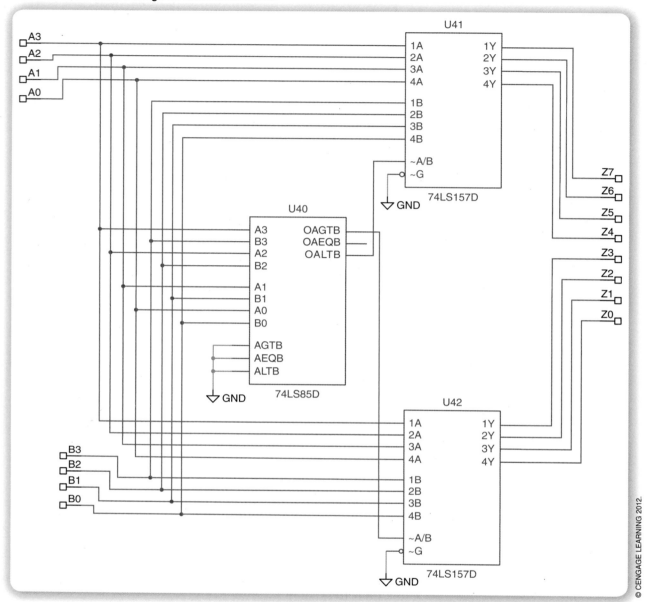

in *Figure 8.30*. Create a block for this circuit and add it to the calculator design. The design with all calculation modules is shown in *Figure 8.31*.

Input and Output Blocks

We have drawn each component with the exception of the input and output blocks. It appears that the input is designed by default: the inputs as they exist meet the design specifications. In other words, we probably did not need a separate module for the input.

The final output is a selection of one of the outputs from our individual components. However, we cannot simply tie these together; circuitry is needed to determine which functional block is to send a value to the output of the system. In other words, if the user selects "addition" by setting inputs our **choice** inputs to C = 00, and we are not concerned about the output of either the **maxmin** or **multiplier** block, we want the output of the system to be the output of the **AddSub** block.

FIGURE 8.31 Calculator with Input and Three Modules Shown

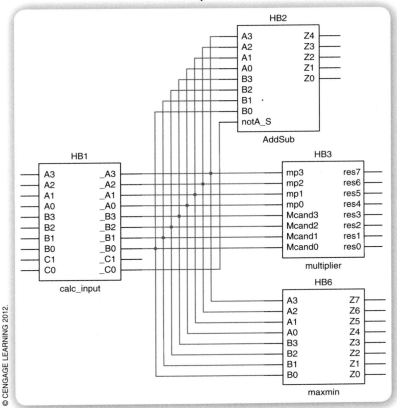

FIGURE 8.32 Inputs to the calcout Module

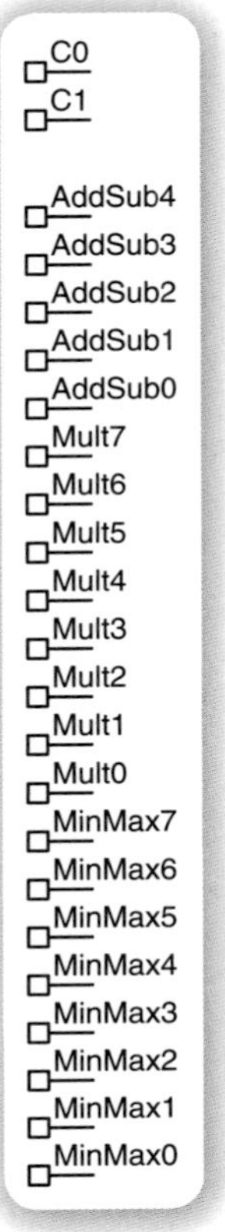

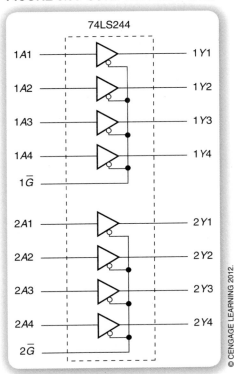

The process of designing an output block called **calcout** begins with a new hierarchical block. The inputs to this block will be the outputs of our other blocks: **AddSub**, **multiplier**, and **maxmin**, in addition to the C inputs, because these will determine which block's outputs are sent to the outputs of the system.

Start a new hierarchical block and add 24 inputs (and eight outputs). Each bit of the final calculator output is one of three possible inputs to the device **calcout**. For example, our least significant output bit, Z_0, will be either AddSub0, mult0, or MaxMin0 from *Figure 8.32*. Quad multiplexer devices could be used to select; however, because the outputs are selected from three different sources instead of two, we would need to use a combination of two devices to select the proper output. Also, each device gives only four bits; therefore, we would need to double the number, requiring four devices.

A more typical solution is to use tristate buffers in a configuration as shown in *Figure 8.33*. Figure 8.33 shows a series of 8 individual tristate buffers (two sets of four) found in a 74LS244 component; the select line (in this case, ~G) determines whether the signal on the input is passed to the output or blocked. When we add a 74LS244 in Multisim, we see a set of 4 tristate buffers and one enable line. We will use a pair of these (8 lines total) to allow one of the outputs from one of our modules to pass through to the outputs of **calcout**, and thus, to the outputs of the calculator.

Refer back to Table 8.1. Inspection shows that we want to select the output of **AddSub** whenever C_1 is LOW. If C_1 is HIGH, **multiplier** should be selected if C_0 is LOW ($C_1 = 1$ AND $C_0 = 0$), or **maxmin** if C_0 is HIGH ($C_1 = 1$ AND $C_0 = 1$). *Figure 8.34* shows this implementation. In effect, we are using AND gates and inverters to decode C_1 and C_0 to select the correct

FIGURE 8.33 Octal Tristate Buffer

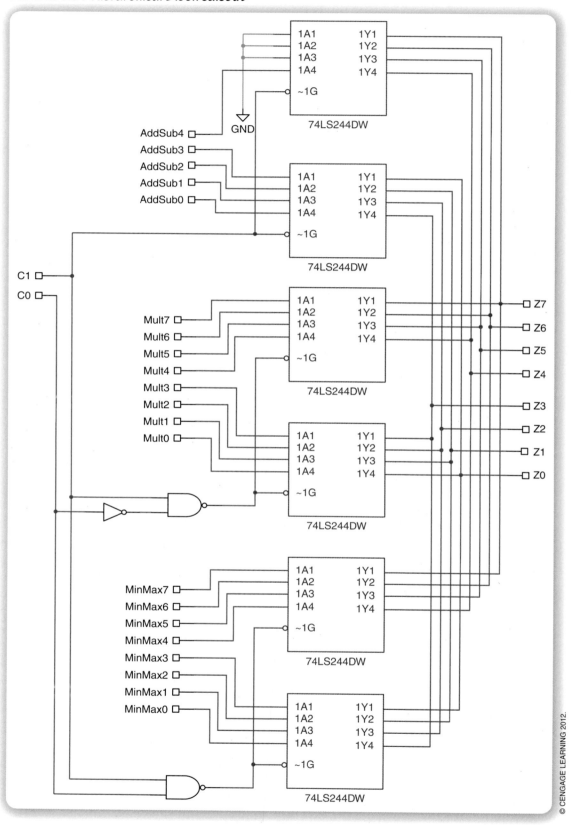

FIGURE 8.34 Hierarchical Block calcout

FIGURE 8.35 *Final Schematic—Calculator*

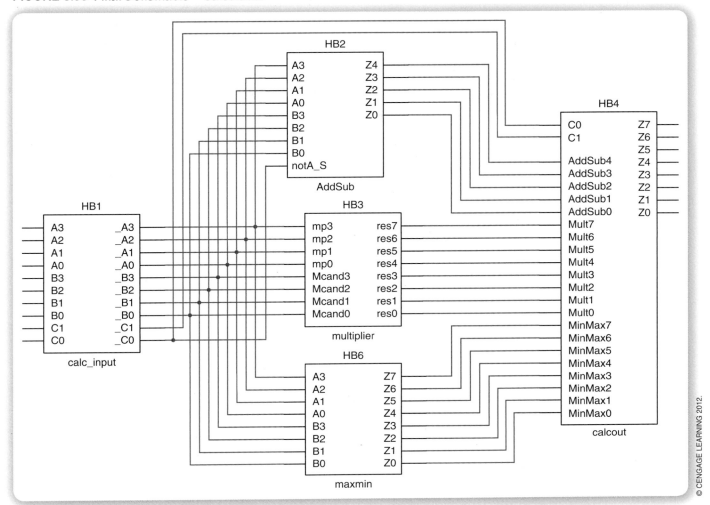

device. Trace through the different possibilities of C to verify that only the correct device is selected. Add the **calcout** block to the final diagram of **calculator**—the final schematic is shown in *Figure 8.35*.

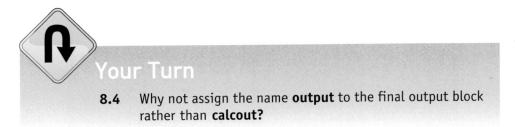

Your Turn

8.4 Why not assign the name **output** to the final output block rather than **calcout?**

8.4 TROUBLESHOOTING

Our overall calculator circuit appears to be a relatively straightforward circuit if we look at the block diagram, and should function correctly if each component was designed and built correctly. If we look inside some of the blocks, individual circuits can appear to be quite complex, with overlapping wires and multiple components.

What happens if the circuit doesn't work correctly? As you might expect, a single wire out of place or missed connection could result in errors that may seem difficult to trace to their source. Methodically reviewing the circuit should lead to clues to these simple errors, rather than erasing and redrawing large circuits.

Troubleshooting—An Example

Let's look at two possible errors. First, suppose we see an error when subtracting two inputs, but other functions, including addition, give correct outputs. We can trace this error to the **AddSub** block. We test the addition function with several different input combinations and our actual output always matches our expected output. However, when we test subtraction, we see errors including:

$$8 - 1: \quad \begin{array}{r} 1000 \\ -0001 \\ \hline 11001 \end{array}$$
(2's complement of 7, or -7)

$$1 - 8: \quad \begin{array}{r} 0001 \\ -1000 \\ \hline 10111 \end{array}$$
(positive 7)

After reviewing different input combinations (including those mentioned), it appears that we are getting the 2's complement of our expected answer. In this example, we open the **AddSub** block shown in *Figure 8.36* and see

FIGURE 8.36 AddSub Module with Inputs A and B Reversed

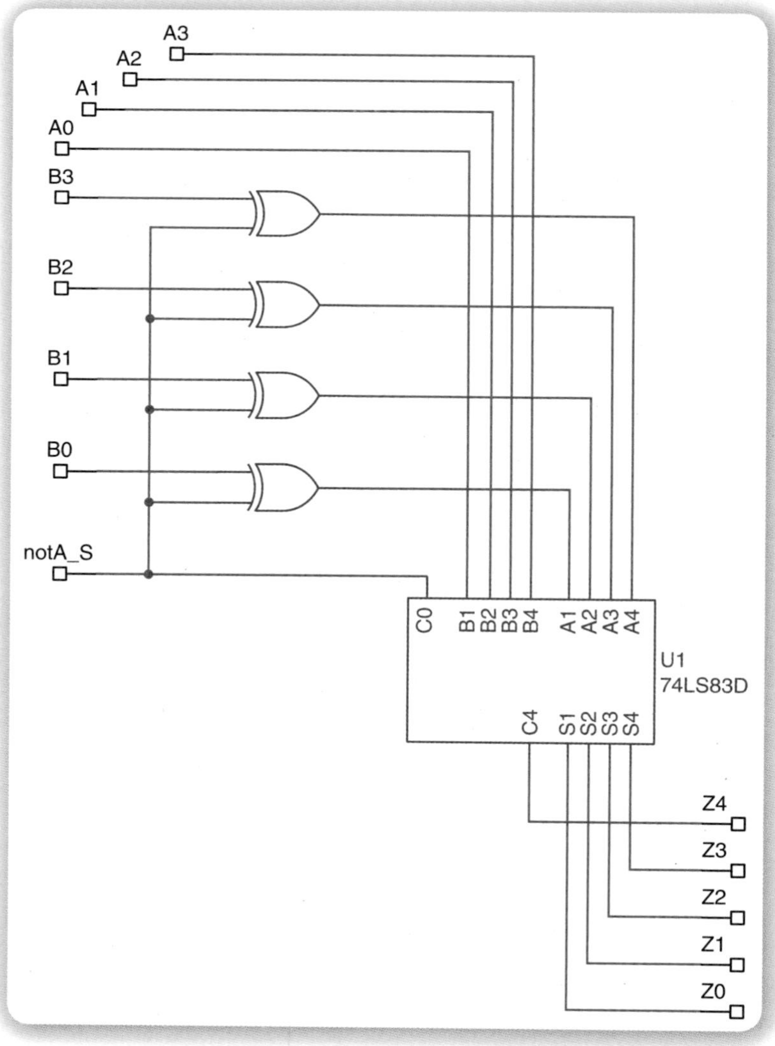

that the A and B inputs are reversed, and we have built a circuit giving us B − A instead of A − B. Fixing this error gives us a correctly functioning circuit.

A second example illustrates a wiring error. In this case, we find errors in both addition and subtraction. A few examples of errors we encounter when testing the circuit are shown in *Table 8.3*. Inspecting these results shows erroneous results whenever we add and input bit B3 is HIGH *or* if we subtract and input bit B3 is LOW. Once we notice this, we can look closely at the portion of the circuit involved with input bit B3, and indeed, we see a wiring error in our circuit, shown in *Figure 8.37*. Once this error is fixed, our entire circuit works correctly.

The purpose of these two examples is to show that, if an error is encountered, it pays to examine the correct and incorrect outputs for patterns that may be a clue for a simple wiring mistake. Time is usually better spent examining the outputs rather than erasing large portions of the circuit and starting over.

FIGURE 8.37 AddSub Module with Wiring Mistake near Input B2

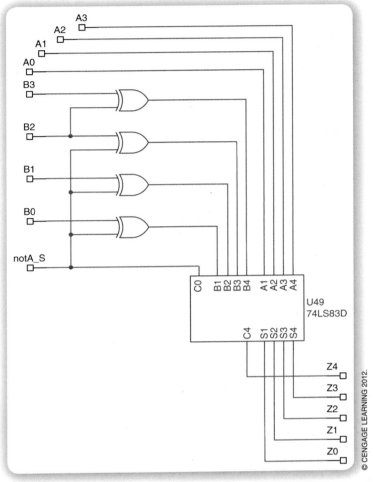

© CENGAGE LEARNING 2012.

TABLE 8.3   Expected and Actual Values for Error Shown in Figure 8.37

		Expected Outputs	Actual Outputs
1 + 1	0001 + 0001	0010	0010
3 + 3	0011 + 0011	1001	1001
7 + 2	0111 + 0010	1001	1001
2 + 7	0010 + 0111	1001	0001
0 + 4	0000 + 0100	0100	1100
9 − 5	1001 − 0101	0100	0100
12 − 12	1100 − 1100	0000	0000
10 − 8	1010 − 1000	0010	1010
3 − 9	0011 − 1001	1010	0011

© CENGAGE LEARNING 2012.

SUMMARY

1. Combinational logic functions such as adders, multiplexers, and decoders are useful on their own and in combination as part of more complex circuitry.

2. Ensuring that a problem is well understood is one of the first steps in solving a word problem or designing a circuit.

3. Understanding the inputs and outputs to a system or to a block within a block diagram of a system is an important step in finding the solution.

4. Hierarchical blocks are useful in breaking up a large problem into smaller, more manageable components.

5. **Control lines** are signals input to or output from a device that functions to control the device rather than determining the outputs. Typical examples of control lines include *enable* signals.

6. **Glue logic** is circuitry (or gates) whose purpose is to connect more complex blocks of a larger circuit.

7. A parallel binary adder may be used to perform other mathematical functions such as subtraction or multiplication.

8. Traditional methods of designing a digital circuit such as the use of truth tables or K-maps become much more difficult to use when a system exceeds four inputs.

9. Designs in Multisim are often best approached by breaking the large design into smaller components. These smaller designs can be built in separate **hierarchical blocks**, then saved and added to larger schematics.

10. Numbering schemes for inputs and outputs, such as A_3, A_2, A_1, and A_0 may not match specific devices: some devices are shown with a least significant bit (LSB) of A_0, whereas others are shown with LSBs of A_1. This can vary by manufacturer or source.

11. Binary multiplication is typically implemented as a series of repeated shifted additions.

12. Many larger combinational circuits involve a final process of selecting certain outputs or the need to enable or disable its outputs. Tristate buffers typically are used for this purpose. In some cases, it may be advantageous to use multiplexers instead.

BRING IT HOME

8.1 Problem-Solving Techniques

8.1 List the specific inputs and outputs of the two-bit multiplier shown in Example 8.2.

8.2 Write a word problem description for the problem given in Example 8.2; be sure to specify the requirements of the circuit.

8.2 Sample Application: A Small Calculator

8.3 List the outputs of the calculator circuit if seven-segment displays were to display the output values instead of the output Z.

8.3 Components of the Calculator

8.4 How many XOR gates would be required to build a 12-bit parallel adder/subtractor?

8.5 If the OAEQB output from the 7485 were connected to the SEL line on the 74157 in the **maxmin** circuit, what would be output from the 74157?

EXTRA MILE

8.3 Components of the Calculator

8.6 Design a parallel 5-bit adder using 7483 components using Multisim. What specific device output is the most significant bit of the result?

8.7 Add seven-segment display drivers to the **calcout** module, changing the output Z to outputs driving two active-LOW seven-segment displays.

8.8 Draw the schematic for the **calcout** component using one or more decoders to enable the proper tristate outputs.

CHAPTER 9
Introduction to Sequential Logic

Menu	START LOCATION	DISTANCE	END LOCATION

Chapter Objectives

Upon successful completion of this chapter, you will be able to:

1 Explain the difference between combinational and sequential circuits.

2 Define the SET and RESET functions of an SR latch.

3 Draw circuits, function tables, and timing diagrams of NAND and NOR latches.

4 Explain the effect of each possible input combination to a NAND and a NOR latch, including SET, RESET, and no change functions, and the ambiguous or forbidden input condition.

5 Design circuit applications that employ NAND and NOR latches.

6 Describe the use of the *ENABLE* input of a gated SR or D latch as an enable/inhibit function and as a synchronizing function.

7 Outline the problems involved with using a level-sensitive *ENABLE* input on a gated SR or D latch.

8 Explain the concept of edge-triggering and why it is an improvement over level-sensitive enabling.

9 Draw circuits, function tables, and timing diagrams of edge-triggered D, and JK T flip-flops.

10 Describe the toggle function of a JK flip-flop and a T flip-flop.

11 Describe the operation of the asynchronous preset and clear functions of D, and JK T flip-flops and be able to draw timing diagrams showing their functions.

12 Use Multisim to create simple circuits and simulations with D latches and D, and JK T flip-flops.

13 Calculate the rise time and fall time of a rising edge or falling edge of a digital pulse waveform.

PHOTO: © ISTOCKPHOTO.COM/XJBEN.

Many of the digital devices we use every day have to consider timing: a digital alarm clock counts the seconds until your alarm sounds, an MP3 player keeps time of the music and remembers the order of songs. A stopwatch keeps time. Your computer runs operations millions, billions, or trillions times every second. All of these devices have a clock, or timer, somewhere inside. What other devices do you use every day that must contain a timer?

The digital circuits studied to this point have all been combinational circuits, that is, circuits whose outputs are functions only of their present inputs. A particular set of input states will always produce the same output state in a combinational circuit.

This chapter will introduce a new category of digital circuitry: the sequential circuit. The output of a sequential circuit is a function both of the present input conditions and the previous conditions of the inputs and/or outputs. In other words, the output "remembers" its previous state and its next state depends on the sequence in which the inputs are applied.

We will begin our study of sequential circuits by examining the two most basic sequential circuit elements: the latch and the flip-flop, both of which are used to store a single bit of information. The difference between a latch and a flip-flop is the condition under which the stored bit is allowed to change.

Latches and flip-flops are also used as integral parts of more complex devices, such as programmable logic devices (PLDs) and microprocessors. In these applications, the flip-flops are often bundled together in groups called registers. These registers can be used to store bytes inside the PLD or microprocessor or to store the input or output values of these devices.

9.1 LATCHES

All of the circuits we have seen up to this point have been combinational circuits. That is, their present outputs depend only on their present inputs. The output state of a combinational circuit results only from a combination of input logic states.

The other major class of digital circuits is the **sequential circuit**. The present outputs of a sequential circuit depend not only on its present inputs, but also on its past input states, so that the circuit can act as a simple memory device.

The simplest sequential circuit is the SR **latch**, whose logic symbol is shown in *Figure 9.1a*. The latch has two inputs, *SET (S)* and *RESET (R)*, and two complementary outputs, *Q* and $\overline{Q}$. If the latch is operating normally, the outputs are always in opposite logic states.

FIGURE 9.1 SR Latch (Active-HIGH Inputs) © CENGAGE LEARNING 2012.

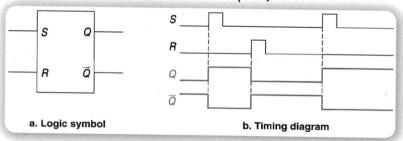

a. Logic symbol   b. Timing diagram

The latch operates like a momentary-contact pushbutton with START and STOP functions, shown in *Figure 9.2*. A momentary-contact switch operates only when it is held down. When released, a spring returns the switch to its rest position.

Suppose the switch in Figure 9.2 is used to control a motor starter. When you push the START button, the motor begins to run. Releasing the START switch does not turn the motor off; that can be done only by pressing the STOP button. If the motor is running, pressing the START button again has no effect, except continuing to let the motor run. If the motor is not running, pressing the STOP switch has no effect, because the motor is already stopped.

There is a conflict if we press both switches simultaneously. In such a case we are trying to start and stop the motor at the same time. We will come back to this point later.

Another important point to note is that if the motor is not nearby where we can see or hear if it is running and if neither START nor STOP is pressed, there is no way to tell if the motor is running or if it stopped. In other words, *not* pressing

FIGURE 9.2 Industrial Pushbutton (e.g., Motor Starter)

© CENGAGE LEARNING 2012.

the pushbuttons does not change the state of the motor, and we *cannot know* what its current state is unless we know whether START or STOP was pressed last. The state of the motor depends on what it was doing after the last pushbutton was pressed.

The latch *SET* input is like the START button in Figure 9.2. The *RESET* input is like the STOP button.

The latch in Figure 9.1 has active-HIGH *SET* and *RESET* inputs. To set the latch, make $R = 0$ and make $S = 1$. This makes $Q = 1$ until the latch is actively reset, as shown in the timing diagram in Figure 9.1b. To activate the reset function, make $S = 0$ and make $R = 1$. The latch is now reset ($Q = 0$) until the set function is next activated.

Combinational circuits produce an output by *combining inputs*. In sequential circuits, it is more accurate to think in terms of *activating functions*. In the latch described, S and R are not *combined* by a Boolean function to produce a particular result at the output. Rather, the set function is *activated* by making $S = 1$, and the reset function is *activated* by making $R = 1$, much as we would activate the START or STOP function of a motor starter by pressing the appropriate pushbutton.

The timing diagram in Figure 9.1b shows that the inputs need not remain active after the set or reset functions have been selected. In fact, the S or R input *must* be inactive before the opposite function can be applied, to avoid conflict between the two functions.

Example 9.1

Latches can have active-HIGH or active-LOW inputs, but in each case $Q = 1$ after the set function is applied and $Q = 0$ after reset. For each latch shown in *Figure 9.3*, complete the timing diagram shown. Q is initially LOW in both cases. (The state of Q before the first active *SET* or *RESET* is unknown unless specified, because the present state depends on previous history of the circuit.)

Solution

The Q and $\overline{Q}$ waveforms are shown in Figure 9.3. Note that the outputs respond only to the first set or reset command in a sequence of several pulses.

FIGURE 9.3 Example 9.1: SR Latch © CENGAGE LEARNING 2012.

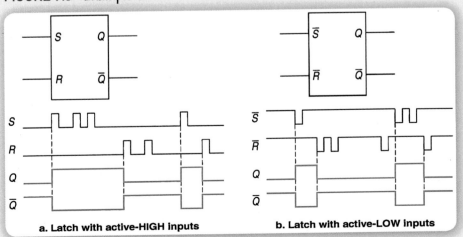

a. Latch with active-HIGH inputs   b. Latch with active-LOW inputs

Example 9.2

Multisim Example

Multisim File: 09.01 SR Latch (Active-HIGH).ms11

Open the Multisim file for this example. The design, shown in *Figure 9.4,* is a Multisim version of the latch with active-HIGH inputs of Figure 9.3a. Run the file as a simulation and operate the Normally Closed (NC) pushbuttons to try the set and reset functions of the latch.

FIGURE 9.4 **Example 9.2: SR Latch in Multisim (Active-HIGH inputs)** © CENGAGE LEARNING 2012.

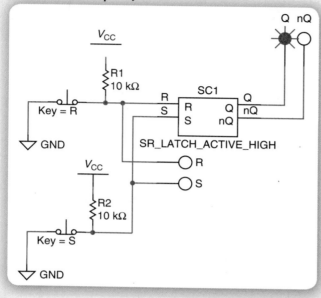

a. What are the logic levels applied to the S and R inputs when the switches are in their rest positions? What happens to the S and R inputs when the S or R key on the keyboard is pressed during a simulation?

b. What action of the switches will make the latch set?

c. What action of the switches will make the latch reset?

d. How does the latch respond to more than one press of the set or reset pushbutton?

Solution

a. The S and R inputs are both in the LOW state (i.e., inactive) when their respective pushbuttons are in their rest positions. During a simulation, when an S or R key is pressed on the PC keyboard, the corresponding input goes HIGH briefly. This can be seen on the logic probe connected to the input.

b. The latch sets when the S switch is pressed.

c. The latch resets when the R switch is pressed.

d. If the S pushbutton is pressed more than once, the latch sets the first time it is pressed and remains set when the S pushbutton is pressed again. If the R pushbutton is pressed more than once, the latch resets the first time it is pressed and remains reset when the R pushbutton is pressed again.

Example 9.3

Figure 9.5 shows a latching HOLD circuit for an office desk telephone. When HIGH, the *HOLD* output allows you to replace the handset without disconnecting a call in progress.

FIGURE 9.5 **Example 9.3: Latching HOLD Button**

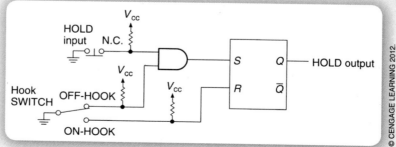

© CENGAGE LEARNING 2012.

continued...

The two-position switch is the telephone's hook switch (the switch the handset pushes down when you hang up), shown in the off-hook (in-use) position. The normally closed pushbutton is a momentary-contact switch used as a HOLD button. The circuit is such that the HOLD button does not need to be held down to keep the HOLD active. The latch "remembers" that the switch was pressed, until told to "forget" by the reset function.

Describe the sequence of events that will place a caller on hold and return the call from hold. Also draw timing diagrams showing the waveforms at the HOLD input, hook switch inputs, S input, and HOLD output for one hold-and-return sequence. (HOLD output = 1 means the call is on hold.)

Solution

To place a call on hold, we must set the latch. We can do so if we press and hold the HOLD switch, then the hook switch. This combines two HIGHs—one from the HOLD switch and one from the on-hook position of the hook switch—into the AND gate, making $S = 1$ and $R = 0$. Note the sequence of events: press HOLD, hang up, release HOLD. The S input is HIGH only as long as the HOLD button is pressed. The handset can be kept on-hook and the HOLD button released. The latch stays set, as $S = R = 0$ (neither SET nor RESET active) as long as the handset is on-hook.

To restore a call, lift the handset. This places the hook switch into the off-hook position and now $S = 0$ and $R = 1$, which resets the latch and turns off the HOLD condition.

Figure 9.6 shows the timing diagram for the sequence described.

FIGURE 9.6 Example 9.3: HOLD Timing Diagram © CENGAGE LEARNING 2012.

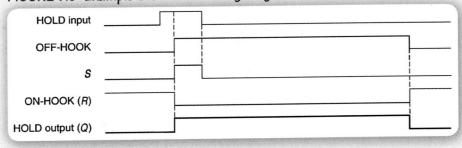

Your Turn

9.1 A latch with active-HIGH S and R inputs is initially set. R is pulsed HIGH three times, with $S = 0$. Describe how the latch responds.

9.2 NAND/NOR LATCHES

An SR latch is easy to build with logic gates. *Figure 9.7* shows two such circuits, one made from NOR gates and one from NANDs. The NAND gates in the second circuit are drawn in DeMorgan equivalent form.

FIGURE 9.7 SR Latch Circuits

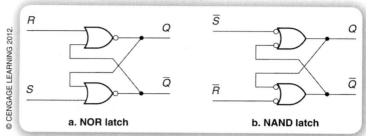

a. NOR latch   b. NAND latch

Both circuits have the following three features:

1. OR-shaped gates
2. Logic level inversion between the gate input and output
3. Feedback from the output of one gate to an input of the opposite gate

During our examination of the NAND and NOR latches, we will discover why these features are important.

A significant difference between the NAND and NOR latches is the placement of *SET* and *RESET* inputs with respect to the Q and $\overline{Q}$ outputs. Once we define which output is Q and which is $\overline{Q}$, the locations of the *SET* and *RESET* inputs are automatically defined.

In a NOR latch, the gates have active-HIGH inputs and active-LOW outputs. When the input to the Q gate is HIGH, Q = 0, because either input HIGH makes the output LOW. Therefore, this input must be the *RESET* input. By default, the other is the *SET* input.

In a NAND latch, the gate inputs are active-LOW (in DeMorgan equivalent form) and the outputs are active-HIGH. A LOW input on the Q gate makes Q = 1. This, therefore, is the *SET* input, and the other gate input is *RESET*.

Because the NAND and NOR latch circuits have two binary inputs, there are four possible input states. *Table 9.1* summarizes the action of each latch for each input combination. The functions are the same for each circuit, but they are activated by opposite logic levels.

We will examine the NAND latch circuit for each of the input conditions in Table 9.1. The analysis of a NOR latch is similar and will be left as an exercise.

TABLE 9.1 NOR and NAND Latch Functions © CENGAGE LEARNING 2012.

S	R	Action (NOR Latch)	$\overline{S}$	$\overline{R}$	Action (NAND Latch)
0	0	Neither *SET* nor *RESET* active; output does not change from previous state	0	0	Both *SET* and *RESET* active forbidden condition
0	1	*RESET* input active	0	1	*SET* input active
1	0	*SET* input active	1	0	*RESET* input active
1	1	Both *SET* and *RESET* active; forbidden condition	1	1	Neither *SET* nor *RESET* active; output does not change from previous state

NAND Latch Operation

Figure 9.8 shows a NAND latch in its two possible stable states. In each case, the inputs $\overline{S}$ and $\overline{R}$ are both HIGH (inactive).

FIGURE 9.8 NAND Latch Stable States

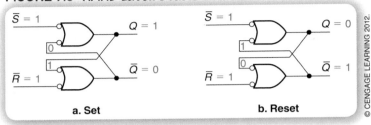

a. Set   b. Reset

Figure 9.8a shows a latch stable in its *SET* condition ($Q = 1$). Note the following characteristics of this state:

▶ The upper gate in Figure 9.8a has a LOW on its "inner" input. For a NAND gate either input LOW makes the output HIGH. Therefore, the output Q is HIGH.

▶ The HIGH at Q feeds back to the "inner" input of the lower gate. Both inputs of this gate are HIGH, making the output $\overline{Q}$ LOW.

▶ The LOW at $\overline{Q}$ feeds back to the "inner" input of the upper gate. This returns us to our starting point with the same logic level as we had to begin with. Therefore, we find a consistent path of logic levels with stable output states.

The stable *RESET* state ($Q = 0$), shown in Figure 9.8b, is a mirror image of the *SET* condition. It has the following characteristics:

▶ The lower gate in Figure 9.8b has a LOW on its "inner" input. Either input LOW makes output HIGH, so this makes $\overline{Q}$ HIGH.

▶ The HIGH on $\overline{Q}$ feeds back to the upper gate. Both inputs of this gate are HIGH, so output Q is LOW.

▶ The LOW on Q feeds back to the "inner" input of the lower gate. Because this is the same logic level as our starting point, we find that the logic levels are consistent throughout the path and the latch has stable output states.

The stability of the latch depends on the feedback connections between the two gates, which supply the logic levels to the "inner" inputs of the latch gates. Notice that in the two stable states, only one of the four inputs to the two NAND gates has a 0. The difference between *SET* and *RESET* states has to do with the placement of the logic 0 on one of these "inner" inputs. Whichever gate has the 0 input will have the HIGH output; the other gate will have a LOW output. When we change the state of the latch, we are moving this 0 from one side of the latch to the other. Thus, because the output state of the latch can be determined solely by these "inner" inputs, it is possible for the same input values of $\overline{S} = 1$ and $\overline{R} = 1$ to yield two different sets of output values for Q and $\overline{Q}$.

What function do $\overline{S}$ and $\overline{R}$ have then? They are used to change the state of the latch.

Figure 9.9 shows the transition of the latch from the *RESET* state to the *SET* state. The following actions occur in the transition:

▶ The latch begins in the stable *RESET* state ($Q = 0$), as shown in Figure 9.8b and Figure 9.9a. In this state, $\overline{S} = 1$ and $\overline{R} = 1$, the no change condition.

▶ To set the latch, we make $\overline{S} = 0$ (Figure 9.9b).

▶ This change propagates through the upper gate of the latch circuit. Because either input LOW makes output HIGH, $Q = 1$ (Figure 9.9c).

FIGURE 9.9 RESET-to-SET Transition © CENGAGE LEARNING 2012.

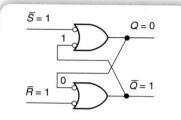

a. Stable in the RESET condition.
SET and RESET inputs inactive.

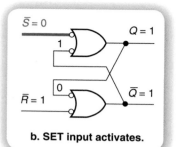

b. SET input activates.

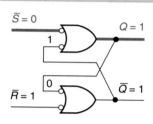

c. Change propagates through upper gate.
(Either input LOW makes output HIGH.)

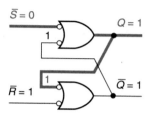

d. HIGH transfers across feedback line to lower
gate, removing active input condition.

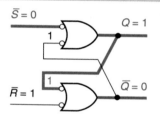

e. Change propagates through lower gate.
(Both inputs HIGH, therefore output LOW.)

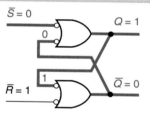

f. Feedback transfers LOW to upper gate,
completing change to new state.

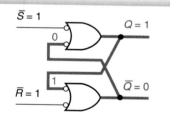

g. S input goes back to inactive state. SET state
held by LOW at inner input of upper gate.

- The HIGH on Q transfers across to the lower gate via the feedback line, removing its active input condition (Figure 9.9d).
- The lower gate changes state. Both inputs are now HIGH, making $\overline{Q} = 0$ (Figure 9.9e).
- The feedback line carries the LOW to the inner input of the upper gate. At this point the *RESET-to-SET* transition is complete (Figure 9.9f).
- Because only one LOW input is required to hold the output of the upper gate in the HIGH state, we can remove the LOW on input $\overline{S}$. The latch is now stable in the *SET* condition (Figure 9.9g).

FIGURE 9.10 **SET-to-RESET Transition** © CENGAGE LEARNING 2012.

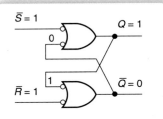

a. Stable in the SET condition.
   SET and RESET inputs inactive.

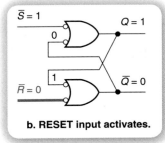

b. RESET input activates.

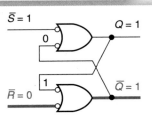

c. Change propagates through lower gate.
   (Either input LOW makes output HIGH.)

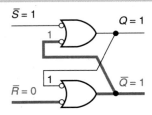

d. HIGH transfers across feedback line to upper
   gate, removing active input condition.

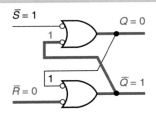

e. Change propagates through upper gate.
   (Both inputs HIGH, therefore output LOW.)

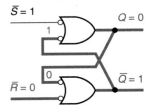

f. Feedback line transfers LOW to lower gate,
   completing transition to RESET state.

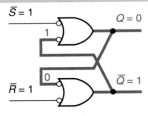

g. R goes back to inactive level. Latch is stable in
   new state, held by the 0 on inner input of lower gate.

A similar action occurs when the latch makes a transition from the *SET* state to the *RESET* state. This is shown in *Figure 9.10*.

- The latch begins in the stable *SET* condition ($Q = 1$), as shown in Figure 9.8a and Figure 9.10a. $\overline{S} = 1$ and $\overline{R} = 1$ (no change).
- To reset the latch, we make $\overline{R} = 0$ (Figure 9.10b).
- This change propagates through the lower gate of the circuit. Either input LOW makes output HIGH, so $\overline{Q} = 1$ (Figure 9.10c).
- The HIGH on $\overline{Q}$ transfers across the feedback line to the inner input of the upper gate, removing its active input condition (Figure 9.10d).
- Both inputs on the upper gate are now HIGH, so output $Q = 0$ (Figure 9.10e).

▶ The 0 on *Q* transfers across the feedback line to the inner input of the lower gate, thus completing the transition from *RESET* to *SET* (Figure 9.10f).

▶ Because only one LOW input on the lower gate is required to maintain the *RESET* state, we can remove the 0 from the $\overline{R}$ input. The latch is now stable in the *RESET* state (Figure 9.10g).

Note that for each of these cases, the "outer" inputs of the circuit (i.e., $\overline{S}$ and $\overline{R}$) are used to *change* the state of the latch, whereas the "inner" inputs (i.e., the feedback connections) are used to *maintain* the present state of the latch.

Also note that the transition between states is not complete until the change initiated at $\overline{S}$ or $\overline{R}$ propagates through both gates; the circuit is not stable until the 0 transfers to the inner input of the opposite gate. If the set or reset pulse is shorter than the time required for the change to propagate through the gates, the latch output will oscillate between states. In practice this is not a huge problem, because the total delay through the latch is only on the order of 3 ns to 20 ns. Any manual input to the latch, such as a pushbutton, will be far longer than this. Electronic inputs, such as logic gate outputs, will have to account for this delay, but as they, too, are subject to their own delay times, this seldom presents a practical problem.

Figure 9.11 shows a NAND latch with $\overline{S} = \overline{R} = 0$. This implies that both *SET* and *RESET* functions are active. Because a NAND gate requires at least one input LOW to make the output HIGH, both outputs respond by going HIGH. This condition is not unstable in and of itself, but instability can result when the inputs change.

There are three possible results when the inputs go back to the HIGH state.

1. The *SET* input goes HIGH before the *RESET* input. In this case, the latch resets, as *RESET* is the last input active. This is shown in the timing diagram of *Figure 9.12*.

2. The *RESET* input goes HIGH before *SET*. In this case, the latch sets, as shown in *Figure 9.13*.

3. The *SET* and *RESET* inputs go HIGH at the same time. This is an unstable case. *Figure 9.14* shows how the latch will oscillate under this condition. When the inputs *S* and *R* are both LOW (Figure 9.14a), both outputs of the latch are HIGH. When *S* and *R* go HIGH (Figure 9.14b), all gate inputs in the circuit are HIGH. This makes both outputs LOW (Figure 9.14c). The LOWs transfer across the latch to the opposite gates (Figure 9.14d) and, after a delay, make

FIGURE 9.11 NAND Latch Forbidden State

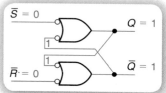

$\overline{S} = 0$

$Q = 1$

1

1

$\overline{R} = 0$

$\overline{Q} = 1$

© CENGAGE LEARNING 2012.

FIGURE 9.12 Transition from Forbidden State to RESET State © CENGAGE LEARNING 2012.

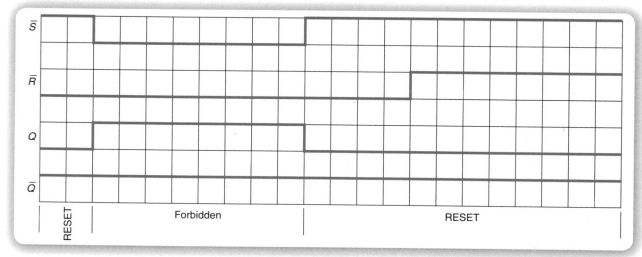

FIGURE 9.13 Transition from Forbidden State to SET State

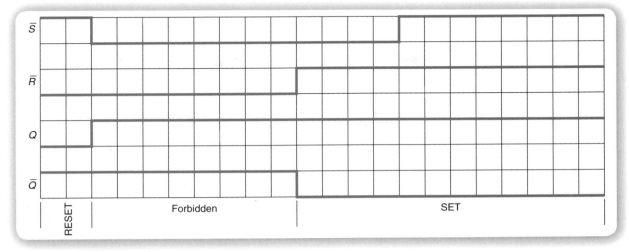

RESET | Forbidden | SET

FIGURE 9.14 Transition from Forbidden State to Oscillation

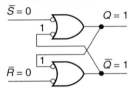
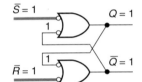
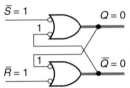

a. Both SET and RESET inputs active. Either input LOW makes output HIGH. Therefore, both outputs HIGH.

b. SET and RESET inputs deactive simultaneously.

c. Change propagate through gates simultaneously.

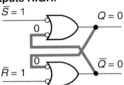
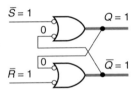
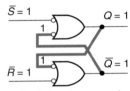

d. New output levels cross circuit via feedback lines.

e. Either input LOW makes output HIGH. Changes on feedback lines propagate through both gates simultaneously.

f. Output logic levels cross circuit via feedback lines. Cycle repeats and circuit oscillates.

both outputs HIGH (Figure 9.14e). At this point, the latch outputs will oscillate until the latch is either set or reset. The waveforms in *Figure 9.15* show how the latch outputs oscillate under these conditions.

In practice, the latch will probably not oscillate for very long. One of the two gates is likely to be slightly faster than the other, which will allow the latch to drop into the set or reset state, as described in case 1 or case 2, above.

The operation of the NAND latch can be summarized in a function table, shown in *Table 9.2*. The notation Q_{t+1} indicates that the column shows the *next* state of Q; that is, the value of Q *after* the specified input is applied. Q_t indicates the present state of the Q input.[1] Thus, the entry for the no change state indicates that after the inputs $\overline{S} = 1$ and $\overline{R} = 1$ are applied, the next state of the output is the same as its present state.

Table 9.3 shows the function table for the NOR latch.

TABLE 9.2 NAND Latch Function Table

$\overline{S}$	$\overline{R}$	Q_{t+1}	$\overline{Q}_{t+1}$	Function
0	0	1	1	Forbidden
0	1	1	0	Set
1	0	0	1	Reset
1	1	Q_t	$\overline{Q}_t$	No change

TABLE 9.3 NOR Latch Function Table

S	R	Q_{t+1}	$\overline{Q}_{t+1}$	Function
0	0	Q_t	$\overline{Q}_t$	No change
0	1	0	0	Reset
1	0	1	0	Set
1	1	0	0	Forbidden

---

[1] Many sources (such as datasheets) use the notation Q_0 to refer to the previous state of Q. We will use the notation indicated (Q_t for present state and Q_{t+1} for next state) so as to be able to reserve Q_0 for the least significant bit of a circuit that requires multiple Q outputs.

FIGURE 9.15 *Transition from Forbidden State to Oscillation*

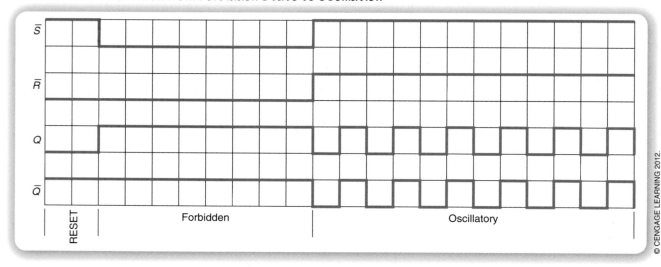

Latch as a Switch Debouncer

Pushbutton or toggle switches are sometimes used to generate pulses for digital circuit inputs, as illustrated in *Figure 9.16*. However, when a switch is operated and contact is made on a new terminal, the contact, being mechanical, will bounce a few times before settling into the new position. Figure 9.16d shows the effect of contact bounce on the waveform for a pushbutton switch. The contact bounce is shown only on the terminal where contact is being made, not broken.

Contact bounce can be a serious problem, particularly when a switch is used as an input to a digital circuit that responds to individual pulses. If the circuit expects to receive one pulse, but gets several from a bouncy switch, it will behave unpredictably.

A latch can be used as a switch debouncer, as shown in *Figure 9.17a*. When the pushbutton is in the position shown, the latch is set, because $\overline{S} = 0$ and $\overline{R} = 1$. (Recall that the NAND latch inputs are active-LOW.) When the pushbutton is pressed, the $\overline{R}$ contact bounces a few times, as shown in *Figure 9.17b*. However, on the first

FIGURE 9.16 *Switches as Pulse Generators*

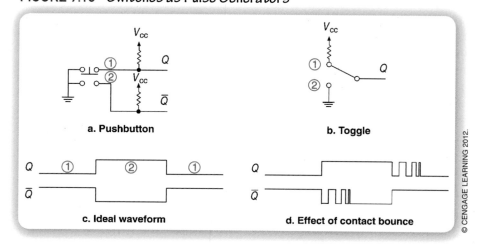

FIGURE 9.17 NAND Latch as a Switch Debouncer

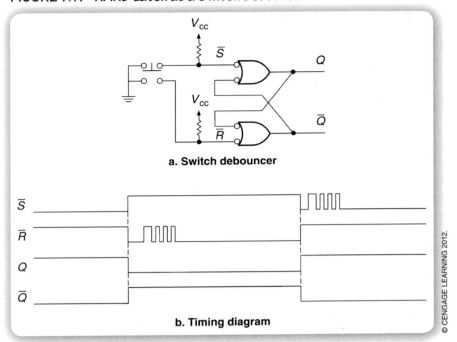

a. Switch debouncer

b. Timing diagram

© CENGAGE LEARNING 2012.

bounce, the latch is reset. Any further bounces are ignored, because the resulting input state is either $\overline{S} = \overline{R} = 1$ (no change) or $\overline{S} = 1, \overline{R} = 0$ (reset).

Similarly, when the pushbutton is released, the $\overline{S}$ input bounces a few times, setting the latch on the first bounce. The latch ignores any further bounces, as they either do not change the latch output ($\overline{S} = \overline{R} = 1$) or set it again ($\overline{S} = 0, \overline{R} = 1$). The resulting waveforms at Q and $\overline{Q}$ are free of contact bounce and can be used reliably as inputs to digital sequential circuits.

Example 9.4

A NOR latch can be used as a switch debouncer, but not in the same way as a NAND latch. *Figure 9.18* shows two NOR latch circuits, only one of which works as a switch debouncer. Draw a timing diagram for each circuit, showing R, S, Q, and $\overline{Q}$, to prove that the circuit in Figure 9.18b eliminates switch contact bounce but the circuit in Figure 9.18a does not.

FIGURE 9.18 Example 9.4: NOR Latch Circuits © CENGAGE LEARNING 2012.

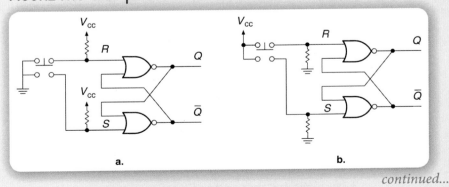

a.

b.

continued...

Solution

Figure 9.19 shows the timing diagrams of the two NOR latch circuits. In the circuit in Figure 9.18a, contact bounce causes the latch to oscillate in and out of the forbidden state of the latch ($S = R = 1$). This causes one of the two outputs to bounce for each contact closure. (Use the function table of the NOR latch to examine each part of the timing diagram to see that this is so.)

By making the resistors pull down rather than pull up, as in Figure 9.18b, the latch oscillates in and out of the no change state ($S = R = 0$) as a result of contact bounce. The first bounce on the *SET* terminal sets the latch, and other oscillations are disregarded. The first bounce on the *RESET* input resets the latch, and further pulses on this input are ignored.

The principle illustrated here is that a closed switch must present the active input level to the latch, because switch bounce is only a problem on contact closure. Thus, a closed switch must make the input of a NOR latch HIGH or the input of a NAND latch LOW to debounce the switch waveform.

FIGURE 9.19 Example 9.4: Timing Diagrams for NOR Latch Circuits of Figure 9.18

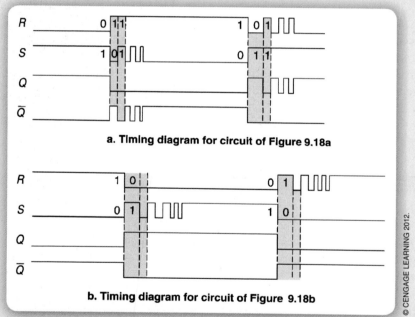

a. Timing diagram for circuit of Figure 9.18a

b. Timing diagram for circuit of Figure 9.18b

© CENGAGE LEARNING 2012.

Your Turn

9.2 Why is the input state $S = R = 1$ considered forbidden in the NOR latch? Why is the same state in the NAND latch the no change condition?

9.3   GATED LATCHES

KEY TERMS

Gated SR latch An SR latch whose ability to change states is controlled by an extra input called the *ENABLE* input.

Steering gates Logic gates, controlled by the *ENABLE* input of a gated latch, that steer a *SET* or *RESET* pulse to the correct input of an SR latch circuit.

Transparent latch (gated D latch) A latch whose output follows its data input when its *ENABLE* input is active.

Gated SR Latch

It is not always desirable to allow a latch to change states at random times. The circuit shown in *Figure 9.20,* called a **gated SR latch**, regulates the times when a latch is allowed to change state. Note that the *S* and *R* inputs are active-HIGH in Figure 9.20.

The gated SR latch has two distinct subcircuits. One pair of gates is connected as an SR latch. A second pair, called the **steering gates**, can be enabled or inhibited

FIGURE 9.20   *Gated SR Latch*

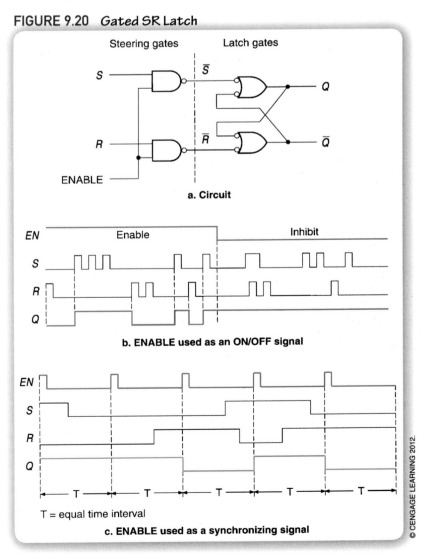

a. Circuit

b. ENABLE used as an ON/OFF signal

T = equal time interval

c. ENABLE used as a synchronizing signal

by a control signal, called *ENABLE,* allowing one or the other of these gates to pass a *SET* or *RESET* signal to the latch gates.

The *ENABLE* input can be used in two principal ways: (1) as an ON/OFF signal, and (2) as a synchronizing signal. This latter technique is also called "strobing."

Figure 9.20b shows the *ENABLE* input functioning as an ON/OFF signal. When *ENABLE* = 1, the circuit acts as an active-HIGH latch. The upper gate converts a HIGH at *S* to a LOW at $\overline{S}$, setting the latch. The lower gate converts a HIGH at *R* to a LOW at $\overline{R}$, thus resetting the latch.

When *ENABLE* = 0, the steering gates are inhibited and do not allow *SET* or *RESET* signals to reach the latch gate inputs. In this condition, the latch outputs cannot change.

Figure 9.20c shows the *ENABLE* input as a synchronizing or "strobe" signal. A periodic pulse waveform is present on the *ENABLE* line. The *S* and *R* inputs are free to change at random, but the latch outputs will change only when the *ENABLE* input is active. Because the *ENABLE* pulses are equally spaced in time, changes to the latch output can occur only at fixed intervals. The outputs can change out of synchronization if *S* or *R* change when *ENABLE* is HIGH. We can minimize this possibility by making the *ENABLE* pulses as short as possible.

Table 9.4 represents the function table for a gated SR latch.

TABLE 9.4 *Gated SR Latch Function Table*

EN	S	R	Q_{t+1}	$\overline{Q}_{t+1}$	Function
1	0	0	Q_t	$\overline{Q}_t$	No change
1	0	1	0	1	Reset
1	1	0	1	0	Set
1	1	1	1	1	Forbidden
0	X	X	Q_t	$\overline{Q}_t$	Inhibited

© CENGAGE LEARNING 2012.

Example 9.5

Figure 9.21 shows a timing diagram for two gated latches with the same *S* and *R* input waveforms but different *ENABLE* waveforms. EN_1 has a 50% duty cycle. EN_2 has a duty cycle of 16.67%.

Draw the output waveforms, Q_1 and Q_2. Describe how the length of the *ENABLE* pulse affects the output of each latch, assuming that the intent of each circuit is to synchronize the output changes to the beginning of the *ENABLE* pulse.

FIGURE 9.21 **Example 9.5: Effect of *ENABLE* Pulse Width** © CENGAGE LEARNING 2012.

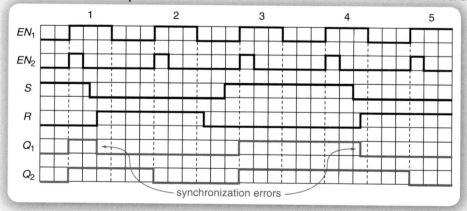

synchronization errors

continued...

Transparent Latch (Gated D Latch)

Figure 9.22 shows the equivalent circuit of a gated D ("data") latch, or **transparent latch**. This circuit has two modes. When the *ENABLE* input is HIGH, the latch is *transparent* because the output Q goes to the level of the data input, D. (We say, "Q follows D.") When the *ENABLE* input is LOW, the latch *stores* the data that was present at D when *ENABLE* was last HIGH. In this way, the latch acts as a simple memory circuit.

FIGURE 9.22 Transparent Latch

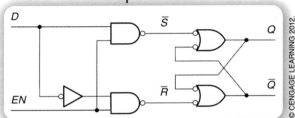

© CENGAGE LEARNING 2012.

The latch in Figure 9.22 is a modification of the gated SR latch, configured so that the *S* and *R* inputs are always opposite. Under these conditions, the states $S = R = 0$ (no change) and $S = R = 1$ (forbidden) can never occur. However, the equivalent of the no change state happens when the *ENABLE* input is LOW, when the latch steering gates are inhibited.

Figure 9.23 shows the operation of the transparent latch in the inhibit (no change), set, and reset states. When the latch is inhibited, the steering gates block any LOW pulses to the latch gates; the latch does not change states, regardless of the logic level at D.

FIGURE 9.23 Operation of Transparent Latch © CENGAGE LEARNING 2012.

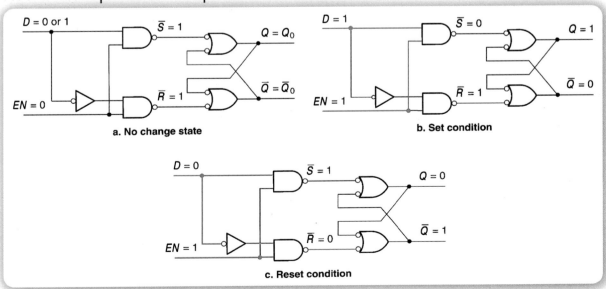

If $EN = 1$, Q follows D. When $D = 1$, the upper steering gate transmits a LOW to the SET input of the latch and $Q = 1$. When $D = 0$, the lower steering gate transmits a LOW to the $RESET$ input of the output latch and $Q = 0$.

Table 9.5 shows the function table for a transparent latch.

TABLE 9.5 *Function Table of a Transparent Latch*

EN	D	Q_{t+1}	$\overline{Q}_{t+1}$	Function	Comment
0	X	Q_t	$\overline{Q}_t$	No change	Store
1	0	0	1	Reset	Transparent: $(Q = D)$
1	1	1	0	Set	

© CENGAGE LEARNING 2012.

Implementing D Latches in Multisim

Figure 9.24 shows a Multisim design for testing the operation of a D latch. The *ENABLE* input is driven by a 50 Hz, 5-volt pulse source, with a duty cycle of 10%. The *D* input is driven by a manually controlled logic switch, operated by the *D* key on a PC keyboard. Notice that the latch has active-LOW inputs labeled *SET* and *RESET* on the top and bottom of the device. These are *SET* and *RESET* inputs that act independently of the *ENABLE* input. We will see how they work later in the chapter, but for now they are both disabled by tying them HIGH.

Both inputs and the Q output are monitored by virtual oscilloscopes in Multisim. The frequency of the ENABLE pulse source is chosen so as to be able to comfortably observe how the output responds to changes on D and ENABLE when we run the simulation of the file in Multisim.

Figure 9.25 shows the three traces on the two virtual oscilloscopes. Notice that the beginning of each of the ENABLE pulses (the edge of the pulse that goes from

FIGURE 9.25 *Oscilloscope Traces for D Latch in Multisim* © CENGAGE LEARNING 2012.

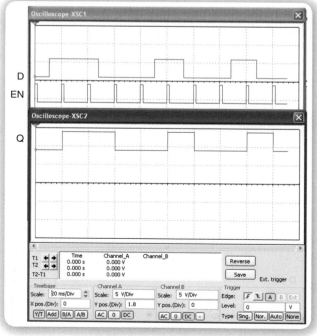

FIGURE 9.24 **D Latch Testing in Multisim**

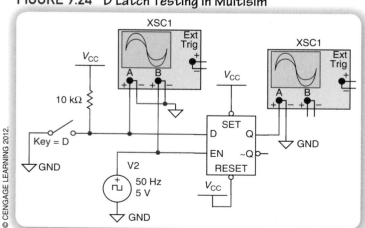

© CENGAGE LEARNING 2012.

0 to 1, or the rising edge) is aligned with the oscilloscope grid lines. These are also the places where Q changes to follow D. If *ENABLE* is LOW, a change on D has no effect on the output. If D remains at its new level, the Q output changes to follow D as soon as *ENABLE* is HIGH, that is, on a grid line.

Example 9.6

Figure 9.26 shows a set of input and output traces for the latch test circuit of Figure 9.24. Briefly explain why the output trace looks the way it does and why the two pulses on D are ignored, that is, not transferred to Q.

Solution

The changes in the Q output trace are aligned to the grid markings because that is where the *ENABLE* input goes HIGH. The two pulses on D are ignored because they occur during times when *ENABLE* is LOW, which is when the latch outputs cannot change.

FIGURE 9.26   Example 9.6: Oscilloscope Traces for D Latch in Multisim   © CENGAGE LEARNING 2012.

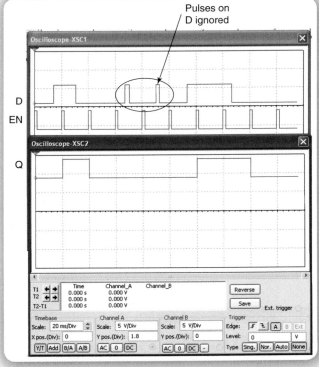

Multiple-Bit D Latches

D latches can also have multiple inputs and outputs, as shown in *Figure 9.27*. The ENABLE input is usually common to all latches in the group, allowing a multiple-bit Q to follow a multiple-bit D when the latches are enabled. For example, if $D_3 D_2 D_1 D_0 = 0101$, then $Q_3 Q_2 Q_1 Q_0$ will equal 0101 as soon as $EN = 1$ for the latch. *Figure 9.28* shows how four latches can be connected in Multisim to make a 4-bit latch subcircuit.

FIGURE 9.27   4-Bit D Latch Circuit

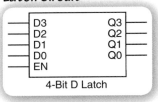

4-Bit D Latch

© CENGAGE LEARNING 2012.

FIGURE 9.28   4-Bit D Latch as a Multisim Subcircuit

© CENGAGE LEARNING 2012.

Example 9.7

Multisim Example

Multisim File: 09.05 4-Bit D Latch.ms11

Figure 9.29 shows a Multisim circuit that can be used to test a 4-bit *D* latch.

   a. In the latch circuit shown, the *D* inputs are not the same as the *Q* outputs. Why not?
   b. How can the *Q* outputs be made the same as the *D* inputs?
   c. Open the Multisim file for this example and make the inputs and outputs look the same as in Figure 9.29. Then make the *Q* outputs the same as the *D* inputs shown in Figure 9.29.

continued...

Solution

a. The latch outputs are not the same as the inputs because the *ENABLE* input is LOW and there has been a change on *D* since the last time that *ENABLE* was HIGH.

b. *Q* follows *D* when *ENABLE* is HIGH. Make the *ENABLE* input HIGH to make the output the same as the input.

c. To reproduce the diagram in Figure 9.29, run the Multisim simulation, set *ENABLE* HIGH and the *D* inputs to 1110. Next, set the *ENABLE* input LOW and change the *D* inputs to 1010 (i.e., make $D_2 = 0$).

FIGURE 9.29 Example 9.7: Testing a 4-Bit D Latch © CENGAGE LEARNING 2012.

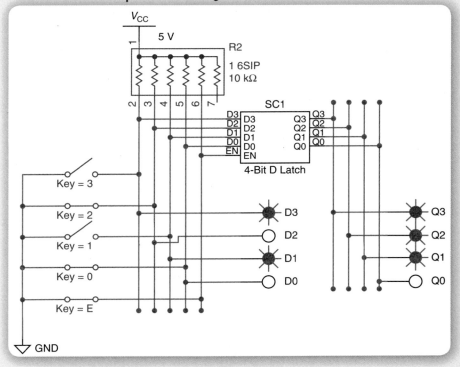

Example 9.8

A system for monitoring automobile traffic is set up at an intersection, with four sensors, placed as shown in *Figure 9.30*. Each sensor monitors traffic for a particular direction. When a car travels over a sensor, it produces a logic HIGH. The status of the sensor system is captured for later analysis by a set of D latches, as shown in *Figure 9.31*. A timing pulse enables the latches once every 5 seconds and thus stores the system status as a "snapshot" of the traffic pattern. This technique of data capture is known as strobing.

Figure 9.32 shows the timing diagram of a typical traffic pattern at the intersection. The *D* inputs show the cars passing through the intersection in the various lanes. Complete this timing diagram by drawing the *Q* outputs of the latches.

How should we interpret the *Q* output waveforms?

FIGURE 9.30 Example 9.8: Sensor Placement in a Traffic Intersection

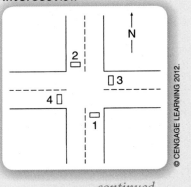

© CENGAGE LEARNING 2012.

continued...

FIGURE 9.31 Example 9.8: D Latch Collection of Data

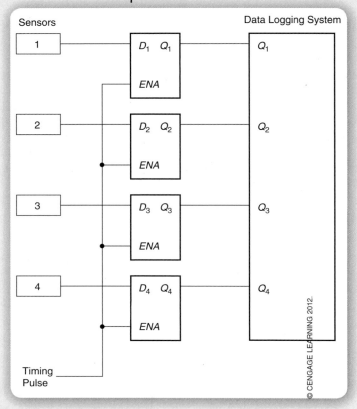

FIGURE 9.32 Example 9.8: Latch Timing Diagram

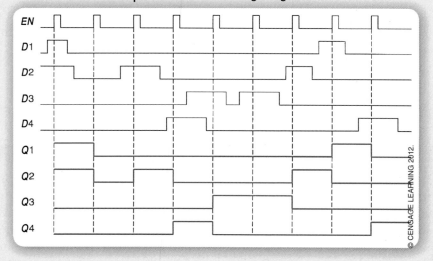

Solution

Figure 9.32 shows the completed timing diagram. The *ENABLE* input synchronizes the random sensor pattern to a 5-second standard interval. A *HIGH* on any Q output indicates a car over a sensor at the beginning of the interval. For example, at the beginning of the first interval, there is a car in the northbound lane (Q1) and one in the southbound lane (Q2). Similar interpretations can be made for each interval. This timing diagram is not completely realistic; the car intervals are spaced too closely in opposing directions. However, the diagram is a good illustration of the principle of data collection.

9.4 EDGE-TRIGGERED D FLIP-FLOPS

In Example 9.5, we saw how a shorter pulse width at the *ENABLE* input of a gated latch increased the chance of the output being synchronized to the *ENABLE* pulse waveform. This is because a shorter *ENABLE* pulse gives less chance for the *SET* and *RESET* inputs to change while the latch is enabled.

A logical extension of this idea is to enable the latch for such a small time that the width of the *ENABLE* pulse is almost zero. The best approximation we can make to this is to allow changes to the circuit output only when an enabling, or *CLOCK*, input receives the **edge** of an input waveform. An edge is the part of a waveform that is in transition from LOW to HIGH (positive edge) or HIGH to LOW (negative edge), as shown in *Figure 9.33*. We can say that a device enabled by an edge is **edge-triggered** or **edge-sensitive**. As shown in Figure 9.33, the *CLOCK* input of a device is sometimes labeled with its abbreviation: *CLK*.

The *CLOCK* input enables a circuit only while in transition, so we can refer to it as a "dynamic" input. This is in contrast to the *ENABLE* input of a gated latch, which is **level-sensitive** or "static," and will enable a circuit for the entire time it is at its active level.

FIGURE 9.33 Edges of a CLOCK Waveform

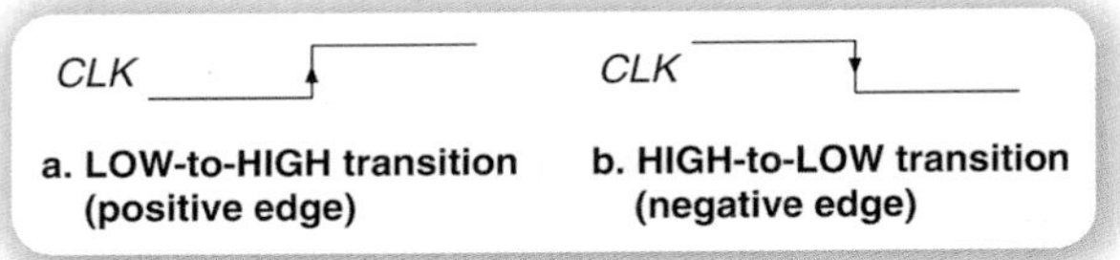

CLK		CLK	
a. LOW-to-HIGH transition (positive edge)		**b. HIGH-to-LOW transition (negative edge)**	

© CENGAGE LEARNING 2012.

Latches versus Flip-Flops

A gated latch with a clock input is called a **flip-flop**. Although the distinction is not always understood, we will define a *latch* as a circuit with a *level-sensitive enable* (e.g., gated D latch) or *no enable* (e.g., NAND latch) and a *flip-flop* as a circuit with an *edge-triggered clock* (e.g., D flip-flop). A NAND or NOR latch is sometimes called an SR flip-flop. By our definition this is not correct, as neither of these circuits has a clock input. (An SR flip-flop would be like the gated SR latch of Figure 9.20 with a clock instead of an enable input.)

The symbol for the D, or data, flip-flop is shown in *Figure 9.34*. The D flip-flop has the same behavior as a gated D latch, except that the outputs change only on the positive edge of the clock waveform, as opposed to the HIGH state of the enable input. The triangle on the *CLK* (clock) input of the flip-flop indicates that the device is edge-triggered.

FIGURE 9.34 D Flip-Flop Logic Symbol

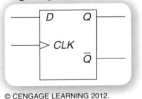

© CENGAGE LEARNING 2012.

TABLE 9.6 Function Table for a Positive Edge-Triggered D Flip-Flop

CLK	D	Q_{t+1}	$\overline{Q}_{t+1}$	Function
↑	0	0	1	Reset
↑	1	1	0	Set
0	X	Q_t	$\overline{Q}_t$	Inhibited
1	X	Q_t	$\overline{Q}_t$	Inhibited
↓	X	Q_t	$\overline{Q}_t$	Inhibited

Table 9.6 shows the function table of a positive edge-triggered D flip-flop. The upward arrow (↑) in the **CLK** column indicates a positive clock edge. The downward arrow (↓) indicates a negative edge.

Figure 9.35 shows the equivalent circuit of a positive edge-triggered D flip-flop. The circuit is the same as the transparent latch of Figure 9.22, except that the enable input (called *CLK* [i.e., "clock"] in the flip-flop) passes through an **edge detector**, a circuit that converts a positive edge to a brief positive-going pulse. (A negative edge detector converts a negative edge to a positive-going pulse.)

FIGURE 9.35 D Flip-Flop Equivalent Circuit

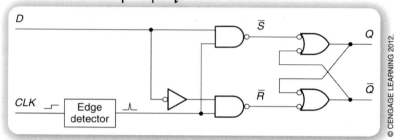

FIGURE 9.36 Positive Edge Detector

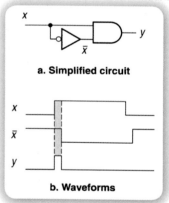

a. Simplified circuit

b. Waveforms

Figure 9.36 shows a circuit that acts as a simplified positive edge detector. Edge detection depends on the fact that a gate output does not switch immediately when its input switches. There is a delay of about 3 ns to 10 ns from input change to output change, called *propagation delay*.

When input *x*, shown in the timing diagram of Figure 9.36, goes from LOW to HIGH, the inverter output, $\overline{x}$, goes from HIGH to LOW after a short delay. This delay causes both *x* and $\overline{x}$ to be HIGH for a short time, producing a high-going pulse at the circuit output immediately following the positive edge at *x*.

When *x* returns to LOW, $\overline{x}$ goes HIGH after a delay. However, there is no time in this sequence when both AND inputs are HIGH. Therefore, the circuit output stays LOW after the negative edge of the input waveform.

Figure 9.37 shows how the D flip-flop circuit operates. When *D* = 0 and the edge detector senses a positive edge at the *CLK* input, the output of the lower

FIGURE 9.37 Operation of a D Flip-Flop

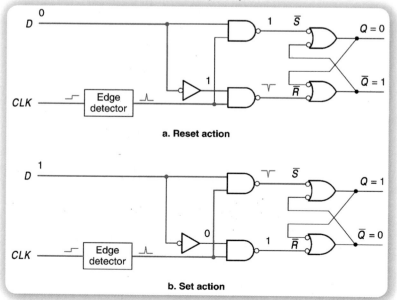

a. Reset action

b. Set action

NAND gate steers a low-going pulse to the *RESET* input of the latch, thus storing a 0 at Q. When $D = 1$, the upper NAND gate is enabled. The edge detector sends a high-going pulse to the upper steering gate, which transmits a low-going *SET* pulse to the output latch. This action stores a 1 at Q.

Example 9.9

Multisim Example

Multisim Files: 09.06 D Latch and DFF with Manual D Input.ms11

Figure 9.38 shows a Multisim circuit for comparing the operation of a D latch and a D flip-flop.

FIGURE 9.38   Example 9.9: Testing a D Latch and a D Flip-Flop in Multisim © CENGAGE LEARNING 2012.

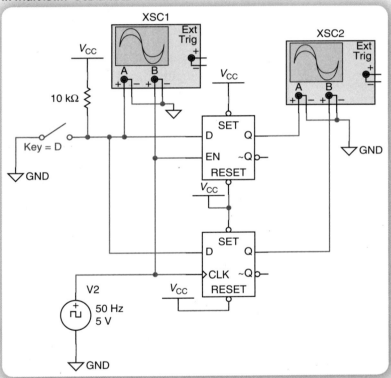

The *D* inputs of both devices are connected to the same switch so that both *D* inputs can be manually changed at the same time. The latch enable and flip-flop clock are both connected to a pulse generator that runs slowly enough to be observed on the virtual oscilloscopes in the circuit. Oscilloscope XSC1 measures *D* on channel A and *CLK/EN* on channel B. Oscilloscope XSC2 measures the output waveforms: from the latch on channel A and the flip-flop on channel B. An example of the input and output traces is shown in *Figure 9.39*.

Open the Multisim file for this example. Open the oscilloscopes by double-clicking on the symbols. Align the oscilloscope windows as shown in Figure 9.39. Adjust the timebase measurement on each oscilloscope to

continued...

FIGURE 9.39　Example 9.9: Oscilloscope Traces for D Latch and D Flip-Flop in Multisim © CENGAGE LEARNING 2012.

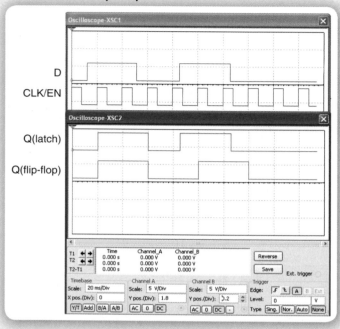

20 ms/division. Adjust the Y position of the traces to make them well-spaced and easy to read when the channel A and B scales are set to 5 V/division. Click once on the Multisim schematic to select that window and run the simulation.

Operate the *D* input by pressing D on your PC keyboard. Try to duplicate the waveforms in Figure 9.39. You can freeze the oscilloscope waveforms by stopping the simulation. When you start the simulation again or when the traces reach the right side of the screens, the oscilloscope screens will clear and start again from the left.

Explain why the latch and flip-flop outputs act as they do in Figure 9.39.

Solution

At the beginning of the oscilloscope trace, the latch and flip-flop outputs both change at the same time. This is because the first two changes on the *D* inputs happen when the enable and clock inputs are LOW. The latch enable goes HIGH at the same time that there is a positive edge on the flip-flop clock, therefore allowing both devices to change output states at the same time.

The next two changes on *D* happen when the latch enable input is HIGH. Because *EN* is HIGH, *Q* follows *D* immediately, both when *D* goes from LOW to HIGH and when it goes from HIGH to LOW. These changes on *D* both happen after a positive edge on the flip-flop clock. The flip-flop output does not change until the next positive edge is available. This difference causes the flip-flop output to be delayed with respect to the latch.

Example 9.10

Multisim Example

Multisim Files: 09.06 D Latch and DFF with Manual D Input.ms11

Three positive edge-triggered D flip-flops are connected as shown in the Multisim drawing of *Figure 9.40*. Inputs D_0 and *CLK* are shown in the timing diagram.

Open the Multisim file for this example. Open the oscilloscopes by double-clicking on the symbols. Align the oscilloscope windows as shown in *Figure 9.41*. Adjust the timebase measurement on each oscilloscope to 20 ms/division. Adjust the Y position of the traces to make them well-spaced and easy to read when the channel A and B scales are set to 5 V/division. Click once on the Multisim schematic to select that window and run the simulation.

Operate the *D* input by pressing D on your PC keyboard. Try to duplicate the waveforms in Figure 9.41. You can freeze the oscilloscope waveforms by

continued...

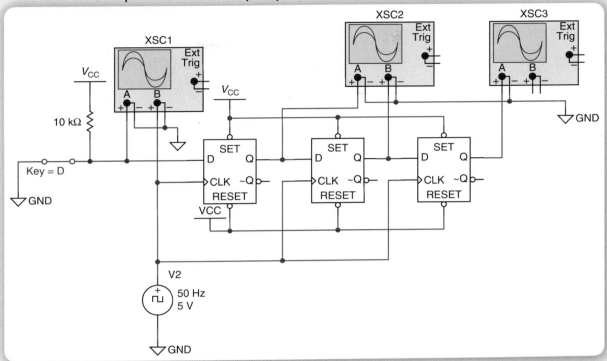

stopping the simulation. When you start the simulation again or when the traces reach the right side of the screens, the oscilloscope screens will clear and start again from the left.

Explain why the latch and flip-flop outputs act as they do in Figure 9.41.

Solution

Figure 9.41 shows the output waveforms. Q_0 follows D_0 at each point where the clock input has a positive edge. $D_1 = Q_0$ and Q_1 follows D_1, so the waveform at Q_1 is the same as at Q_0, but delayed by one clock cycle. If Q_0 changes due to *CLK*, we assume that the value of D_1 is the same as Q_0 just *before* the clock pulse. This is because delays within the circuitry of the flip-flops ensure that their outputs will not change for several nanoseconds after an applied clock pulse. Therefore, the level at D_1 remains constant long enough for it to be clocked into the second flip-flop. Data transfers to D_2, and then Q_2 in a similar way.

The data entering the circuit at D_0 are moved, or shifted, from one flip-flop to the next. This type of data movement, called "serial shifting," is frequently used in data communication and digital arithmetic circuits.

FIGURE 9.41   Example 9.10: Oscilloscope Traces for 3-Bit D Flip-Flop Circuit in Multisim

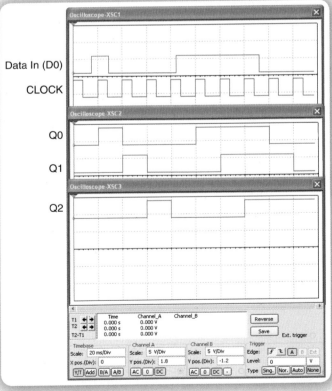

© CENGAGE LEARNING 2012.

9.3 Which part of a D flip-flop accounts for the difference in operation between a D flip-flop and a D latch? How does it work?

9.5 EDGE-TRIGGERED JK FLIP-FLOPS

Toggle Alternate between opposite binary states with each applied clock pulse.

FIGURE 9.42 Edge-Triggered JK Flip-Flops

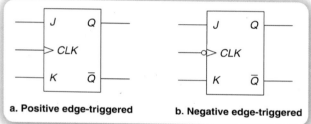

a. Positive edge-triggered

b. Negative edge-triggered

© CENGAGE LEARNING 2012.

A versatile and widely used sequential circuit is the JK flip-flop.

Figure 9.42 shows the logic symbols of a positive- and a negative-edge triggered JK flip-flop. *J* acts as a *SET* input and *K* acts as a *RESET* input, with the output changing on the active clock edge in response to *J* and *K*. When *J* and *K* are both HIGH, the flip-flop will **toggle** between opposite logic states with each applied clock pulse. The function tables of the devices in Figure 9.42 are shown in *Table 9.7*.

TABLE 9.7 Function Tables for Edge-Triggered JK Flip-Flops © CENGAGE LEARNING 2012.

Positive Edge-Triggered						Negative Edge-Triggered					
CLK	J	K	Q_{t+1}	$\overline{Q}_{t+1}$	Function	CLK	J	K	Q_{t+1}	$\overline{Q}_{t+1}$	Function
↑	0	0	Q_t	$\overline{Q}_t$	No change	↓	0	0	Q_t	$\overline{Q}_t$	No change
↑	0	1	0	1	Reset	↓	0	1	0	1	Reset
↑	1	0	1	0	Set	↓	1	0	1	0	Set
↑	1	1	$\overline{Q}_t$	Q_t	Toggle	↓	1	1	$\overline{Q}_t$	Q_t	Toggle
0	X	X	Q_t	$\overline{Q}_t$	Inhibited	0	X	X	Q_t	$\overline{Q}_t$	Inhibited
1	X	X	Q_t	$\overline{Q}_t$	Inhibited	1	X	X	Q_t	$\overline{Q}_t$	Inhibited
↓	X	X	Q_t	$\overline{Q}_t$	Inhibited	↑	X	X	Q_t	$\overline{Q}_t$	Inhibited

Example 9.11

The *J*, *K*, and *CLK* inputs of a negative edge-triggered JK flip-flop are as shown in the timing diagram in *Figure 9.43*. Complete the timing diagram by drawing the waveforms for *Q* and $\overline{Q}$. Indicate which function (no change, set, reset, or toggle) is performed at each clock pulse. The flip-flop is initially reset.

FIGURE 9.43 Example 9.11: Timing Diagram (Negative
Edge-Triggered JK Flip-Flop) © CENGAGE LEARNING 2012.

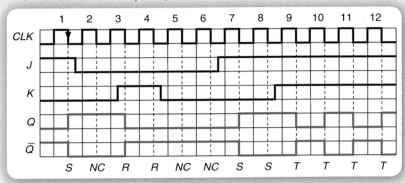

Solution

The completed timing diagram is shown in Figure 9.43, with the function labels (NC, R, S, and T) shown under the waveform grid. The outputs change only on the negative edges of the CLK waveform. Note that the same output sometimes results from different inputs. For example, the function at clock pulse 4 is reset and the function at pulses 5 and 6 is no change, but the Q waveform is LOW in each case.

Example 9.12

The toggle function of a JK flip-flop is often used to generate a desired output sequence from a series of flip-flops. The circuit shown in *Figure 9.44* is configured so that all flip-flops are permanently in toggle mode.

FIGURE 9.44 Example 9.12: Flip-Flop Circuit

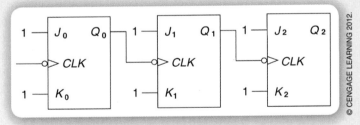

© CENGAGE LEARNING 2012.

Assume that all flip-flops are initially reset. Draw a timing diagram showing the CLK, Q_0, Q_1, and Q_2 waveforms when 8 clock pulses are applied. Make a table showing each combination of Q_2, Q_1, and Q_0. What pattern do the outputs form over the period shown on the timing diagram?

Solution

The circuit timing diagram is shown in *Figure 9.45*. All flip-flops are in toggle mode. Each time a negative clock edge is applied to the flip-flop CLK input, the Q output will change to the opposite state.

For flip-flop 0, this happens with every clock pulse, as it is clocked directly by the CLK waveform. Each of the other flip-flops is clocked by the Q output waveform of the previous stage. Flip-flop 1 is clocked by the negative edge of the Q_0 waveform. Flip-flop 2 toggles when Q_1 goes from HIGH to LOW.

TABLE 9.8 *Sequence of Outputs for Circuit in Figure 9.45*

Clock Pulse	Q_2	Q_1	Q_0
0	0	0	0
1	0	0	1
2	0	1	0
3	0	1	1
4	1	0	0
5	1	0	1
6	1	1	0
7	1	1	1
8	0	0	0

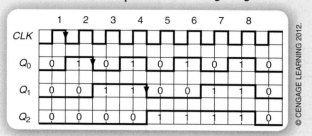

FIGURE 9.45 Example 9.12: Timing Diagram

Table 9.8 shows the flip-flop outputs after each clock pulse. The outputs form a 3-bit number, in the order Q_2 Q_1 Q_0, that counts from 000 to 111 in binary sequence, then returns to 000 and repeats.

This flip-flop circuit is called a 3-bit asynchronous counter. Counters will be explored further in a later chapter.

Synchronous versus Asynchronous Circuits

KEY TERMS

Asynchronous Not synchronized to the system clock.

Synchronous Synchronized to the system clock.

FIGURE 9.46 Detail of a Timing Diagram for a 3-Bit Asynchronous Counter

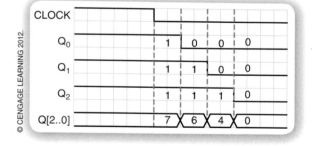

The **asynchronous** counter in Figure 9.44 has the advantage of being simple to construct and analyze. However, because it is asynchronous (that is, not synchronized to a single clock), it is seldom used in modern digital designs. The main problem with this and other asynchronous circuits is that their outputs do not change at the same time, due to delays in the flip-flops. This yields intermediate states that are not part of the desired output sequence.

Figure 9.46 shows a detail of a timing diagram of a circuit similar to that in Figure 9.44. The outputs are shown separately, and also as a group labeled $Q[2..0]$ that shows the combined binary value of the outputs. Figure 9.46 shows the timing diagram at the point where the output goes from 7 to 0 (111 to 000). The circuit output is initially 111. A negative clock edge, applied to flip-flop 0, makes Q_0 toggle after a short delay. The output is now 110 ($=6_{10}$). The resulting negative edge on Q_0 clocks flip-flop 1, making it toggle, and yields a new output of 100 ($=4_{10}$). The negative edge on Q_1 clocks flip-flop 2, making the output equal to 000 after a short delay.

Thus, the output goes through two very short intermediate states that are not in the desired output sequence. Instead of going directly from 111 to 000, as in Figure 9.45, the output goes in the sequence 111-110-100-000. We see in Figure 9.46 that the counter output goes through one or more intermediate transitions after each negative edge of the Q_0 waveform. In other words, intermediate states arise whenever a change propagates through more than one flip-flop. This happens because the flip-flops are clocked from different sources.

Figure 9.47 shows the circuit of a 3-bit **synchronous** counter. Unlike the circuit in Figure 9.44, the flip-flops in this circuit are clocked from a common source. Therefore, flip-flop delays do not add up through the circuit, and all the outputs change at the same time. *Figure 9.48* shows a timing diagram of the circuit of Figure 9.47. Detail of the timing diagram is shown in *Figure 9.49*. The output

FIGURE 9.47 3-Bit Synchronous Counter © CENGAGE LEARNING 2012.

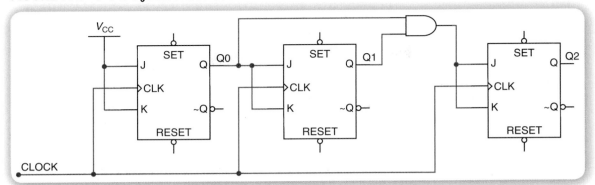

changes are delayed for a short time after the positive edge on CLOCK, but all outputs are delayed equally. Therefore, the outputs progress in a binary sequence, and with no intermediate states.

Also notice that the flip-flops in the synchronous counter circuit are positive edge-triggered. We could use either positive or negative edge-triggered flip-flops for this circuit since they are all clocked at the same time and do not depend on one another for the state of their outputs. This is not true of the asynchronous counter in Figure 9.44, which requires the flip-flops to be negative edge-triggered.

The synchronous counter circuit works as follows:

1. Flip-flop 0 is configured for toggle mode ($J_0K_0 = 11$). Because the flip-flops in Figure 9.47 are positive edge-triggered, Q_0 toggles on each positive clock edge.

2. Q_0 is connected to inputs J_1 and K_1. These inputs are tied together, so only two states are possible: no change ($JK = 00$) or toggle ($JK = 11$). If $Q_0 = 1$, Q_1 toggles. Otherwise, it does not change. This results in a Q_1 waveform that toggles at half the rate of Q_0.

3. J_2 and K_2 are both tied to the output of an AND gate. The AND gate output is HIGH if *both* Q_1 and Q_0 are HIGH. This makes Q_2 toggle, because $J_2K_2 = 11$. In all other cases, there is no change on Q_2. The result of this is that Q_2 toggles every fourth clock pulse, the only times when Q_1 and Q_0 are both HIGH.

FIGURE 9.48 Timing Diagram for a 3-Bit Synchronous Counter

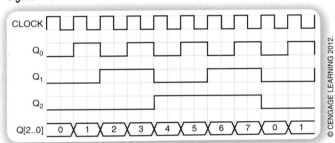

FIGURE 9.49 Detail of a Timing Diagram for a 3-Bit Synchronous Counter

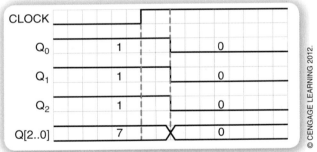

© CENGAGE LEARNING 2012.

Asynchronous Inputs (Preset and Clear)

KEY TERMS

Synchronous inputs The inputs of a flip-flop that do not affect the flip-flop's Q outputs unless a clock pulse is applied. Examples include D, J, and K inputs.

Asynchronous inputs The inputs of a flip-flop that change the flip-flop's Q outputs immediately, without waiting for a pulse at the CLK input. Examples include preset and clear inputs.

Preset An asynchronous set function.

Clear An asynchronous reset function.

FIGURE 9.50 Active-LOW Preset and Clear Inputs © CENGAGE LEARNING 2012.

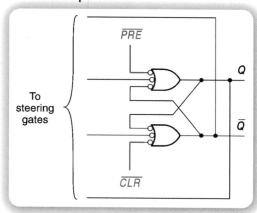

The *D*, *J*, and *K* inputs of the flip-flops examined so far are called **synchronous inputs**. This is because any effect they have on the flip-flop outputs is synchronized to the *CLK* input.

Another class of inputs is also provided on many flip-flops. These inputs, called **asynchronous inputs**, do not need to wait for a clock pulse to make a change at the output. The two functions usually provided are **preset**, an asynchronous set function, and **clear**, an asynchronous reset function. These functions are generally active-LOW, and are abbreviated $\overline{PRE}$ and $\overline{CLR}$.

Figure 9.50 shows the output circuit of a flip-flop with $\overline{PRE}$ and $\overline{CLR}$ inputs. The $\overline{PRE}$ and $\overline{CLR}$ inputs have direct access to the latch gates of the flip-flop and thus are not affected by the *CLK* input. They act exactly the same as the *SET* and *RESET* inputs of an SR latch and will override any synchronous input functions currently active.

Example 9.13

The waveforms for the *CLK*, *J*, *K*, $\overline{PRE}$, and $\overline{CLR}$ inputs of a negative edge-triggered JK flip-flop are shown in the timing diagram of *Figure 9.51*. Complete the diagram by drawing the waveform for output *Q*. Assume that *Q* is initially HIGH.

FIGURE 9.51 Example 9.13: JK Flip-Flop Waveforms Showing Synchronous and Asynchronous Functions © CENGAGE LEARNING 2012.

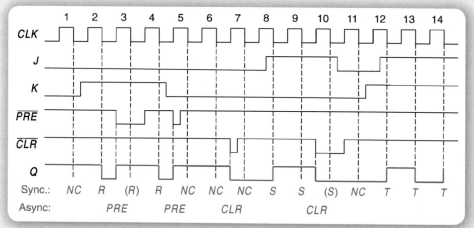

Solution

The *Q* waveform is shown in Figure 9.51. The asynchronous inputs cause an immediate change in *Q*, whereas the synchronous inputs must wait for the next negative clock edge. If asynchronous and synchronous inputs are simultaneously active, the asynchronous inputs have priority. This occurs in two places: pulse 3 (*K*, $\overline{PRE}$) and pulse 10 (*J*, $\overline{CLR}$).

The diagram shows the synchronous function (no change, reset, set, and toggle) at each clock pulse and the asynchronous functions (preset and clear) at the corresponding transition points.

The function table of a negative edge-triggered JK flip-flop with preset and clear functions is shown in *Table 9.9*.

TABLE 9.9 Function Table of a Negative Edge-Triggered JK Flip-Flop with Preset and Clear Functions © CENGAGE LEARNING 2012.

	$\overline{PRE}$	$\overline{CLR}$	CLK	J	K	Q_{t+1}	$\overline{Q}_{t+1}$	Function
Synchronous Functions	1	1	↓	0	0	Q_t	$\overline{Q}_t$	No change
	1	1	↓	0	1	0	1	Reset
	1	1	↓	1	0	1	0	Set
	1	1	↓	1	1	$\overline{Q}_t$	Q_t	Toggle
Asynchronous Functions	0	1	X	X	X	1	0	Preset
	1	0	X	X	X	0	1	Clear
	0	0	X	X	X	1	1	Forbidden
	1	1	0	X	X	Q_t	$\overline{Q}_t$	Inhibited
	1	1	1	X	X	Q_t	$\overline{Q}_t$	Inhibited
	1	1	↑	X	X	Q_t	$\overline{Q}_t$	Inhibited

X = Don't care ↓ = HIGH-to-LOW transition
Q_t = Present state of Q ↑ = LOW-to-HIGH transition
Q_{t+1} = Next state of Q

Note . . .

If preset and clear functions are not used, they should be disabled by connecting them to logic HIGH (for active-LOW inputs). This prevents them from being activated inadvertently by circuit noise. The synchronous functions of some flip-flops will not operate properly unless $\overline{PRE}$ and $\overline{CLR}$ are HIGH.

Using Asynchronous Reset in a Synchronous Circuit

KEY TERM

Master reset An asynchronous reset input used to set a sequential circuit to a known initial state.

Figure 9.52 shows an application of asynchronous clear inputs in a 3-bit synchronous counter. An input called *RESET* is tied to the asynchronous $\overline{CLR}$ inputs of all flip-flops. The counter output is set to 000 when the *RESET* line goes LOW.

FIGURE 9.52 3-Bit Synchronous Counter with Asynchronous Reset © CENGAGE LEARNING 2012.

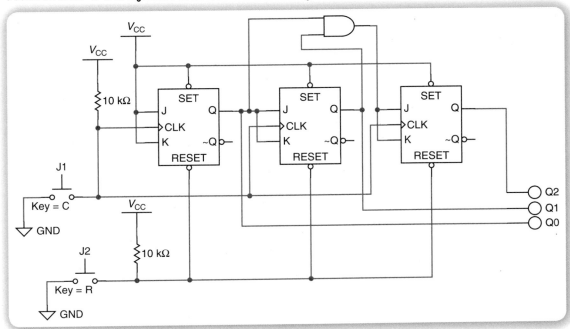

FIGURE 9.53 Timing Diagram for a 3-Bit Synchronous Counter with Asynchronous Reset © CENGAGE LEARNING 2012.

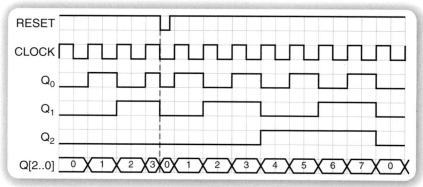

Figure 9.53 shows a timing diagram that illustrates the asynchronous clear function. When *RESET* is HIGH, the count proceeds normally. The third positive clock edge drives the counter to state 011. The reset pulse before the next positive clock edge sets the counter to 000 as soon as it goes LOW. On the next clock edge, the count proceeds from 000.

The function that sets all flip-flops in a circuit to a known initial state is sometimes called **master reset**.

Your Turn

9.4 What is the main difference between synchronous and asynchronous circuits, such as the two counters in Figure 9.44 and Figure 9.47? What disadvantage is there to an asynchronous circuit?

9.6 IDEAL AND NONIDEAL PULSES

Pulse Waveforms

KEY TERMS

Pulse A momentary variation of voltage from one logic level to the opposite level and back again.

Rising edge The part of a signal where the logic level is in transition from a LOW to a HIGH.

Falling edge The part of a signal where the logic level is in transition from a HIGH to a LOW.

Amplitude The instantaneous voltage of a waveform. Often used to mean maximum amplitude, or peak voltage, of a pulse.

Pulse width (t_w) Elapsed time from the 50% point of the leading edge of a pulse to the 50% point of the trailing edge.

Rise time (t_r) Elapsed time from the 10% point to the 90% point of the rising edge of a signal.

Fall time (t_f) Elapsed time from the 90% point to the 10% point of the falling edge of a signal.

Leading edge The edge of a pulse that occurs earliest in time.

Trailing edge The edge of a pulse that occurs latest in time.

Figure 9.54 shows the forms of both an ideal and a nonideal **pulse**. The **rising and falling edges** of an ideal pulse are vertical. That is, the transitions between logic HIGH and LOW levels are instantaneous. There is no such thing as an ideal pulse in a real digital circuit; an **edge** in a real signal is never absolutely vertical. Circuit capacitance, inductance, and other factors make the rising and falling edges of the pulse more like those on the nonideal pulses in Figure 9.54b and Figure 9.54c.

FIGURE 9.54 Ideal and Nonideal Pulses © CENGAGE LEARNING 2012.

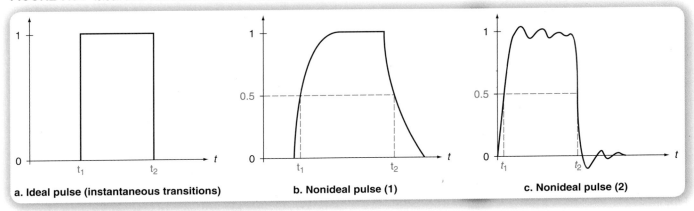

a. Ideal pulse (instantaneous transitions)   b. Nonideal pulse (1)   c. Nonideal pulse (2)

Pulses can be either positive-going or negative-going, as shown in *Figure 9.55*. In a positive-going pulse, the measured logic level is normally LOW, goes HIGH for the duration of the pulse, and returns to the LOW state. A negative-going pulse acts in the opposite direction.

Nonideal pulses are measured in terms of several timing parameters. *Figure 9.56* shows the 10%, 50%, and 90% points on the rising and falling edges of a nonideal pulse. (100% is the maximum **amplitude** of the pulse.)

FIGURE 9.56 Pulse Width, Rise Time, Fall Time © CENGAGE LEARNING 2012.

FIGURE 9.55 Pulse Edges © CENGAGE LEARNING 2012.

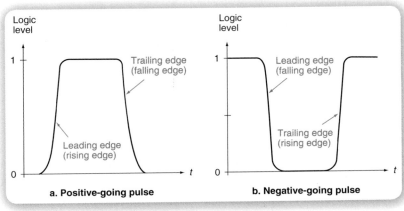

a. Positive-going pulse   b. Negative-going pulse

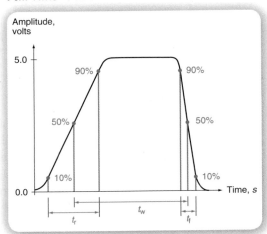

The 50% points are used to measure **pulse width** because the edges of the pulse are not vertical. Without an agreed reference point, the pulse width cannot be determined. The 10% and 90% points are used as references for the **rise and fall times** because the edges of a nonideal pulse are not straight lines. Most of the distortion of the rising or falling edge is below the 10% or above the 90% point.

Example 9.14

Calculate the pulse width, rise time, and fall time of the pulse shown in *Figure 9.57*.

Solution

From the graph in Figure 9.57, read the times corresponding to the 10%, 50%, and 90% values of the pulse on both the **leading and trailing edges**.

Leading edge:	10%:	2 μs	*Trailing edge:*	90%:	20 μs
	50%:	5 μs		50%:	25 μs
	90%:	8 μs		10%:	30 μs

Pulse width: 50% of leading edge to 50% of trailing edge.

$$t_w = 25\ \mu s - 5\ \mu s = 20\ \mu s$$

Rise time: 10% of rising edge to 90% of rising edge.

$$t_r = 8\ \mu s - 2\ \mu s = 6\ \mu s$$

Fall time: 90% of falling edge to 10% of falling edge.

$$t_f = 30\ \mu s - 20\ \mu s = 10\ \mu s$$

FIGURE 9.57 Example 9.14: Pulse

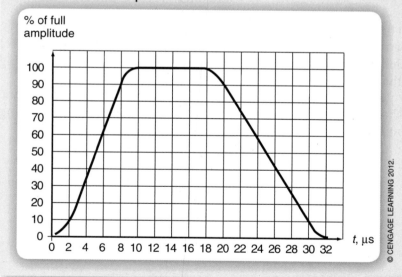

Although there are no perfectly vertical edges in digital circuitry, the time for a signal to rise or fall is usually negligible for most applications. We can almost always treat these edges as vertical, and not be concerned with the rise time or fall time. Rise and fall time may be important to consider if we have circuits that are designed to run with high clock frequencies or if we have inputs changing very close to the clock edge. Most devices will specify times around the clock edge in which the inputs must remain stable.

SUMMARY

1. A combinational circuit combines inputs to generate a particular output logic level that is always the same, regardless of the order in which the inputs are applied. A sequential circuit can generate different outputs for the same inputs, depending on the sequence in which the inputs were applied.

2. An SR latch is a sequential circuit with *SET* (*S*) and *RESET* (*R*) inputs and complementary outputs (*Q* and $\overline{Q}$). By definition, a latch is set when $Q = 1$ and reset when $Q = 0$.

3. A latch sets when its *S* input activates. When *S* returns to the inactive state, the latch remains in the set condition until explicitly reset by activating its *R* input.

4. A latch can have active-HIGH inputs (designated *S* and *R*) or active-LOW inputs (designated $\overline{S}$ and $\overline{R}$).

5. Two basic SR latch circuits are the NAND latch and the NOR latch, each consisting of two gates with cross-coupled feedback. In the NAND form, we draw the gates in their DeMorgan equivalent form so that each circuit has OR-shaped gates, inversion from input to output, and feedback to the opposite gate.

6. A NOR latch has active-HIGH inputs. It is described by the following function table:

S	R	Q_{t+1}	$\overline{Q}_{t+1}$	Function
0	0	Q_t	$\overline{Q}_t$	No change
0	1	0	1	Reset
1	0	1	0	Set
1	1	0	0	Forbidden

© CENGAGE LEARNING 2012.

7. A NAND latch has active-LOW inputs and is described by the following function table:

$\overline{S}$	$\overline{R}$	Q_{t+1}	$\overline{Q}_{t+1}$	Function
0	0	1	1	Forbidden
0	1	1	0	Set
1	0	0	1	Reset
1	1	Q_t	$\overline{Q}_t$	No change

© CENGAGE LEARNING 2012.

8. The transition from the forbidden state of a NAND or NOR latch to the no change state is not always predictable. If the latch inputs do not change at the same time, the latch will take the state represented by the last input to change. If both inputs change simultaneously, one of the latch gates will be slightly faster than the other, causing the latch to drop into the SET or RESET state. However, it cannot be determined beforehand which state will prevail.

9. A NAND latch can be used as a switch debouncer for a switch with a grounded common terminal, a normally open, and a normally closed contact. When the switch operates, one contact closes, resetting the latch on the first bounce. Further bounces are ignored. When the switch returns to its normal position, it sets the latch on the first bounce and further bounces are ignored.

10. A NOR latch can also be used as a debouncer, but the logic switch to be debounced must use pull-down resistors, rather than pull-up, and the common terminal must be connected to V_{CC}, not ground. This configuration is seldom

continues...

continued...

used because of its tendency to draw unacceptably large amounts of idle current from the power supply.

11. A gated SR latch controls the times when a latch can switch. The circuit consists of a pair of latch gates and a pair of steering gates. The steering gates are enabled or inhibited by a control signal called *ENABLE*. When the steering gates are enabled, they can direct a set or reset pulse to the latch gates. When inhibited, the steering gates block any set or reset pulses to the latch gates so the latch output cannot change.

12. A gated D ("data") latch can be constructed by connecting opposite logic levels to the *S* and *R* inputs of an SR latch. Because *S* and *R* are always opposite, the D latch has no forbidden state. The no change state is provided by the inhibit property of the *ENABLE* input.

13. In a gated D latch (or transparent latch), *Q* follows *D* when *ENABLE* is active. This is the transparent mode of the latch. When *ENABLE* is inactive, the latch stores the last value of *D*.

14. A flip-flop is like a gated latch that responds to the edge of a pulse applied to an enable input called *CLOCK*. A flip-flop output will change only when the input makes a transition from LOW to HIGH (for a positive edge-triggered device) or HIGH to LOW (for a negative edge-triggered device).

15. In a positive edge-triggered D flip-flop, *Q* follows *D* when there is a positive edge on the clock input.

16. A JK flip-flop has two synchronous inputs, called *J* and *K*. *J* acts as an active-HIGH set input. *K* acts as an active-HIGH reset function. When both inputs are made HIGH, the flip-flop toggles between 0 and 1 with each applied clock pulse.

17. The toggle function in a JK flip-flop is implemented with additional cross-coupled feedback from the latch gate outputs to the steering gate inputs.

18. A chain of JK flip-flops can implement an asynchronous binary counter if the *Q* of each flip-flop is connected to the clock input of the next and each flip-flop is configured to toggle. Although this is an easy way to create a counter, it is seldom used because internal flip-flop delays result in unwanted intermediate states in the count sequence.

19. JK flip-flops can be combined with a network of logic gates to make a synchronous binary counter. The gates are connected in such a way that each flip-flop toggles when all previous bits are HIGH: otherwise the flip-flops are in a no-change state. Although more complex than an asynchronous counter, a synchronous counter is free of unwanted intermediate states.

20. Many flip-flops are provided with asynchronous preset (set) and clear (reset) functions. Because these functions are connected directly to the latch gates of a flip-flop, they act immediately, without waiting for the clock. In most cases, these functions are active-LOW.

21. Asynchronous inputs, such as preset and clear, are usually designed so that they will override the synchronous inputs, such as *D* or *JK*.

22. Unused asynchronous inputs should be disabled by tying them to a logic HIGH (for an active-LOW input).

23. Pulse waveforms are measured by pulse width (t_w: time from 50% of leading edge to 50% of trailing edge).

24. Signal edges, including edges on clock signals and pulse waveforms, have rise times (t_r: time from 10% to 90% of rising edge) and fall times (t_f: time from 90% to 10% of falling edge). Usually, rise times and fall times are negligible, and rising and falling edges are treated as ideal (instantaneous rise and fall times).

9.1 Latches

9.1 Complete the timing diagram in *Figure 9.58* for the active-HIGH latch shown. The latch is initially set.

FIGURE 9.58 Problem 9.1: Timing Diagram

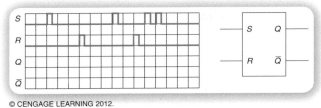

© CENGAGE LEARNING 2012.

9.2 Repeat Problem 9.1 for the timing diagram shown in *Figure 9.59*.

FIGURE 9.59 Problem 9.2: Timing Diagram

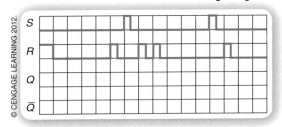

9.3 Complete the timing diagram in *Figure 9.60* for the active-LOW latch shown.

FIGURE 9.60 Problem 9.3: Timing Diagram

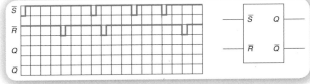

© CENGAGE LEARNING 2012.

9.4 Multisim Example
Multisim File: 09.02 SR Latch (Active-LOW) .ms11

Open the Multisim file for this example. The design, shown in *Figure 9.61*, is a Multisim version of the latch with active-HIGH inputs of Figure 9.3b. Run the file as a simulation and operate the Normally Open (NO) pushbuttons to try the set and reset functions of the latch.

a. What are the logic levels applied to the $\overline{S}$ and $\overline{R}$ inputs when the switches are in their rest positions? What happens to the $\overline{S}$ and $\overline{R}$ inputs when the S or R key on the keyboard is pressed during a simulation?

FIGURE 9.61 Problem 9.4: SR Latch in Multisim (Active-LOW Inputs) © CENGAGE LEARNING 2012.

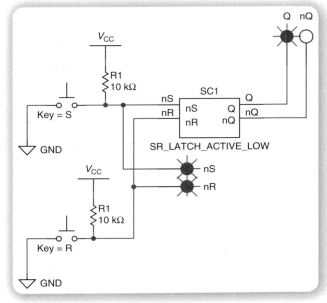

b. What action of the switches will make the latch set?

c. What action of the switches will make the latch reset?

d. How does the latch respond to more than one press of the set or reset pushbutton?

9.2 NAND/NOR Latches

9.5 Draw a NAND latch, correctly labeling the inputs and outputs. Describe the operation of a NAND latch for all four possible combinations of $\overline{S}$ and $\overline{R}$.

9.6 Draw a NOR latch, correctly labeling the inputs and outputs. Describe the operation of a NOR latch for all four possible combinations of S and R.

9.7 The timing diagram in *Figure 9.62* shows the input waveforms of a NAND latch. Complete the diagram by showing the output waveforms.

FIGURE 9.62 Problem 9.7: Input Waveforms to a NAND Latch © CENGAGE LEARNING 2012.

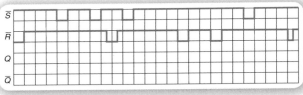

continues...

continued...

9.8 *Figure 9.63* shows the input waveforms to a NOR latch. Draw the corresponding output waveforms.

FIGURE 9.63 Problem 9.8: Input Waveforms to a NOR Latch © CENGAGE LEARNING 2012.

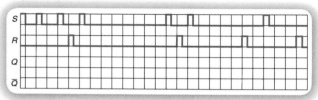

9.3 Gated Latches

9.9 Complete the timing diagram for the gated latch shown in *Figure 9.64*.

9.10 Complete the timing diagram for the gated latch shown in *Figure 9.65*.

9.11 The *S* and *R* waveforms in *Figure 9.66* are applied to two different gated latches. The *ENABLE* waveforms for the latches are shown as EN_1 and EN_2. Draw the output waveforms Q_1 and Q_2, assuming that *S*, *R*, and *EN* are all active-HIGH. Which output is least prone to synchronization errors? Why?

FIGURE 9.64 Problem 9.9: Gated Latch © CENGAGE LEARNING 2012.

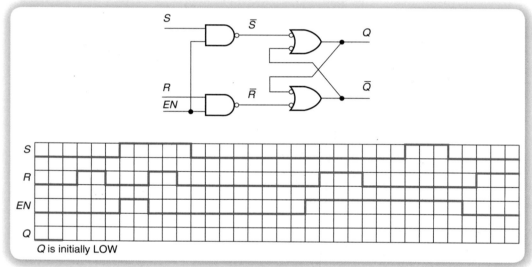

Q is initially LOW

FIGURE 9.65 Problem 9.10: Gated Latch © CENGAGE LEARNING 2012.

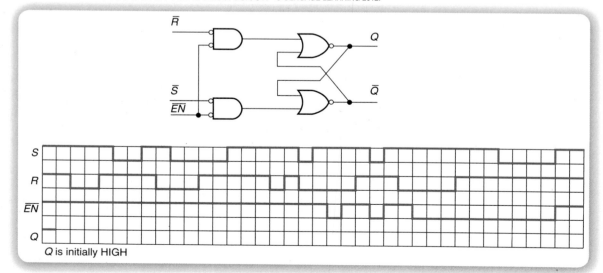

Q is initially HIGH

FIGURE 9.66 Problem 9.11: Waveforms

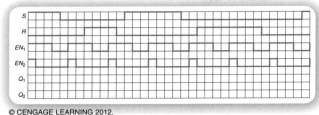

© CENGAGE LEARNING 2012.

9.12 *Figure 9.67* represents the waveforms of the *EN* and *D* inputs of a 4-bit transparent latch. Complete the timing diagram by drawing the waveforms for Q_1 to Q_4.

FIGURE 9.67 Problem 9.12: 4-Bit Transparent Latch and Waveforms © CENGAGE LEARNING 2012.

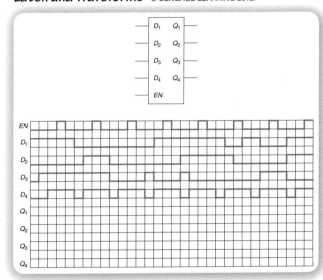

9.4 Edge-Triggered D Flip-Flops

9.13 The waveforms in *Figure 9.68* are applied to the inputs of a positive edge-triggered D flip-flop and a gated D latch. Complete the timing diagram where Q_1 is the output of the flip-flop and Q_2 is the output of the gated latch. Account for any differences between the Q_1 and Q_2 waveforms.

FIGURE 9.68 Problem 9.13: Waveforms for a D Latch and a D Flip-Flop © CENGAGE LEARNING 2012.

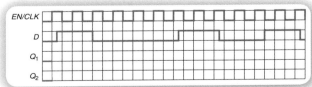

9.14 Complete the timing diagram for a positive edge-triggered D flip-flop if the waveforms shown in *Figure 9.69* are applied to the flip-flop inputs.

FIGURE 9.69 Problem 9.14: Waveforms for a D Flip-Flop © CENGAGE LEARNING 2012.

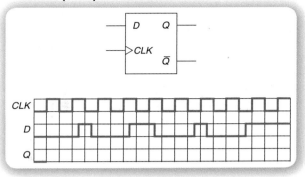

9.15 Repeat Problem 9.14 for the waveforms shown in *Figure 9.70*.

FIGURE 9.70 Problem 9.15: Waveforms for a D Flip-Flop © CENGAGE LEARNING 2012.

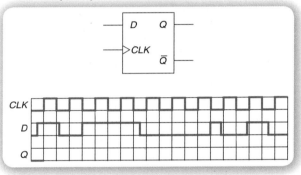

9.16 Repeat Problem 9.14 for the waveforms shown in *Figure 9.71*.

FIGURE 9.71 Problem 9.16: Waveforms for a D Flip-Flop © CENGAGE LEARNING 2012.

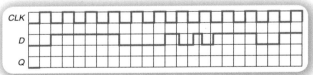

9.5 Edge-Triggered JK Flip-Flops

9.17 The waveforms in *Figure 9.72* are applied to a negative edge-triggered JK flip-flop. Complete the timing diagram by drawing the Q waveform.

FIGURE 9.72 Problem 9.17: JK Flip-Flop Input Waveforms © CENGAGE LEARNING 2012.

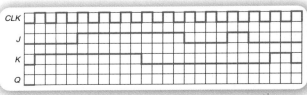

continues...

continued...

9.18 Repeat Problem 9.17 for the waveforms in *Figure 9.73*.

FIGURE 9.73 Problem 9.18: JK Flip-Flop Input Waveforms © CENGAGE LEARNING 2012.

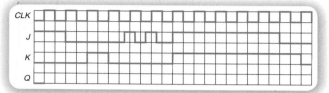

9.19 Given the inputs *x*, *y*, and *z* to the circuit in *Figure 9.74*, draw the waveform for output Q.

FIGURE 9.74 Problem 9.19: JK Flip-Flop Input Waveforms © CENGAGE LEARNING 2012.

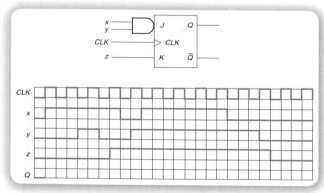

9.20 The waveforms shown in *Figure 9.75* are applied to a negative edge-triggered JK flip-flop. The flip-flop's Preset and Clear inputs are active-LOW. Complete the timing diagram by drawing the output waveforms.

FIGURE 9.75 Problem 9.20: JK Flip-Flop Input Waveforms © CENGAGE LEARNING 2012.

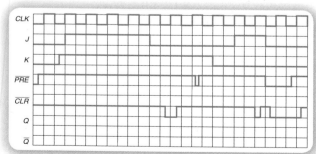

9.21 Repeat Problem 9.20 for the waveforms in *Figure 9.76*.

FIGURE 9.76 Problem 9.21: JK Flip-Flop Input Waveforms © CENGAGE LEARNING 2012.

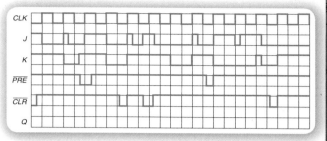

9.6 Ideal and Nonideal Pulses

9.22 Calculate the pulse width, rise time, and fall time of the pulse shown in *Figure 9.77*.

FIGURE 9.77 Problem 9.22: Positive-Going Pulse

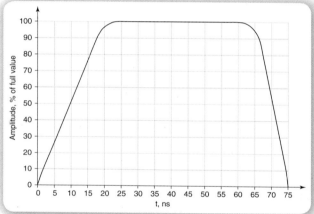

© CENGAGE LEARNING 2012.

9.23 Calculate the pulse width, rise time, and fall time of the pulse shown in *Figure 9.78*.

FIGURE 9.78 Problem 9.23: Negative-Going Pulse

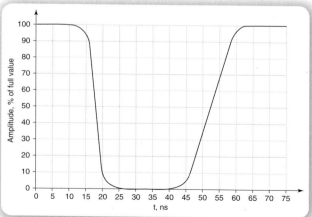

© CENGAGE LEARNING 2012.

9.1 Latches

9.24 Multisim Problem
Multisim File: 09.01 SR Latch (Active-HIGH) .ms11

Figure 9.79 shows the Multisim version of an active-HIGH latch used to control the HOLD function of a desk telephone, as described in Example 9.3. Open the Multisim file for this example and modify the circuit to be like the one in Figure 9.79. Run the simulation of the file and try various operations to make the circuit behave as described in Example 9.3.

FIGURE 9.79 Problem 9.24: Telephone Hold Circuit in Multisim © CENGAGE LEARNING 2012.

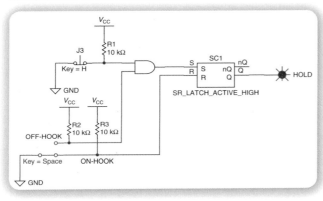

9.25 *Figure 9.80* shows an active-LOW latch used to control a motor starter. The motor runs when $Q = 1$ and stops when $Q = 0$.

FIGURE 9.80 Problem 9.25: Latch for Motor Starter © CENGAGE LEARNING 2012.

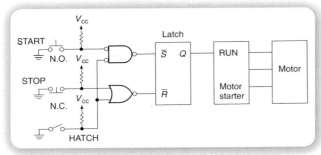

The motor is housed in a safety enclosure that has an access hatch for service. A safety interlock prevents the motor from running when the hatch is open. The *HATCH* switch opens when the hatch opens, supplying a logic *HIGH* to the circuit. The *START* switch is a normally open momentary-contact pushbutton (LOW when pressed). The *STOP* switch is a normally closed momentary-contact pushbutton (HIGH when pressed).

Draw the timing diagram of the circuit, showing *START, STOP, HATCH*, $\bar{S}$, $\bar{R}$, and Q for the following sequence of events:

a. *START* is pressed and released.
b. The hatch cover is opened.
c. *START* is pressed and released.
d. The hatch cover is closed.
e. *START* is pressed and released.
f. *STOP* is pressed and released.

Briefly describe the functions of the three switches and how they affect the motor operation.

9.26 A pump motor can be started at two different locations with momentary-contact pushbuttons S_1 and S_2. It can be stopped by momentary-contact pushbuttons ST_1 and ST_2. As in Problem 9.25, a *RUN* input on the motor controller must be kept HIGH to keep the motor running. After the motor is stopped, a timer prevents the motor from starting for 5 minutes.

Draw a circuit block diagram showing how an SR latch and some additional gating logic can be used in such an application. The timer can be shown as a block activated by the *STOP* function. Assume that the timer output goes HIGH for 5 minutes when activated.

9.2 NAND/NOR Latches

9.27 *Figure 9.81* represents two input waveforms to a latch circuit.

FIGURE 9.81 Problem 9.27: Input Waveforms to a Latch © CENGAGE LEARNING 2012.

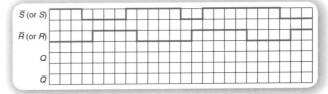

continues...

continued...

a. Draw the outputs Q and $\overline{Q}$ if the latch is a NAND latch.

b. Draw the output waveforms if the latch is a NOR latch.

(Note that in each case, the waveforms will produce the forbidden state at some point. Even under this condition, it is still possible to produce unambiguous output waveforms. Refer to Table 9.2 and Table 9.3 for guidance.)

9.28 a. Draw a timing diagram for a NAND latch showing each of the following sequences of events:

    i. $\overline{S}$ and $\overline{R}$ are both LOW; $\overline{S}$ goes HIGH before $\overline{R}$

    ii. $\overline{S}$ and $\overline{R}$ are both LOW; $\overline{R}$ goes HIGH before $\overline{S}$

    iii. $\overline{S}$ and $\overline{R}$ are both LOW; $\overline{S}$ and $\overline{R}$ go HIGH simultaneously.

b. State why $\overline{S} = \overline{R} = 0$ is a forbidden state for the NAND latch.

c. Briefly explain what the final result is for each of the above transitions.

9.29 a. Draw a timing diagram for a NOR latch showing each of the following sequences of events:

    i. S and R are both HIGH; S goes LOW before R.

    ii. S and R are both HIGH; R goes LOW before S.

    iii. S and R are both HIGH; S and R go LOW simultaneously.

b. Briefly explain what the final result is for each of the transitions listed in part **a** of this question.

c. State why $S = R = 1$ is a forbidden state for the NOR latch.

9.30 *Figure 9.82* shows the effect of mechanical bounce on the switching waveforms of a single-pole double-throw (SPDT) switch.

a. Briefly explain how this effect arises.

b. Draw a NAND latch circuit that can be used to eliminate this mechanical bounce, and briefly explain how it does so.

c. Briefly explain why we would not want to use a NOR latch as a switch debouncer.

FIGURE 9.82 Problem 9.30: Effect of Mechanical Bounce on an SPDT Switch

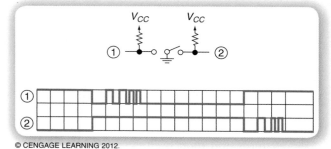

© CENGAGE LEARNING 2012.

9.3 Gated Latches

9.31 *Figure 9.83* shows a set of input and output traces for the latch test circuit of Figure 9.24. Briefly explain why the output trace looks the way it does.

FIGURE 9.83 Problem 9.31: D Latch Test Waveforms in Multisim © CENGAGE LEARNING 2012.

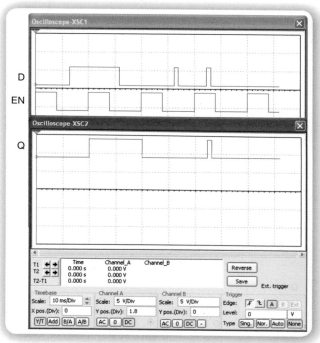

9.32 *Figure 9.84* shows a set of input and output traces for the latch test circuit of Figure 9.24. Briefly explain why the output trace looks the way it does.

FIGURE 9.84 Problem 9.32: D Latch Test Waveforms in Multisim © CENGAGE LEARNING 2012.

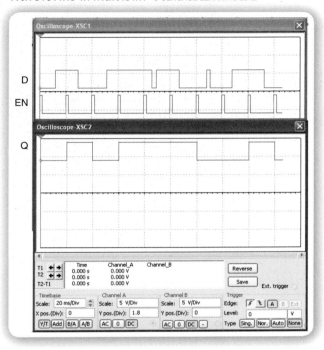

9.33 An electronic direction finder aboard an aircraft uses a 4-bit number to distinguish 16 different compass points as follows:

Direction	Degrees	Output Code
N	0/360	0000
NNE	22.5	0001
NE	45	0011
ENE	67.5	0010
E	90	0110
ESE	112.5	0111
SE	135	0101
SSE	157.5	0100
S	180	1100
SSW	202.5	1101
SW	225	1111
WSW	247.5	1110
W	270	1010
WNW	295.5	1011
NW	315	1001
NNW	337.5	1000

The output of the direction finder is stored in a 4-bit latch so that the aircraft flight path can be logged by a computer. The latch is periodically updated by a continuous pulse on the latch enable line.

Figure 9.85 shows a sample reading of the direction finder's output as presented to the latch.

FIGURE 9.85 Problem 9.33: Direction Finder and Sample Output © CENGAGE LEARNING 2012.

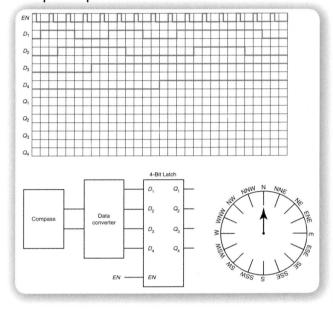

a. Complete the timing diagram by filling in the data for the *Q* outputs.

b. Based on the completed timing diagram of Figure 9.85 make a rough sketch of the aircraft's flight path for the monitored time.

9.4 Edge-Triggered D Flip-Flops

9.34 *Figure 9.86* shows a Multisim circuit for testing a positive edge-triggered D flip-flop. *Figure 9.87* shows a set of input and output traces for the flip-flop test circuit. Briefly explain why the output trace looks the way it does.

FIGURE 9.86 Problem 9.34: D Flip-Flop Testing in Multisim © CENGAGE LEARNING 2012.

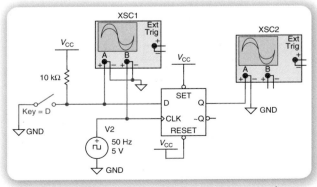

continues...

continued...

FIGURE 9.87 Problem 9.34: D Flip-Flop Test Waveforms in Multisim © CENGAGE LEARNING 2012.

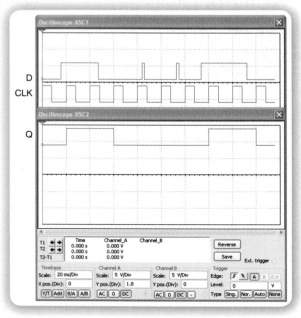

9.5 Edge-Triggered JK Flip-Flops

9.35 *Figure 9.88* and *Figure 9.89* show a Multisim circuit for testing a negative edge-triggered JK flip-flop. Both figures show the same Q output value, but with different input states. One figure shows the flip-flop before a clock pulse is applied and one shows it after the clock pulse. Which figure represents which case? Briefly explain your answer.

FIGURE 9.88 Problem 9.35: JK Flip-Flop Test Circuit in Multisim © CENGAGE LEARNING 2012.

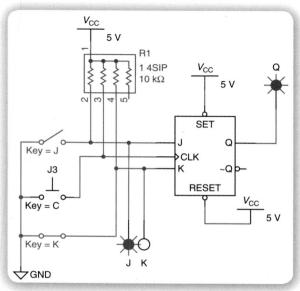

FIGURE 9.89 Problem 9.35: JK Flip-Flop Test Circuit in Multisim © CENGAGE LEARNING 2012.

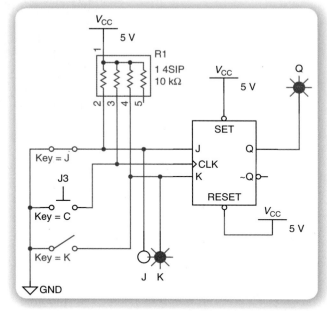

9.36 Assume that all flip-flops in *Figure 9.90* are initially set. Draw a timing diagram showing the CLK, Q_0, Q_1, and Q_2 waveforms when 8 clock pulses are applied. Make a table showing each combination of Q_2, Q_1, and Q_0. What pattern do the outputs form over the period shown on the timing diagram?

FIGURE 9.90 Problem 9.36: Flip-Flops

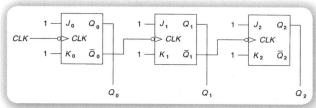

© CENGAGE LEARNING 2012.

9.37 Refer to the JK flip-flop circuit in *Figure 9.91*. Is the circuit synchronous or asynchronous? Explain your answer.

9.38 Assume all flip-flops in the circuit in Figure 9.91 are reset. Analyze the operation of the circuit when 16 clock pulses are applied by making a table showing the sequence of states of $Q_3 Q_2 Q_1 Q_0$, beginning at 0000.

9.39 Draw a timing diagram showing the sequence of states from the table derived in Problem 9.38.

9.40 Draw a logic diagram of a D flip-flop configured for toggle mode. (Hint: The D input must always be the opposite of the Q output.)

FIGURE 9.91 Problem 9.37: Flip-Flop Circuit © CENGAGE LEARNING 2012.

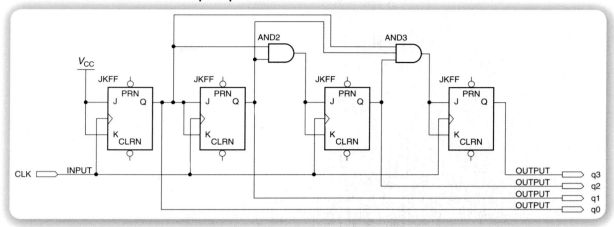

9.41 The term *asynchronous* is sometimes used to refer to the configuration of a circuit (e.g., a 3-bit asynchronous counter) and sometimes to a type of input to a device (e.g., an asynchronous clear input). Briefly explain how these two usages are similar and how they are different.

9.42 *Figure 9.92* shows a 3-bit synchronous counter drawn in Multisim.

FIGURE 9.92 Problem 9.42: 3-Bit Counter in Multisim © CENGAGE LEARNING 2012.

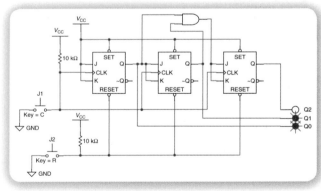

a. Why are the *SET* inputs of the flip-flops connected to V_{CC}?

b. Given the state of the counter in Figure 9.92, what will happen to the flip-flop outputs when the next clock pulse is applied? Explain your answer by referring to the behavior of each flip-flop when it transitions to the next state and what this means for the binary output of the counter.

c. What happens to the flip-flop outputs when the *RESET* button is pressed? Is this action synchronous or asynchronous?

9.6 Ideal and Nonideal Pulses

9.43 *Figure 9.93* shows a positive-going pulse on a Multisim oscilloscope. Estimate the pulse width, rise time, and fall time of one of the pulses shown in Figure 9.93.

FIGURE 9.93 Problem 9.43: Pulse Shown on Multisim Oscilloscope © CENGAGE LEARNING 2012.

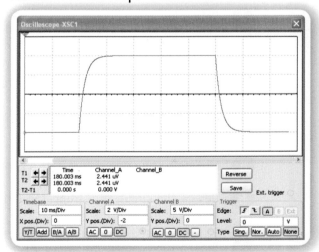

CHAPTER 10
Digital Counters

Menu		START LOCATION	DISTANCE	END LOCATION

Chapter Objectives

Upon successful completion of this chapter, you will be able to:

1 Determine the modulus of a counter.

2 Determine the maximum modulus of a counter, given the number of circuit outputs.

3 Draw the count sequence table, state diagram, and timing diagram of a counter.

4 Determine the count sequence, state diagram, timing diagram, and modulus of any synchronous counter.

5 Complete the state diagram of a synchronous counter to account for unused states.

6 Design the circuit of a truncated-sequence synchronous counter, using flip-flops and logic gates.

7 Draw a timing diagram showing the operation of a serial and parallel-load shift register.

8 Design and compare shift registers, ring counters, and Johnson counters.

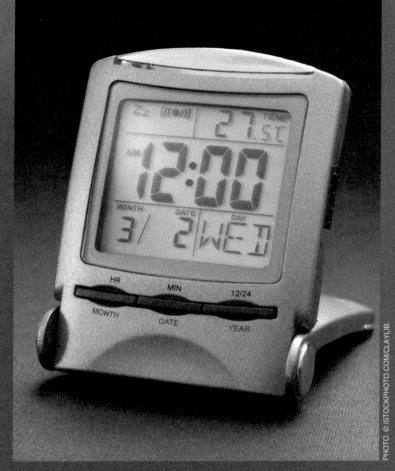

PHOTO: © ISTOCKPHOTO.COM/CLAYLIB

A special type of sequential digital circuit is a counter, a circuit that counts (the name seems to fit!) Some counters will count up, some will count down, some will count so far then start over. One of the most common digital counters is an alarm clock; once it counts to 12:59:59, we can be sure that it will go to 1:00:00 next. If it decided to count in a different order, there's a good chance someone would oversleep and miss school.

Counters and shift registers are two important classes of sequential circuits. In the simplest terms, a counter is a circuit that counts pulses. As such, it is used in many circuit applications, such as event counting and sequencing, timing, frequency division, and control. A basic counter can be enhanced to include functions such as synchronous or asynchronous parallel loading, synchronous or asynchronous clear, count enable, directional control, and output decoding. In this chapter, we will design counters from simple up or down counters to those with the above features.

Shift registers are circuits that store and move data. They can be used in serial data transfer, serial/parallel conversion, arithmetic functions, and delay elements. As with counters, many shift registers have additional functions such as parallel load, clear, and directional control.

10.1 BASIC CONCEPTS OF DIGITAL COUNTERS

KEY TERMS

Counter A sequential digital circuit whose output progresses in a predictable repeating pattern, advancing by one state for each clock pulse.

Recycle To make a transition from the last state of the count sequence to the first state.

Count sequence The specific series of output states through which a counter progresses.

Modulus The number of states through which a counter sequences before repeating.

UP counter A counter with an ascending sequence.

DOWN counter A counter with a descending sequence.

Modulo-*n* (or mod-*n*) counter A counter with a modulus of *n*.

State diagram A diagram showing the progression of states of a sequential circuit.

Modulo arithmetic A closed system of counting and adding, whereby a sum greater than the largest number in a sequence "rolls over" and starts from the beginning. For example, on a clock face, four hours after 10 A.M. is 2 P.M., so in a mod-12 system, 10 + 4 = 2.

The simplest definition of a **counter** is "a circuit that counts pulses." Knowing only this, let us look at an example of how we might use a counter circuit.

Example 10.1

Figure 10.1 shows a 10-bit binary counter that can be used to count the number of people passing by an optical sensor. Every time the sensor detects a person passing by, it produces a pulse. Briefly describe the counter's operation. What is the maximum number of people it can count? What happens if this number is exceeded?

Solution

The counter has a 10-bit output, allowing a binary number from 00 0000 0000 to 11 1111 1111 (0 to 1023) to appear at its output. The sensor causes the counter to advance by one binary number for every pulse applied to the counter's clock (*CLK*) input. If the counter is allowed to register *no people* (i.e., 00 0000 0000), then the circuit can count 1023 people because there are 1024 unique binary combinations of a 10-bit number, including 0. (This is because $2^{10} = 1024$.) When the 1024th pulse is applied to the clock input, the counter rolls over to 0 (or **recycles**) and starts counting again.

FIGURE 10.1 *Example 10.1: 10-Bit Counter*

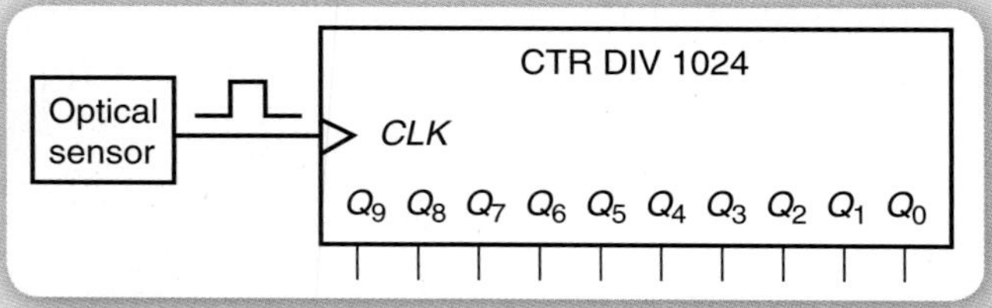

© CENGAGE LEARNING 2012.

continued...

(After this point, the counter would not accurately reflect the number of people counted.)

The counter is labeled CTR DIV 1024 to indicate that one full cycle of the counter requires 1024 clock pulses (i.e., the frequency of the MSB output signal, Q_9, is the clock frequency divided by 1024).

A counter is a digital circuit that has a number of binary outputs whose states progress through a fixed sequence. This **count sequence** can be ascending, descending, or nonlinear. The output sequence of a counter is usually defined by its **modulus**, that is, the number of states through which the counter progresses. An **UP counter** with a modulus of 12 counts through 12 states from 0000 up to 1011 (0 to 11 in decimal), recycles to 0000, and continues. A **DOWN counter** with a modulus of 12 counts from 1011 down to 0000, recycles to 1011, and continues downward. Both types of counter are called **modulo-12**, or just **mod-12** counters, because they both have sequences of 12 states.

State Diagram

The states of a counter can be represented by a **state diagram**. *Figure 10.2* compares the state diagram of a mod-12 UP counter to an analog clock face. Each counter state is illustrated in the state diagram by a circle containing its binary value. The progression is shown by a series of directional arrows. With each clock pulse, the counter progresses by one state, from its present position in the state diagram to the next state in the sequence.

Both the clock face and the state diagram represent a closed system of counting. In each case, when we reach the end of the count sequence, we start over from the beginning of the cycle.

For instance, if it is 10:00 A.M. and we want to meet a friend in 4 hours, we know we should turn up for the appointment at 2:00 P.M. We arrive at this figure by starting at 10 on the clock face and counting 4 digits forward in a "clockwise" circle. This takes us two digits past 12, the "recycle point" of the clock face.

Similarly, if we want to know the 8th state after 0111 in a mod-12 UP counter, we start at state 0111 and count 8 positions in the direction of the arrows. This brings us to state 0000 (the recycle point) in 5 counts and then on to state 0011 in another 3 counts. This closed system of counting and adding is known as **modulo arithmetic**.

FIGURE 10.2 Mod-12 State Diagram and Analog Clock Face

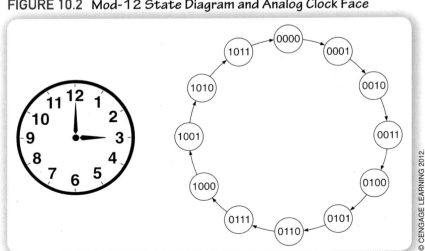

© CENGAGE LEARNING 2012.

Number of Bits and Maximum Modulus

FIGURE 10.3 *State Diagram of a Mod-16 Counter*

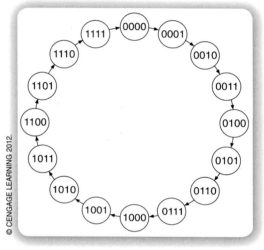

© CENGAGE LEARNING 2012.

The state diagram of Figure 10.2 represents the states of a mod-12 counter as a series of 4-bit numbers. Counter states are always written with a fixed number of bits, because each bit represents the logic level of a physical location in the counter circuit. A mod-12 counter requires four bits because its highest count value is a 4-bit number: 1011.

The **maximum modulus** of a 4-bit counter is 16 ($= 2^4$). The count sequence of a mod-16 UP counter is from 0000 to 1111 (0 to 15 in decimal), as illustrated in the state diagram of *Figure 10.3*.

In general, an n-bit counter has a maximum modulus of 2^n and a count sequence from 0 to ($2^n - 1$) (i.e., all 0's to all 1's). Because a mod-16 counter has a modulus of 2^n ($= m_{max}$), we say that it is a **full-sequence counter**. We can also call this a **binary counter** if it generates the sequence in binary order. A counter, such as a mod-12 counter, whose modulus is less than 2^n, is called a **truncated-sequence counter**.

Count-Sequence Table and Timing Diagram

TABLE 10.1 Mod-16 Count-Sequence Table

Q_3	Q_2	Q_1	Q_0
0	0	0	0
0	0	0	1
0	0	1	0
0	0	1	1
0	1	0	0
0	1	0	1
0	1	1	0
0	1	1	1
1	0	0	0
1	0	0	1
1	0	1	0
1	0	1	1
1	1	0	0
1	1	0	1
1	1	1	0
1	1	1	1

© CENGAGE LEARNING 2012.

Two ways to represent a count sequence other than a state diagram are by a **count-sequence table** and by a timing diagram. The count sequence table is simply a list of counter states in the same order as the count sequence. *Table 10.1* and *Table 10.2* show the count-sequence tables of a mod-16 UP counter and a mod-12 UP counter, respectively.

We can derive timing diagrams from each of these tables. We know that each counter advances by one state with each applied clock pulse. The mod-16 count sequence shows us that the Q_0 waveform changes state with each clock pulse. Q_1 changes with every two clock pulses, Q_2 with every four, and Q_3 with every eight. *Figure 10.4* shows this pattern for the mod-16 UP counter, assuming the counter is a positive edge-triggered device.

A divide-by-two ratio relates the frequencies of adjacent outputs of a binary counter. For example, if the clock frequency is $f_c = 16$ MHz, the frequencies of the output waveforms are: 8 MHz ($f_0 = f_c/2$); 4 MHz ($f_1 = f_c/4$); 2 MHz ($f_2 = f_c/8$); 1 MHz ($f_3 = f_c/16$).

We can construct a similar timing diagram, illustrated in *Figure 10.5*, for a mod-12 UP counter. The changes of state can be monitored by noting where Q_0 (the least significant bit) changes. This occurs on each positive edge of the *CLK*

FIGURE 10.4 Mod-16 Timing Diagram © CENGAGE LEARNING 2012.

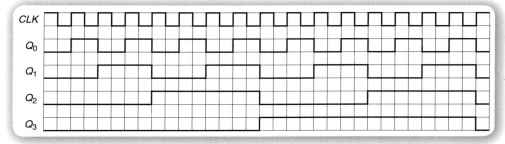

FIGURE 10.5 Mod-12 Timing Diagram

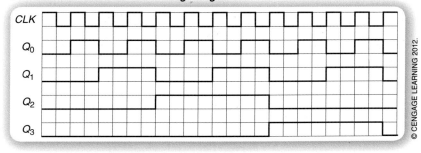

© CENGAGE LEARNING 2012.

TABLE 10.2 Mod-12 Count-Sequence Table

Q_3	Q_2	Q_1	Q_0
0	0	0	0
0	0	0	1
0	0	1	0
0	0	1	1
0	1	0	0
0	1	0	1
0	1	1	0
0	1	1	1
1	0	0	0
1	0	0	1
1	0	1	0
1	0	1	1

© CENGAGE LEARNING 2012.

waveform. The sequence progresses by 1 with each *CLK* pulse until the outputs all go to 0 on the first *CLK* pulse after state $Q_3Q_2Q_1Q_0 = 1011$.

The output waveform frequencies of a truncated sequence counter do not necessarily have a simple relationship to one another as do binary counters. For the mod-12 counter the relationships between clock frequency, f_c, and output frequencies are: $f_0 = f_c/2$; $f_1 = f_c/4$; $f_2 = f_c/12$; $f_3 = f_c/12$. Note that both Q_2 and Q_3 have the same frequencies (f_2 and f_3), but are out of phase with one another.

Example 10.2

Draw the state diagram, count-sequence table, and timing diagram for a mod-12 DOWN counter.

Solution

Figure 10.6 shows the state diagram for the mod-12 DOWN counter. The states are identical to those of a mod-12 UP counter, but progress in the opposite direction. *Table 10.3* shows the count-sequence table of this circuit.

The timing diagram of this counter is illustrated in *Figure 10.7*. The output starts in state $Q_3Q_2Q_1Q_0 = 1011$ and counts DOWN until it reaches 0000. On the next pulse, it recycles to 1011 and starts over.

FIGURE 10.6 Example 10.2: State Diagram of a Mod-12 DOWN Counter

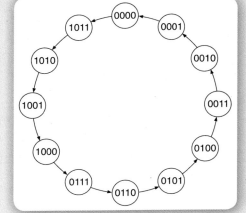

© CENGAGE LEARNING 2012.

continued...

TABLE 10.3 Count-Sequence Table for a Mod-12 DOWN Counter

Q_3	Q_2	Q_1	Q_0
1	0	1	1
1	0	1	0
1	0	0	1
1	0	0	0
0	1	1	1
0	1	1	0
0	1	0	1
0	1	0	0
0	0	1	1
0	0	1	0
0	0	0	1
0	0	0	0

© CENGAGE LEARNING 2012.

FIGURE 10.7 *Example 10.2: Timing Diagram of a Mod-12 DOWN Counter*

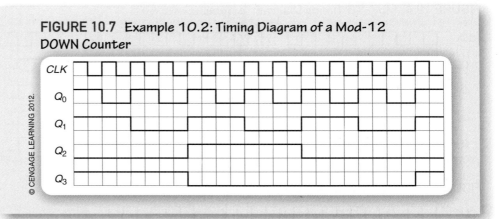

Example 10.3

Multisim Example

Multisim File: 10.03 Mod 12 Down Counter.ms11

Figure 10.8 shows a Multisim version of a Mod-12 DOWN counter (using JK flip-flops). The clock frequency can be changed by opening the file and double-clicking the clock component. The state of the counter is shown on the LED indicators as well as the seven-segment display.

a. Why is this a synchronous counter instead of an asynchronous counter?

b. What are the equations for the J and K inputs on the flip-flop whose output is Q3?

Solution

a. This is a synchronous counter because the same clock signal is used for each flip-flop in the counter.

b. The equation for J_{Q3} and K_{Q3} is the same: $\overline{Q2}\ \overline{Q1}\ \overline{Q0}$.

FIGURE 10.8 *Example 10.3: Mod-12 DOWN Counter*

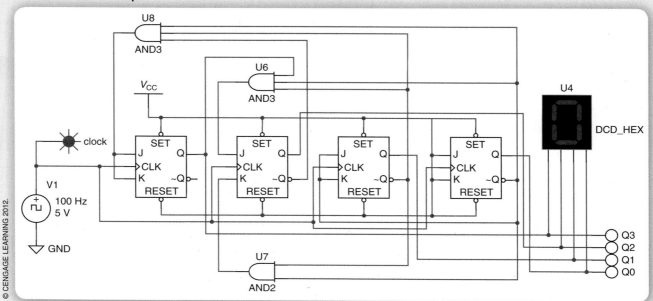

Your Turn

10.1 How many outputs does a mod-24 counter require? Is this a full-sequence or a truncated-sequence counter? Explain your answer.

10.2 SYNCHRONOUS COUNTERS

KEY TERMS

Synchronous counter A counter whose flip-flops are all clocked by the same source and thus change in synchronization with each other.

Memory section A set of flip-flops in a synchronous circuit that holds its present state.

Present state The current state of flip-flop outputs in a synchronous sequential circuit.

Control section The combinational logic portion of a synchronous circuit that determines the next state of the circuit.

Next state The desired future state of flip-flop outputs in a synchronous sequential circuit after the next clock pulse is applied.

Status lines Signals that communicate the present state of a synchronous circuit from its memory section to its control section.

Command lines Signals that connect the control section of a synchronous circuit to its memory section and direct the circuit from its present to its next state.

In an earlier chapter, we briefly examined the circuits of a 3-bit and a 4-bit **synchronous counter**. A synchronous counter is a circuit consisting of flip-flops and control logic, whose outputs progress through a regular predictable sequence, driven by a clock signal. The counter is synchronous because all flip-flops are clocked at the same time.

Figure 10.9 shows the block diagram of a synchronous counter, which consists of a **memory section** to keep track of the **present state** of the counter and a **control section** to direct the counter to its **next state**. The memory section is a sequential circuit (flip-flops) and the control section is combinational (gates). They communicate through a set of **status lines** that go from the Q outputs of the flip-flops to the control gate inputs and **command lines** that connect the control gate outputs to the synchronous inputs (J, K, D, or T) of the flip-flops. Outputs can be tied directly to the status lines or can be decoded to give a sequence other than that of the flip-flop output states. The circuit might have inputs to implement one or more control functions, such as changing the count direction, clearing the counter, or presetting the counter to a specific value.

FIGURE 10.9 Synchronous Counter Block Diagram

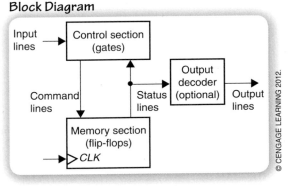

© CENGAGE LEARNING 2012.

Analysis of Synchronous Counters

A 3-bit synchronous binary counter based on JK flip-flops is shown in *Figure 10.10* and available in Multisim file **10.10 3 bit binary counter.ms11**. Let us analyze its count sequence in detail so that we can see how the J and K inputs are affected by the Q outputs and how transitions between states are made. Later we will look at

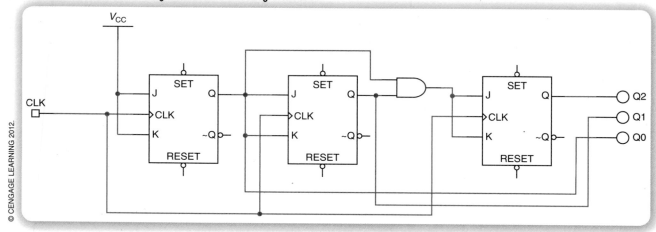

FIGURE 10.10 3-Bit Synchronous Binary Counter

TABLE 10.4 Function Table of a JK Flip-Flop

J	K	Q_{t+1}	Function
0	0	Q_t	No change
0	1	0	Reset
1	0	1	Set
1	1	$\overline{Q}_t$	Toggle

the function of truncated-sequence counter circuits and counters that are made from flip-flops other than JK.

The synchronous input equations are given by:

$$J_2 = K_2 = Q_1 \cdot Q_0$$
$$J_1 = K_1 = Q_0$$
$$J_0 = K_0 = 1$$

For reference, the JK flip-flop function table is shown in *Table 10.4*.

Q_t indicates the state of Q before a clock pulse is applied (i.e, the present state). Q_{t+1} indicates the state of Q after the clock pulse (i.e., the next state).

Assume that the counter output is initially $Q_2 Q_1 Q_1 = 000$. Before any clock pulses are applied, the J and K inputs are at the following states:

$$J_2 = K_2 = Q_1 \cdot Q_0 = 0 \cdot 0 = 0 \quad \text{(No change)}$$
$$J_1 = K_1 = Q_0 = 0 \quad \text{(No change)}$$
$$J_0 = K_0 = 1 \text{ (Constant)} \quad \text{(Toggle)}$$

The transitions of the outputs after the clock pulse are:

$$Q_2: 0 \rightarrow \quad \text{(No change)}$$
$$Q_1: 0 \rightarrow 0 \quad \text{(No change)}$$
$$Q_0: 0 \rightarrow 1 \quad \text{(Toggle)}$$

The output goes from $Q_2 Q_1 Q_1 = 000$ to $Q_2 Q_1 Q_1 = 001$ (see *Figure 10.11*). The transition is defined by the values of J and K *before* the clock pulse, because the propagation delays of the flip-flops prevent the new output conditions from changing the J and K values until after the transition.

FIGURE 10.11 Timing Diagram for a Synchronous 3-Bit Binary Counter

The new conditions of the J and K inputs are:

$$J_2 = K_2 = Q_1 \cdot Q_0 = 0 \cdot 1 = 0 \quad \text{(No change)}$$
$$J_1 = K_1 = Q_0 = 1 \quad \text{(Toggle)}$$
$$J_0 = K_0 = 1 \text{ (Constant)} \quad \text{(Toggle)}$$

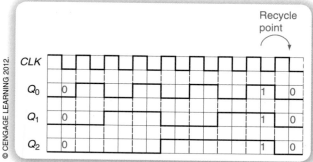

The transitions of the outputs generated by the second clock pulse are:

$$Q_2: 0 \rightarrow 0 \quad \text{(No change)}$$
$$Q_1: 0 \rightarrow 1 \quad \text{(Toggle)}$$
$$Q_0: 1 \rightarrow 0 \quad \text{(Toggle)}$$

The new output is $Q_2 Q_1 Q_0 = 010$, because both Q_0 and Q_1 change and Q_2 stays the same. The J and K conditions are now:

$$J_2 = K_2 = Q_1 \cdot Q_0 = 1 \cdot 0 = 0 \qquad \text{(No change)}$$
$$J_1 = K_1 = Q_0 = 0 \qquad \text{(No change)}$$
$$J_0 = K_0 = 1 \text{ (Constant)} \qquad \text{(Toggle)}$$

The output transitions are:

$$Q_2: 0 \to 0 \qquad \text{(No change)}$$
$$Q_1: 1 \to 1 \qquad \text{(No change)}$$
$$Q_0: 0 \to 1 \qquad \text{(Toggle)}$$

The output is now $Q_2 Q_1 Q_0 = 011$, which results in the JK conditions:

$$J_2 = K_2 = Q_1 \cdot Q_0 = 1 \cdot 1 = 1 \qquad \text{(Toggle)}$$
$$J_1 = K_1 = Q_0 = 1 \qquad \text{(Toggle)}$$
$$J_0 = K_0 = 1 \text{ (Constant)} \qquad \text{(Toggle)}$$

The above conditions result in output transitions:

$$Q_2: 0 \to 1 \qquad \text{(Toggle)}$$
$$Q_1: 1 \to 0 \qquad \text{(Toggle)}$$
$$Q_0: 1 \to 0 \qquad \text{(Toggle)}$$

All of the outputs toggle and the new output state is $Q_2 Q_1 Q_0 = 100$. The J and K values repeat this pattern in the second half of the counter cycle (states 100 to 111). Go through the exercise of calculating the J, K, and Q values for the rest of the cycle. Compare the result with the timing diagram in Figure 10.11.

In the counter we have just analyzed, the combinational circuit generates either a toggle ($JK = 11$) or a no change ($JK = 00$) state at each point through the count sequence. We could use any combination of JK modes (no change, reset, set, or toggle) to make the transitions from one state to the next. For instance, instead of using only the no change and toggle modes, the 000 → 001 transition could also be done by making Q_0 set ($J_0 = 1$, $K_0 = 0$) and Q_1 and Q_2 reset ($J_1 = 0, K_1 = 1$ and $J_2 = 0$, $K_2 = 1$). To do so we would need a different set of combinational logic in the circuit.

The simplest synchronous counter design uses only the no change ($JK = 00$) or toggle ($JK = 11$) modes because the J and K inputs of each flip-flop can be connected together. The no change and toggle modes allow us to make any transition (i.e., not just in a linear sequence), even though for truncated sequence and non-binary counters this is not usually the most efficient design.

There is a simple progression of algebraic expressions for the J and K inputs of a synchronous binary (full-sequence) counter, which uses only the no change and toggle states:

$$J_0 = K_0 = 1$$
$$J_1 = K_1 = Q_0$$
$$J_2 = K_2 = Q_1 \cdot Q_0$$
$$J_3 = K_3 = Q_2 \cdot Q_1 \cdot Q_0$$
$$J_4 = K_4 = Q_3 \cdot Q_2 \cdot Q_1 \cdot Q_0 \text{ and so on}$$

The J and K inputs of each stage are the ANDed outputs of all previous stages. This implies that a flip-flop toggles only when the outputs of *all* previous stages are HIGH. For example, Q_2 doesn't change unless *both* Q_1 AND Q_0 are HIGH (and therefore $J_2 = K_2 = 1$) before the clock pulse. In a 3-bit counter, this occurs only at states 011 and 111, after which Q_2 will toggle, along with Q_1 and Q_0, giving transitions to states 100 and 000 respectively. Look at the timing diagram of Figure 10.11 to confirm this.

Determining the Modulus of a Synchronous Counter

We can use a more formal technique to analyze any synchronous counter, as follows.

1. Determine the equations for the synchronous inputs (JK, D, or T) in terms of the Q outputs for all flip-flops. (For counters other than straight binary full-sequence types, the equations will *not* be the same as the algebraic progressions previously listed.)
2. Lay out a table with headings for the present state of the counter (Q outputs before CLK pulse), each Synchronous Input before CLK pulse, and next state of the counter (Q outputs after the clock pulse).
3. Choose a starting point for the count sequence, usually 0, and enter the starting point in the Present State column.
4. Substitute the Q values of the initial present state into the synchronous input equations and enter the results under the appropriate columns.
5. Determine the action of each flip-flop on the next CLK pulse (e.g., for a JK flip-flop, the output either will not change [$JK = 00$], or will reset [$JK = 01$], set [$JK = 10$], or toggle [$JK = 11$]).
6. Look at the Q values for every flip-flop. Change them according to the function determined in Step 5 and enter them in the column for the counter's next state.
7. Enter the result from Step 6 on the next line of the column for the counter's present state (i.e., this line's next state is the next line's present state).
8. Repeat this process until the result in the next state column is the same as in the initial state.

Example 10.4

Multisim Example

Multisim File: 10.04 Example.ms11

Find the count sequence of the synchronous counter shown in *Figure 10.12*, and from the count sequence table, draw the timing diagram and state diagram. What is the modulus of the counter?

FIGURE 10.12 Example 10.4: *Synchronous Counter of Unknown Modulus*

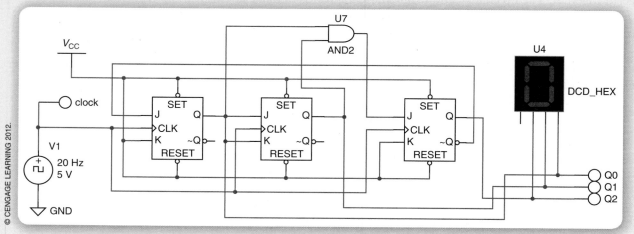

continued...

Solution

J and K equations are:

$$J_2 = Q_1 \cdot Q_0 \qquad J_1 = Q_0 \qquad J_0 = \overline{Q_2}$$
$$K_2 = 1 \qquad\qquad K_1 = Q_0 \qquad K_0 = 1$$

The output transitions can be determined from the values of the J and K functions before each clock pulse, as shown in *Table 10.5*. There are five unique output states, so the counter's modulus is 5.

The timing diagram and state diagram are shown in *Figure 10.13*. Because this circuit produces one pulse on Q_2 for every 5 clock pulses, we can use it as a divide-by-5 circuit.

TABLE 10.5 *State Table for Figure 10.12*

Present State			Synchronous Inputs									Next State		
Q_2	Q_1	Q_0	J_2	K_2		J_1	K_1		J_0	K_0		Q_2	Q_1	Q_0
0	0	0	0	1	(R)	0	0	(NC)	1	1	(T)	0	0	1
0	0	1	0	1	(R)	1	1	(T)	1	1	(T)	0	1	0
0	1	0	0	1	(R)	0	0	(NC)	1	1	(T)	0	1	1
0	1	1	1	1	(T)	1	1	(T)	1	1	(T)	1	0	0
1	0	0	0	1	(R)	0	0	(NC)	0	1	(R)	0	0	0

© CENGAGE LEARNING 2012.

FIGURE 10.13 **Example 10.4: Timing Diagram and State Diagram of a Mod-5 Counter**

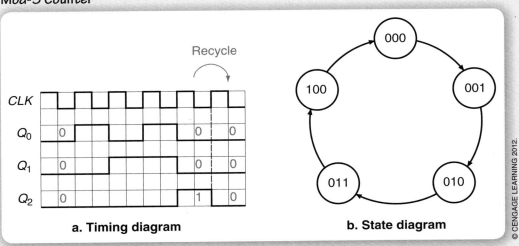

| **a. Timing diagram** | **b. State diagram** |

© CENGAGE LEARNING 2012.

The analysis in Example 10.4 did not account for the counter using only 5 of a possible 8 output states. In any truncated-sequence counter, it is good practice to determine the next state for each unused state to ensure that if the counter powers up in one of these unused states, it will eventually enter the main sequence.

An unused state might be problematic because a flip-flop is not necessarily guaranteed to power up in a particular state. The worst possible situation would be one such as where the counter flip-flops would power up in state 110, then make a transition to 111, then back to 110. The counter would be stuck between these two states and never enter the main count sequence. Therefore, we should check that the counter's combinational logic never lets it get stuck in one or more unused states.

Example 10.5

Extend the analysis of the counter in Example 10.4 to include its unused states. Redraw the counter's state diagram to show how these unused states enter the main sequence (if they do).

Solution

The synchronous input equations are:

$$J_2 = Q_1 \cdot Q_0 \qquad J_1 = Q_0 \qquad J_0 = \overline{Q_2}$$
$$K_2 = 1 \qquad K_1 = Q_0 \qquad K_0 = 1$$

The unused states are $Q_2 Q_1 Q_0 = 101$, 110, and 111. *Table 10.6* shows the transitions made by the unused states. *Figure 10.14* shows the complete state diagram.

TABLE 10.6 *State Table for Mod-5 Counter Including Unused States*

Present State			Synchronous Inputs									Next State		
Q_2	Q_1	Q_0	J_2	K_2		J_1	K_1		J_0	K_0		Q_2	Q_1	Q_0
0	0	0	0	1	(R)	0	0	(NC)	1	1	(T)	0	0	1
0	0	1	0	1	(R)	1	1	(T)	1	1	(T)	0	1	0
0	1	0	0	1	(R)	0	0	(NC)	1	1	(T)	0	1	1
0	1	1	1	1	(T)	1	1	(T)	1	1	(T)	1	0	0
1	0	0	0	1	(R)	0	0	(NC)	0	1	(R)	0	0	0
1	0	1	0	1	(R)	1	1	(T)	0	1	(R)	0	1	0
1	1	0	0	1	(R)	0	0	(NC)	0	1	(R)	0	1	0
1	1	1	1	1	(T)	1	1	(T)	0	1	(R)	0	0	0

© CENGAGE LEARNING 2012.

FIGURE 10.14 Example 10.5: *Complete State Diagram*

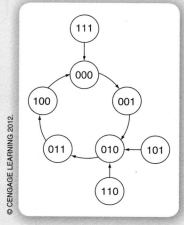

© CENGAGE LEARNING 2012.

Your Turn

10.2 A 4-bit synchronous counter based on JK flip-flops is described by the following set of equations:

$$J_3 = Q_2 Q_1 Q_0 \qquad J_2 = Q_1 Q_0 \qquad J_1 = \overline{Q_3} Q_0 \qquad J_0 = 1$$
$$K_3 = Q_0 \qquad K_2 = Q_1 Q_0 \qquad K_1 = Q_0 \qquad K_0 = 1$$

Assume that the counter output is at 1000 in the count sequence. What will the output be after 1 clock pulse? After 2 clock pulses?

10.3 DESIGN OF SYNCHRONOUS COUNTERS

KEY TERMS

State machine A synchronous sequential circuit.

Excitation table A table showing the required input conditions for every possible transition of a flip-flop output.

A synchronous counter can be designed using established techniques that involve the derivation of Boolean equations for the counter's next state logic. We will leave the state machine design for the following chapter.

Classical Design Technique

Several steps are involved in the classical design of a synchronous counter.

1. Define the problem. Before you can begin design of a circuit, you have to know what its purpose is and what it should do under all possible conditions.
2. Draw a state diagram showing the progression of states under various input conditions and what outputs the circuit should produce, if any.
3. Make a state table which lists all possible present states and the next state for each one. *List the present states in **binary order**.*
4. Use flip-flop excitation tables to determine at what states the flip-flop synchronous inputs must be to make the circuit go from each present state to its next state.
5. The logic levels of the synchronous inputs are Boolean functions of the flip-flop outputs and the control inputs. Simplify the expression for each input and write the simplified Boolean expression.
6. Use the Boolean expressions found in Step 5 to draw the required logic circuit.

Flip-Flop Excitation Tables

In the synchronous counter circuits we examined earlier in this chapter, we used JK flip-flops that were configured to operate only in toggle or no change mode. We can use any type of flip-flop for a synchronous sequential circuit. If we choose to use JK flip-flops, we can use any of the modes (no change, reset, set, or toggle) to make transitions from one state to another.

A flip-flop excitation table shows all possible transitions of a flip-flop output and the synchronous input levels needed to effect these transitions. *Table 10.7 is* the excitation table of a JK flip-flop.

TABLE 10.7 JK Flip-Flop Excitation Table

Transition	Function	JK	
0 → 0	No change or reset	00 01	0X
0 → 1	Toggle or set	11 10	1X
1 → 0	Toggle or reset	11 01	X1
1 → 1	No change or set	00 10	X0

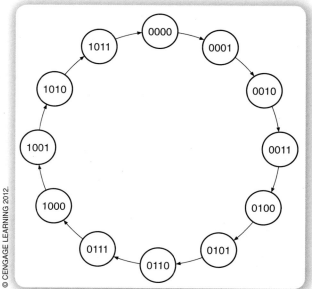

FIGURE 10.15 State Diagram for a Mod-12 Counter

TABLE 10.8 Condensed Excitation Table for a JK Flip-Flop

Transition	JK
0 → 0	0X
0 → 1	1X
1 → 0	X1
1 → 1	X0

If we want a flip-flop to make a transition from 0 to 1, we can use either the toggle function ($JK = 11$) or the set function ($JK = 10$). It doesn't matter what K is, as long as $J = 1$. This is reflected by the variable pair ($JK = 1X$) beside the $0 \rightarrow 1$ entry in Table 10.7. The X is a don't care state, a 0 or 1 depending on which is more convenient for the simplification of the Boolean function of the J or K input affected.

Table 10.8 shows a condensed version of the JK flip-flop excitation table.

Design of a Synchronous Mod-12 Counter

We will follow the procedure previously outlined to design a synchronous mod-12 counter circuit, using JK flip-flops. The aim is to derive the Boolean equations of all J and K inputs and to draw the counter circuit.

1. Define the problem. The circuit must count in binary sequence from 0000 to 1011 and repeat. The output progresses by 1 for each applied clock pulse. The outputs are 4-bit numbers, so we require 4 flip-flops.
2. Draw a state diagram. The state diagram for this problem is shown in *Figure 10.15*.
3. Make a state table showing each present state and the corresponding next state.
4. Use flip-flop excitation tables to fill in the J and K entries in the state table. *Table 10.9* shows the combined result of Steps 3 and 4. Note that all present states are in binary order.

We assume for now that states 1100 to 1111 never occur. If we assign their corresponding next states to be don't care states, they can be used to simplify the J and K expressions that we derive from the state table.

Let us examine one transition to show how the table is completed. The transition from $Q_3Q_2Q_1Q_0 = 0101$ to $Q_3Q_2Q_1Q_0 = 0110$ consists of the following individual flip-flop transitions.

$$Q_3: 0 \rightarrow 0 \quad \text{(No change or reset;} \quad J_3K_3 = 0X)$$
$$Q_2: 1 \rightarrow 1 \quad \text{(No change or set;} \quad J_2K_2 = X0)$$
$$Q_1: 0 \rightarrow 1 \quad \text{(Toggle or set;} \quad J_1K_1 = 1X)$$
$$Q_0: 1 \rightarrow 0 \quad \text{(Toggle or reset;} \quad J_0K_0 = X1)$$

The other lines of the table are similarly completed.

TABLE 10.9   State Table for a Mod-12 Counter

Present State				Next State				Synchronous Inputs							
Q_3	Q_2	Q_1	Q_0	Q_3	Q_2	Q_1	Q_0	J_3	K_3	J_2	K_2	J_1	K_1	J_0	K_0
0	0	0	0	0	0	0	1	0	X	0	X	0	X	1	X
0	0	0	1	0	0	1	0	0	X	0	X	1	X	X	1
0	0	1	0	0	0	1	1	0	X	0	X	X	0	1	X
0	0	1	1	0	1	0	0	0	X	1	X	X	1	X	1
0	1	0	0	0	1	0	1	0	X	X	0	0	X	1	X
0	1	0	1	0	1	1	0	0	X	X	0	1	X	X	1
0	1	1	0	0	1	1	1	0	X	X	0	X	0	1	X
0	1	1	1	1	0	0	0	1	X	X	1	X	1	X	1
1	0	0	0	1	0	0	1	X	0	0	X	0	X	1	X
1	0	0	1	1	0	1	0	X	0	0	X	1	X	X	1
1	0	1	0	1	0	1	1	X	0	0	X	X	0	1	X
1	0	1	1	0	0	0	0	X	1	0	X	X	1	X	1
1	1	0	0	X	X	X	X	X	X	X	X	X	X	X	X
1	1	0	1	X	X	X	X	X	X	X	X	X	X	X	X
1	1	1	0	X	X	X	X	X	X	X	X	X	X	X	X
1	1	1	1	X	X	X	X	X	X	X	X	X	X	X	X

5. Simplify the Boolean expression for each input. Table 10.9 can be treated as eight truth tables, one for each J or K input. We can simplify each function by using a Karnaugh map.

   Figure 10.16 shows K-map simplification for all eight synchronous inputs. These maps yield the following simplified Boolean expressions.

$$J_0 = 1$$
$$K_0 = 1$$
$$J_1 = Q_0$$
$$K_1 = Q_0$$
$$J_2 = \overline{Q_3}Q_1Q_0$$
$$K_2 = Q_1Q_0$$
$$J_3 = Q_2Q_1Q_0$$
$$K_3 = Q_1Q_0$$

6. Draw the required logic circuit. *Figure 10.17* shows the circuit corresponding to the above Boolean expressions. Open Multisim file **10.17 Mod-12 Up Counter.ms11** to try the Mod-12 counter.

We have assumed that states 1100 to 1111 will never occur in the operation of the mod-12 counter. This is normally the case, but when the circuit is powered up, there is no guarantee that the flip-flops will be in any particular state.

If a counter powers up in an unused state, the circuit should enter the main sequence after one or more clock pulses. To test whether or not this happens, let us make a state table, applying each unused state to the J and K equations as

FIGURE 10.16 K-Map Simplification of Table 10.9

Q_3Q_2 \ Q_1Q_0	00	01	11	10
00	0	0	0	0
01	0	0	1	0
11	X	X	X	X
10	X	X	X	X

J_3

Q_3Q_2 \ Q_1Q_0	00	01	11	10
00	X	X	X	X
01	X	X	X	X
11	X	X	X	X
10	0	0	1	0

K_3

Q_3Q_2 \ Q_1Q_0	00	01	11	10
00	0	0	1	0
01	X	X	X	X
11	X	X	X	X
10	0	0	0	0

J_2

Q_3Q_2 \ Q_1Q_0	00	01	11	10
00	X	X	X	X
01	0	0	1	0
11	X	X	X	X
10	X	X	X	X

K_2

Q_3Q_2 \ Q_1Q_0	00	01	11	10
00	0	1	X	X
01	0	1	X	X
11	X	X	X	X
10	0	1	X	X

J_1

Q_3Q_2 \ Q_1Q_0	00	01	11	10
00	X	X	1	0
01	X	X	1	0
11	X	X	X	X
10	X	X	1	0

K_1

Q_3Q_2 \ Q_1Q_0	00	01	11	10
00	1	X	X	1
01	1	X	X	1
11	X	X	X	X
10	1	X	X	1

J_0

Q_3Q_2 \ Q_1Q_0	00	01	11	10
00	X	1	1	X
01	X	1	1	X
11	X	X	X	X
10	X	1	1	X

K_0

FIGURE 10.17 Synchronous Mod-12 Counter

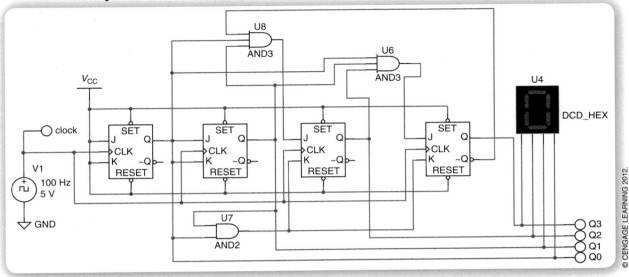

TABLE 10.10 Unused States in a Mod-12 Counter

Present State				Synchronous Inputs								Next State			
Q_3	Q_2	Q_1	Q_0	J_3	K_3	J_2	K_2	J_1	K_1	J_0	K_0	Q_3	Q_2	Q_1	Q_0
1	1	0	0	0	0	0	0	0	0	1	1	1	1	0	1
1	1	0	1	0	0	0	0	1	1	1	1	1	1	1	0
1	1	1	0	0	0	0	0	0	0	1	1	1	1	1	1
1	1	1	1	1	1	0	1	1	1	1	1	0	0	0	0

implemented, to see what the next state is for each case. This analysis is shown in *Table 10.10.*

Figure 10.18 shows the complete state diagram for the designed mod-12 counter. If the counter powers up in an unused state, it will enter the main sequence in no more than 4 clock pulses.

If we want an unused state to make a transition directly to 0000 in 1 clock pulse, we have a couple of options:

1. We could reset the counter asynchronously and otherwise leave the design as is.
2. We could rewrite the state table to specify these transitions, rather than make the unused states don't cares.

Option 1 is simpler and is considered perfectly acceptable as a design practice. Option 2 would yield a more complicated set of Boolean equations and hence a more complex circuit, but might be worthwhile if a direct synchronous transition to 0000 were required.

FIGURE 10.18 Complete State Diagram of Mod-12 Counter in Figure 10.17

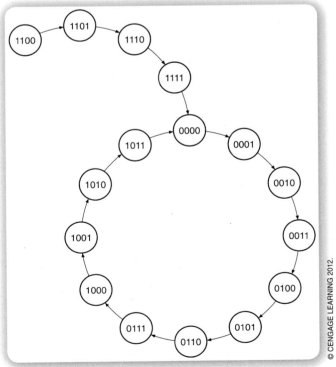

Example 10.6

Derive the synchronous input equations of a 4-bit synchronous binary counter based on D flip-flops. Draw the corresponding counter circuit.

TABLE 10.11 Excitation Table of a D Flip-Flop

Transition	D
$0 \rightarrow 0$	0
$0 \rightarrow 1$	1
$1 \rightarrow 0$	0
$1 \rightarrow 1$	1

© CENGAGE LEARNING 2012.

Solution

The first step in the counter design is to derive the excitation table of a D flip-flop. Recall that Q follows D when the flip-flop is clocked. Therefore, the next state of Q is the same as the input D for any transition. This is illustrated in *Table 10.11*.

Next, we must construct a state table, shown in *Table 10.12*, with present and next states for all possible transitions. Note that the binary value of $D_3 D_2 D_1 D_0$ is the same as the next state of the counter.

This state table yields four Boolean equations, for D_3 through D_0, in terms of the present state outputs. *Figure 10.19* shows four Karnaugh maps used to simplify these functions.

The simplified equations are:

$$D_3 = \overline{Q_3} Q_2 Q_1 Q_0 + Q_3 \overline{Q_2} + Q_3 \overline{Q_1} + Q_3 \overline{Q_0}$$
$$D_2 = \overline{Q_2} Q_1 Q_0 + Q_2 \overline{Q_1} + Q_1 \overline{Q_0}$$
$$D_1 = \overline{Q_1} Q_0 + Q_1 \overline{Q_0}$$
$$D_0 = \overline{Q_0}$$

These equations represent the maximum SOP simplifications of the input functions. However, we can rewrite them to make them more compact. For example, the equation for D_3 can be rewritten using DeMorgan's theorem $(\overline{x} + \overline{y} + \overline{z} = xyz)$ and our knowledge of Exclusive OR (XOR) functions $(\overline{x}y + x\overline{y} = x \oplus y)$.

$$D_3 = \overline{Q_3} Q_2 Q_1 Q_0 + Q_3 \overline{Q_2} + Q_3 \overline{Q_1} + Q_3 \overline{Q_0}$$
$$= \overline{Q_3}(Q_2 Q_1 Q_0) + Q_3(\overline{Q_2} + \overline{Q_1} + \overline{Q_0})$$
$$= \overline{Q_3}(Q_2 Q_1 Q_0) + Q_3(\overline{Q_2 Q_1 Q_0})$$
$$= Q_3 \oplus Q_2 Q_1 Q_0$$

TABLE 10.12 State Table for a 4-Bit Binary Counter

Present State				Next State				Synchronous Inputs			
Q_3	Q_2	Q_1	Q_0	Q_3	Q_2	Q_1	Q_0	D_3	D_2	D_1	D_0
0	0	0	0	0	0	0	1	0	0	0	1
0	0	0	1	0	0	1	0	0	0	1	0
0	0	1	0	0	0	1	1	0	0	1	1
0	0	1	1	0	1	0	0	0	1	0	0
0	1	0	0	0	1	0	1	0	1	0	1
0	1	0	1	0	1	1	0	0	1	1	0
0	1	1	0	0	1	1	1	0	1	1	1
0	1	1	1	1	0	0	0	1	0	0	0
1	0	0	0	1	0	0	1	1	0	0	1
1	0	0	1	1	0	1	0	1	0	1	0
1	0	1	0	1	0	1	1	1	0	1	1
1	0	1	1	1	1	0	0	1	1	0	0
1	1	0	0	1	1	0	1	1	1	0	1
1	1	0	1	1	1	1	0	1	1	1	0
1	1	1	0	1	1	1	1	1	1	1	1
1	1	1	1	0	0	0	0	0	0	0	0

© CENGAGE LEARNING 2012.

We can write similar equations for the other D inputs as follows:

$$D_2 = Q_2 \oplus Q_1 Q_0$$
$$D_1 = Q_1 \oplus Q_0$$
$$D_0 = Q_0 \oplus 1$$

These equations follow a predictable pattern of expansion. Each equation for an input D_n is simply Q_n XORed with the logical product (AND) of all previous Qs.

Figure 10.20 shows the circuit for the 4-bit counter, including an asynchronous reset.

continued...

FIGURE 10.19 Example 10.6: K-Maps for a 4-Bit Counter Based on D Flip-Flops

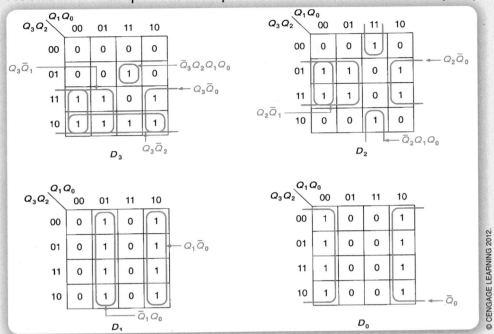

FIGURE 10.20 Example 10.6: 4-Bit Counter Using D Flip-Flops

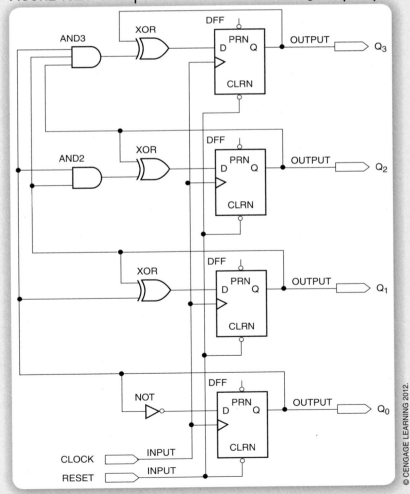

We have previously seen how a D flip-flop could be configured for a switchable toggle function. The flip-flops in Figure 10.20 are similarly configured. Each flip-flop output, except Q_0, is fed back to its input through an Exclusive OR gate. The other input to the XOR controls whether this feedback is inverted (for toggle mode) or not (for no change mode). Recall that $x \oplus 0 = x$ and $x \oplus 1 = \bar{x}$.

For example, Q_3 is fed back to D_3 through an XOR gate. The feedback is inverted only if the 3-input AND gate has a HIGH output. Thus, the Q_3 output toggles only if all previous bits are HIGH ($Q_3Q_2Q_1Q_0 = 0111$ or 1111). The flip-flop toggle mode is therefore controlled by the states of the XOR and AND gates in the circuit.

Your Turn

10.3 A 4-bit synchronous counter must make a transition from state $Q_3Q_2Q_1Q_0 = 1011$ to $Q_3Q_2Q_1Q_0 = 1100$. Write the required states of the synchronous inputs for a set of four JK flip-flops used to implement the counter. Write the required states of the synchronous inputs if the counter is made from D flip-flops.

10.4 CONTROL OPTIONS FOR SYNCHRONOUS COUNTERS

KEY TERMS

Parallel load A function that allows simultaneous loading of binary values into all flip-flops of a synchronous circuit. Parallel loading can be synchronous or asynchronous.

Clear Reset (synchronous or asynchronous).

Count enable A control function that allows a counter to progress through its count sequence when active and disables the counter when inactive.

Bidirectional counter A counter that can count up or down, depending on the state of a control input.

Output decoding A feature in which one or more outputs activate when a particular counter state is detected.

Ripple carry out or ripple clock out (RCO) An output that produces one pulse with the same period as the clock upon terminal count.

Terminal count The last state in a count sequence before the sequence repeats (for example, 1111 is the terminal count of a 4-bit binary UP counter; 0000 is the terminal count of a 4-bit binary DOWN counter).

Presettable counter A counter with a parallel load function.

Synchronous counters can be designed with several features other than just simple counting. Some of the most common features include:

▶ Synchronous or asynchronous **parallel load**, which allows the count to be set to any value whenever a *LOAD* input is asserted

▶ Synchronous or asynchronous **clear** (reset), which sets all of the counter outputs to zero

- **Count enable**, which allows the count sequence to progress when asserted and inhibits the count when deasserted
- **Bidirectional** control, which determines whether the counter counts up or down
- **Output decoding**, which activates one or more outputs when detecting particular states on the counter outputs
- **Ripple carry out or ripple clock out (RCO)**, a special case of output decoding that produces a pulse upon detecting the **terminal count**, or last state, of a count sequence.

Parallel Loading

Figure 10.21 shows the symbol of a 4-bit **presettable counter** (i.e., a counter with a parallel load function). The parallel inputs, *A* to *D*, have direct access to the flip-flops of the counter. When the *~LOAD* input is asserted (LOW), the values at the *A–D* inputs are loaded directly into the counter and appear at the *QA–QD* outputs.

Parallel loading can be synchronous or asynchronous. When *~LOAD* goes LOW, the value on *A–D* is loaded into an asynchronously loading counter immediately. The counter with synchronous load is loaded on the next positive clock edge.

Synchronous Load

The logic diagram of *Figure 10.22* shows the concept of synchronous parallel load. Depending on the status of the *LOAD* input, the flip-flop will either count according to its count logic (the next-state combinational circuit) or load an external value. The flip-flop shown is the most significant bit of a 4-bit binary counter, such as shown in Figure 10.20, but with the count logic represented only by an input pin.

The *LOAD* input selects whether the flip-flop synchronous input will be fed by the count logic or by the parallel input P_3. When *LOAD* = 0, the upper AND gate steers the count logic to the flip-flop, and the count progresses with each clock pulse. When *LOAD* = 1, the lower AND gate loads the logic level at P_3 directly into the flip-flop on the next clock pulse.

Asynchronous Load

The asynchronous load function of a counter makes use of the asynchronous preset and clear inputs of the counter's flip-flops. *Figure 10.23* shows

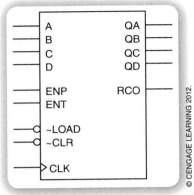
FIGURE 10.22 **Synchronous Count/Load Selection** © CENGAGE LEARNING 2012.

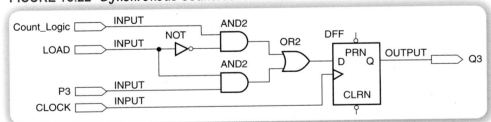

FIGURE 10.23 **Asynchronous Load Element**

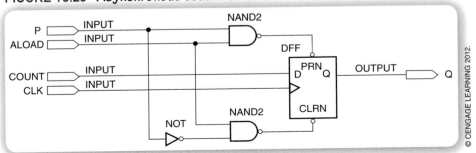

© CENGAGE LEARNING 2012.

the circuit implementation of the asynchronous load function, without any count logic.

When *ALOAD* (Asynchronous LOAD) is HIGH, both NAND gates in Figure 10.23 are enabled. If the *P* input is HIGH, the output of the upper NAND gate goes LOW, activating the flip-flop's asynchronous *PRESET* input, thus setting *Q* = 1. The lower NAND gate has a HIGH output, thus deactivating the flip-flop's *CLEAR* input.

If *P* is LOW the situation is reversed. The upper NAND output is HIGH and the lower NAND has a LOW output, activating the flip-flop's *CLEAR* input, resetting *Q*. Thus, *Q* will be the same value as *P* when the *ALOAD* input is asserted. When *ALOAD* is not asserted (= 0), both NAND outputs are HIGH and thus do not activate either the preset or clear function of the flip-flop.

Figure 10.24 shows the asynchronous load circuit with an asynchronous clear (reset) function added. The flip-flop can be cleared by a logic LOW either from the *P* input (via the lower NAND gate) or the *CLEAR* input pin. The clear function disables the upper NAND gate when it is LOW, preventing the flip-flop from being cleared and preset simultaneously. This extra connection also ensures that the clear function has priority over the load function.

FIGURE 10.24 *Asynchronous Load Element with Asynchronous Clear*

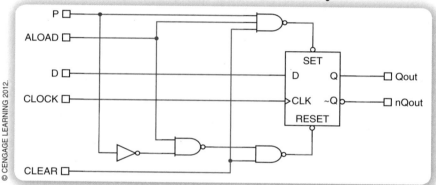

© CENGAGE LEARNING 2012.

Example 10.7

Multisim Example

Multisim File: 10.24 Async Load Element with Async Clr.ms11

Open the Multisim file for the circuit shown in Figure 10.24, a Multisim version of the circuit element with asynchronous load and clear. Run the file as a simulation and operate the switches for inputs P and D and controls A and C.

  a. With A, C, and P all set to LOW (inactive), describe the output Q as D is changed.

  b. Activate A (Asynchronous LOAD). Describe the output as the input P is changed, then as the input D is changed.

  c. Activate C, and describe the output as P and D are changed.

Solution

  a. The output becomes the value of input D on the clock pulse. This is a synchronous output.

  b. The output changes immediately to the value of P if A is active—this is an asynchronous change. D has no effect on the output if A is active.

  c. The output clears immediately (asynchronously) if C is active.

Other Control Lines

Count Enable

If we want to stop (or pause) the count sequence, we must disable the count logic of the counter circuit. Deactivating COUNT ENABLE will leave the outputs in their current state—although the clock signal continues, the output does not change state. Note that the count enable has no effect on the synchronous load and asynchronous reset functions; for details on whether COUNT ENABLE must be active for a synchronous or asynchronous LOAD, the specific counter to be used must be investigated using a datasheet.

Bidirectional Counters

Figure 10.25 shows the logic diagram of a 4-bit synchronous DOWN counter. Its count sequence starts at 1111 and counts backward to 0000, then repeats. The Boolean equations for this circuit will not be derived at this time, but will be left for an exercise in an end-of-chapter problem.

We can intuitively analyze the operation of the counter if we understand that the upper three flip-flops will each toggle when their associated XOR gates have a HIGH input from the rest of the count logic.

Q_0 is set to toggle on each clock pulse. Q_1 toggles whenever Q_0 is LOW (every second clock pulse, at states 1110, 1100, 1010, 1000, 0110, 0100, 0010, and 0000).

FIGURE 10.25 4-Bit Synchronous DOWN Counter

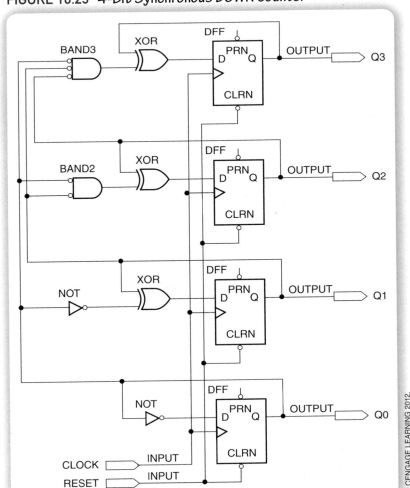

FIGURE 10.26 *Synchronous Counter Element*

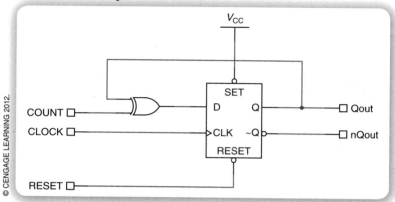

Q_2 toggles when Q_1 AND Q_0 are LOW (1100, 1000, 0100, and 0000). Q_3 toggles when Q_2 AND Q_1 AND Q_0 are LOW (1000 and 0000).

We can create a bidirectional counter by including a circuit to select count logic for an UP or DOWN sequence. *Figure 10.26* shows a basic synchronous counter element that can be used to create a synchronous counter. This is built in the file **10.26 Synchronous Counter Element.ms11**, a Multisim hierarchical block. The element is simply a D flip-flop configured for switchable toggle mode.

Four of these elements can be combined with selectable count logic to make a 4-bit bidirectional counter, as shown in *Figure 10.27*. Each counter element has a pair of AND-shaped gates and an OR gate to steer the count logic to the XOR in

FIGURE 10.27 *4-Bit Bidirectional Counter*

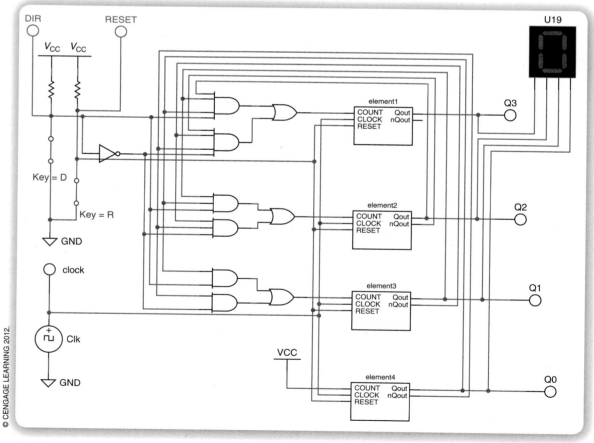

the element. When $DIR = 1$, the upper gate in each pair is enabled and the lower gates disabled, steering the UP count logic to the counter element. When $DIR = 0$, the lower gate in each pair is enabled, steering the DOWN count logic to the counter element. The directional function can also be combined with the load and count enable functions, as was shown for unidirectional UP counters.

Example 10.8

Multisim Example

Multisim File: 10.27 4 Bit Bidirectional Counter.ms11

Open the Multisim file for this example. The design, shown in Figure 10.27, is a Multisim version of a 4-bit bidirectional counter with an active-LOW reset. Run a simulation and operate the pushbuttons to change the direction of the count (if switch D is open, the counter counts up; if D is closed, the counter counts down) and test the RESET function (note that this is an active-LOW RESET).

a. What is the MOD for this counter?
b. What modification would be necessary to let us SET the counter as well as RESET?

Solution

a. This is a MOD-16 counter.
b. We could add an input to the hierarchical block to the SET input of the D flip-flop, then connect that to another switch in the main schematic.

Terminal Count and RCO

A special case of output decoding is a circuit that will detect the **terminal count**, or last state, of a count sequence and activate an output to indicate this state. The terminal count depends on the count sequence. A 4-bit binary UP counter has a terminal count of 1111; a 4-bit binary DOWN counter has a terminal count of 0000. A circuit to detect these conditions must detect the *maximum value* of an UP count and the *minimum value* of a DOWN count.

The decoder shown in *Figure 10.28* fulfills both of these conditions. The directional input *DIR* enables the upper gate when HIGH and the lower gate when

FIGURE 10.28 Terminal Count Decoder for a 4-Bit Bidirectional Counter

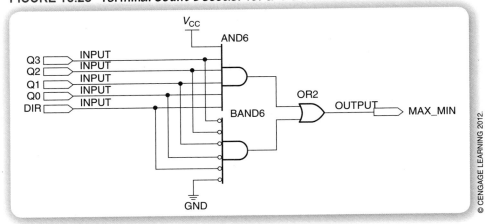

© CENGAGE LEARNING 2012.

FIGURE 10.29 *Counter Expansion Using RCO*

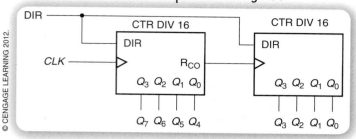

LOW. Thus, the upper gate generates a HIGH output when $DIR = 1$ AND $Q_3Q_2Q_1Q_0 = 1111$. The lower gate generates a HIGH when $DIR = 0$ AND $Q_3Q_2Q_1Q_0 = 0000$.

This function is generally found in counters with a fixed number of bits (i.e., fixed-function counter chips, not PLDs) and is used to asynchronously clock a further counter stage, as in *Figure 10.29*. This allows us to extend the width of the counter beyond the number of bits available in the fixed-function device. This is not necessary when designing synchronous counters using programmable logic, but is included for the sake of completeness.

Your Turn

10.4 *Figure 10.30* shows two presettable counters, one with asynchronous load and clear, the other with synchronous load and clear. The counter with asynchronous functions has a 4-bit output labeled *QA*. The synchronously loaded counter has a 4-bit output labeled *QS*. The load *and* reset inputs to both counters are active-LOW.

Figure 10.31 shows a partial timing diagram for the counters. Complete the diagram.

FIGURE 10.30 *Your Turn Two Presettable Counters*

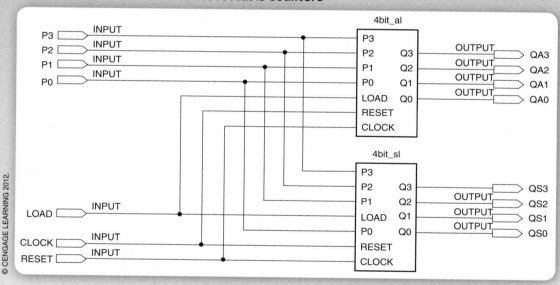

FIGURE 10.31 *Your Turn Problem 10.4: Timing Diagram for Counters in Figure 10.30*

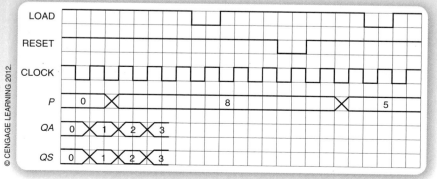

10.5 SHIFT REGISTERS

KEY TERMS

Shift register A synchronous sequential circuit that will store and move n-bit data, either serially or in parallel, in n flip-flops.

SRGn Abbreviation for an n-bit shift register (e.g., SRG4 indicates a 4-bit shift register).

Serial shifting Movement of data from one end of a shift register to the other at a rate of one bit per clock pulse.

Parallel transfer Movement of data into all flip-flops of a shift register at the same time.

Rotation Serial shifting of data with the output(s) of the last flip-flop connected to the synchronous input(s) of the first flip-flop. The result is continuous circulation of the same data.

Right shift A movement of data from the left to the right in a shift register. This is usually defined as data moving from the MSB toward the LSB.

Left shift A movement of data from the right to the left in a shift register. This is usually defined as moving from the LSB toward the MSB.

Bidirectional shift register A shift register that can serially shift bits left or right according to the state of a direction control input.

Parallel-load shift register A shift register that can be preset to any value by directly loading a binary number into its internal flip-flops.

Universal shift register A shift register that can operate with any combination of serial and parallel inputs and outputs (i.e., serial in/serial out, serial in/parallel out, parallel in/serial out, parallel in/parallel out). A universal shift register is often bidirectional, as well.

A **shift register** (abbreviated **SRGn** for an n-bit circuit) is a synchronous sequential circuit used to store or move data. It consists of several flip-flops, connected so that data are transferred into and out of the flip-flops in a standard pattern.

Figure 10.32 represents three types of data movement in three 4-bit shift registers. The circuits each contain four flip-flops, configured to move data in one of the ways shown.

Figure 10.32a shows the operation of **serial shifting**. The stored data are taken in one bit at a time from the input and moved one position toward the output with each applied clock pulse.

Parallel transfer is illustrated in Figure 10.32b. As with the synchronous parallel load function of a presettable counter, data move simultaneously into all flip-flops when a clock pulse is applied. The data are available in parallel at the register outputs.

Rotation, depicted in Figure 10.32c, is similar to serial shifting in that data are shifted one place to the right with each clock pulse. In this operation, however, data are continuously circulated in the shift register by moving the rightmost bit back to the leftmost flip-flop with each clock pulse.

Serial Shift Registers

Figure 10.33 shows the most basic shift register circuit: the serial shift register, so called because data are shifted through the circuit in a linear or serial fashion. The circuit shown consists of four D flip-flops connected in cascade and clocked synchronously.

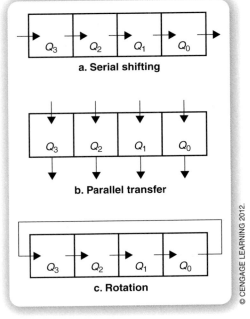

FIGURE 10.32 *Data Movement in a 4-Bit Shift Register*

a. Serial shifting

b. Parallel transfer

c. Rotation

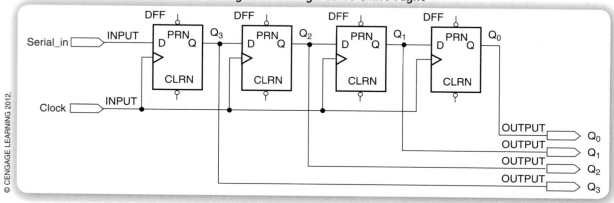

FIGURE 10.33 4-Bit Serial Shift Register Configured to Shift Right

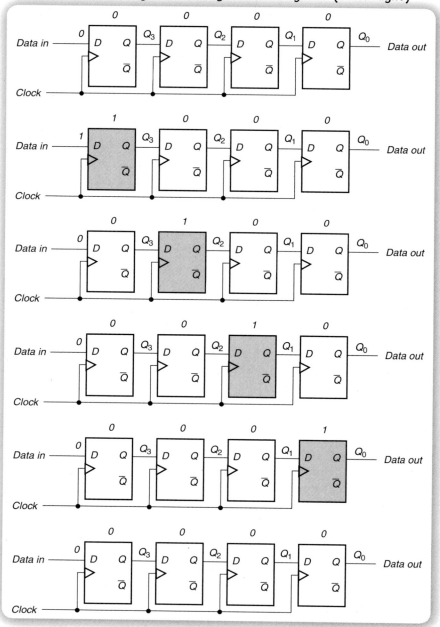

FIGURE 10.34 Shifting a "1" through a Shift Register (Shift Right)

For a D flip-flop, Q follows D. The value of a bit stored in any flip-flop *after* a clock pulse is the same as the bit in the flip-flop to its left *before* the pulse. As a result, when a clock pulse is applied to the circuit, the contents of the flip-flops move one position to the right and the bit at the circuit input is shifted into Q_3. The bit stored in Q_0 is overwritten by the former value of Q_1 and is lost. Because the data move from left to right, we say that the shift register implements a **right shift** function. (Data movement in the other direction, requiring a different circuit connection, is called **left shift**.)

Let us track the progress of data through the circuit in two cases. All flip-flops are initially cleared in each case.

Case 1: A 1 is clocked into the shift register, followed by a string of 0's, as shown in *Figure 10.34*. The flip-flop containing the 1 is shaded.

Before the first clock pulse, all flip-flops are filled with 0's. Data In goes to a 1 and on the first clock pulse, the 1 is clocked into the first flip-flop. After that, the input goes to 0. The 1 moves one position right with each clock pulse, the register filling up with 0's behind it, fed by the 0 at Data In. After four clock pulses, the 1 reaches the Data Out flip-flop. On the fifth pulse, the 0 coming behind overwrites the 1 at Q_0, leaving the register filled with 0's.

Case 2: *Figure 10.35* shows a shift register, initially cleared, being filled with 1's.

FIGURE 10.35 Filling a Shift Register with 1's (Shift Right)

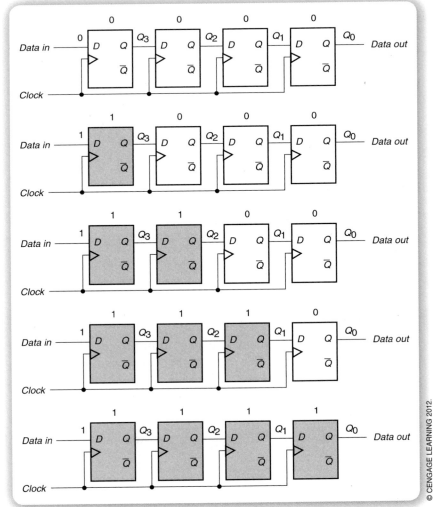

© CENGAGE LEARNING 2012.

Note . . .

Conventions differ about whether the rightmost or leftmost bit in a shift register should be considered the most significant bit. We will follow a convention where the leftmost bit is the MSB. The convention has no physical meaning; the concept of right or left shift only makes sense on a logic diagram. The actual flip-flops may be laid out in any configuration at all in the physical circuit and still implement the right or left shift functions as defined on the logic diagram. (That is to say, wires, circuit board traces, and internal programmable logic connections can run wherever you want; left and right are defined on the logic diagram.)

As before, the initial 1 is clocked into the shift register and reaches the Data Out line on the fourth clock pulse. This time, the register fills up with 1's, not 0's, because the data input remains HIGH.

Example 10.9

Multisim Example

Multisim File: 10.06 Example with Indicators.ms11

Use Multisim to create the logic diagram of a 4-bit serial shift register that shifts left, rather than right.

Solution

Figure 10.36 shows the required logic diagram. File **10.06 example with indicators.ms11** shows the schematic with a clock and input switch controlling the indicators at each output. The flip-flops are laid out the same way as in Figure 10.33, with the MSB (Q_3) on the left. The D input of each flip-flop is connected to the Q output of the flip-flop to its right, resulting in a looped-back connection. A bit at D_0 is clocked into the rightmost flip-flop. Data in the other flip-flops are moved one place to the left. The bit in Q_2 overwrites Q_3. The previous value of Q_3 is lost.

FIGURE 10.36 **Example 10.9: 4-Bit Serial Shift Register Configured to Shift Left**

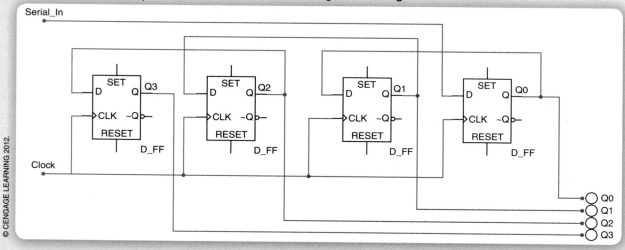

© CENGAGE LEARNING 2012.

Example 10.10

Draw a diagram showing the movement of a single 1 through the register in Figure 10.36. Also draw a diagram showing how the register can be filled up with 1's.

Solution

Figure 10.37 and *Figure 10.38* show the required data movements.

continued...

FIGURE 10.37 Example 10.10: Shifting a "1" through a Shift Register (Shift Left)

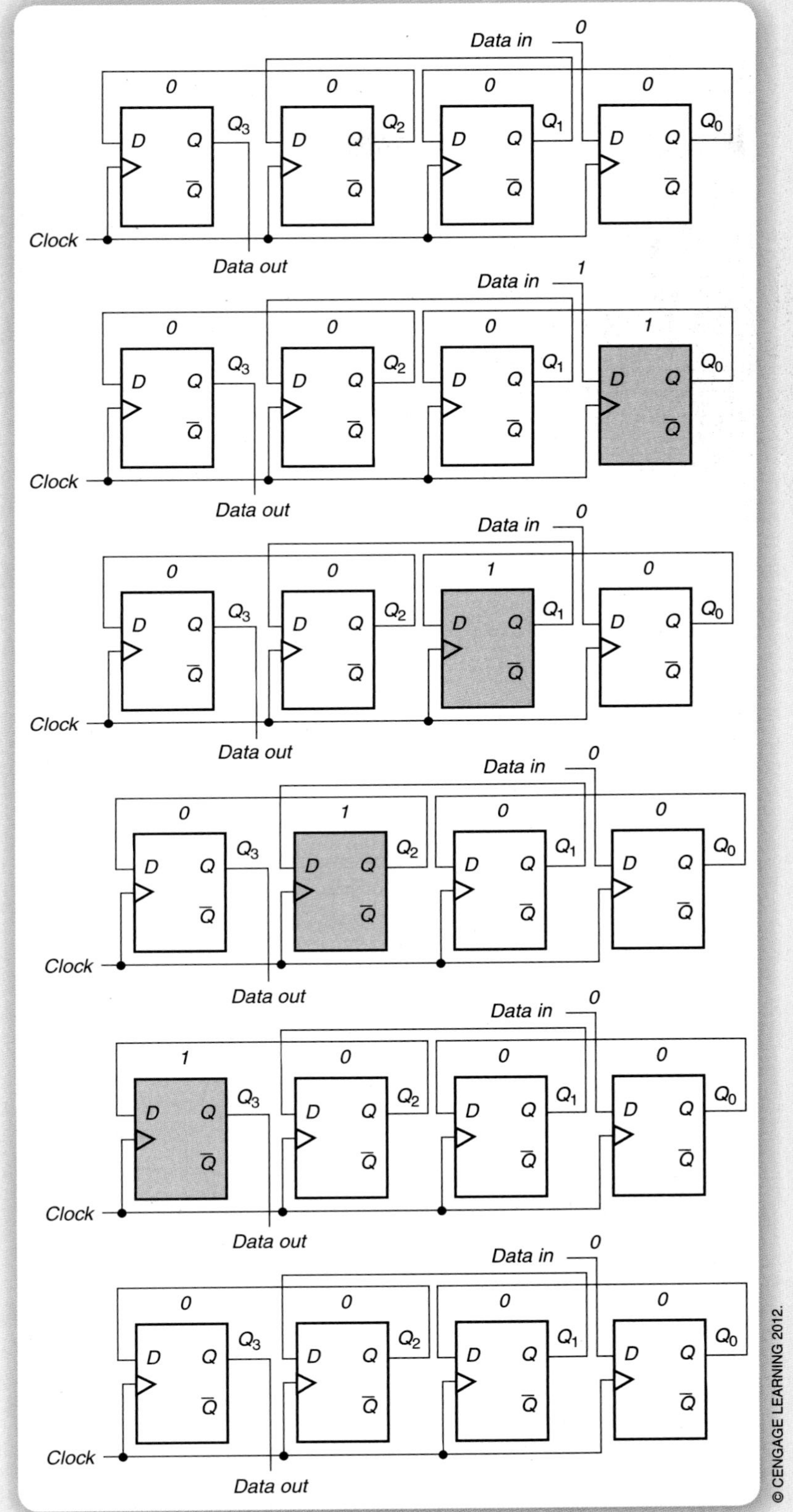

continued...

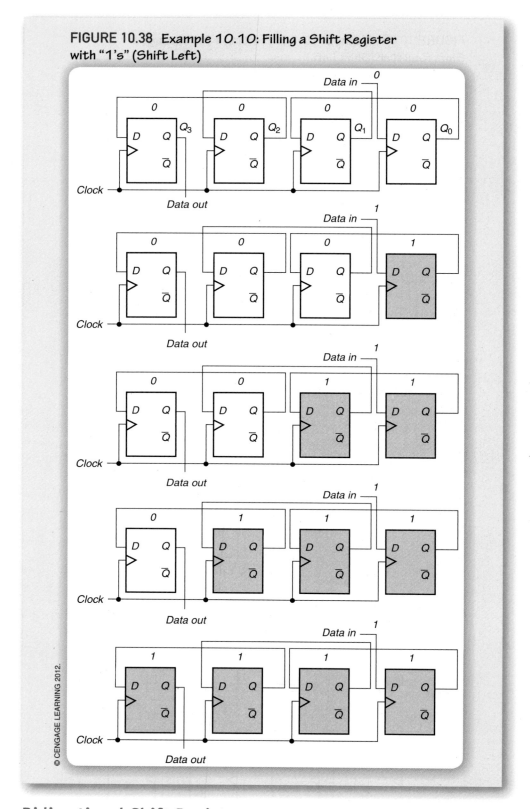

FIGURE 10.38 Example 10.10: Filling a Shift Register with "1's" (Shift Left)

Bidirectional Shift Registers

Figure 10.39 shows the logic diagram of a **bidirectional shift register**. This circuit combines the properties of the right shift and left shift circuits, seen earlier in Figure 10.33 and Figure 10.36. This circuit can serially move data right or left, depending on the state of a control input, called *DIRECTION*.

FIGURE 10.39 **Bidirectional Shift Register**

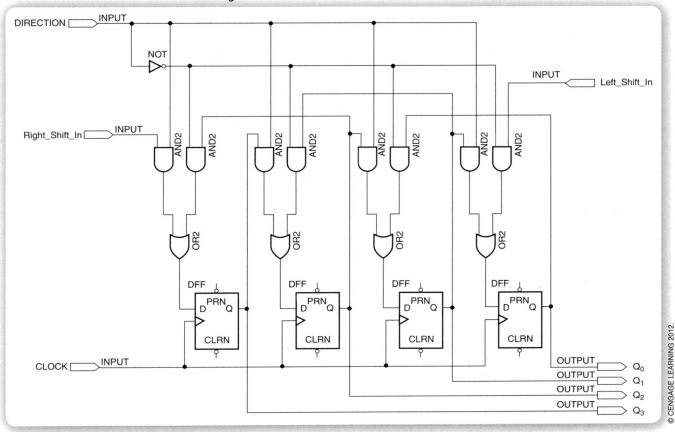

The shift direction is controlled by enabling or inhibiting four pairs of AND-OR circuit paths that direct the bits at the flip-flop outputs to other flip-flop inputs. When $DIRECTION = 0$, the right-hand AND gate in each pair is enabled and the flip-flop outputs are directed to the D inputs of the flip-flops one position left. Thus, the enabled pathway is from $Left\_Shift\_In$ to Q_0, then to Q_1, Q_2, and Q_3.

When $DIRECTION = 1$, the left-hand AND gate of each pair is enabled, directing the data from $Right\_Shift\_In$ to Q_3, then to Q_2, Q_1, and Q_0. Thus, $DIRECTION = 0$ selects left shift and $DIRECTION = 1$ selects right shift.

Shift Register with Parallel Load

Earlier in this chapter, we saw how a counter could be set to any value by synchronously loading a set of external inputs directly into the counter flip-flops. We can implement the same function in a **parallel-load shift register**, as shown in *Figure 10.40*.

The circuit is similar to that of the bidirectional shift register in Figure 10.39. The synchronous input of each flip-flop is fed by an AND-OR circuit that directs one of two signals to the flip-flop: the output of the previous flip-flop (shift function) or a parallel input (load function). The circuit is configured such that the shift function is enabled when $LOAD = 0$ and the load function is enabled when $LOAD = 1$. We could create a **universal shift register** by adding the capability to shift right or left to the shift register shown in Figure 10.40.

FIGURE 10.40 *Serial Shift Register with Parallel Load*

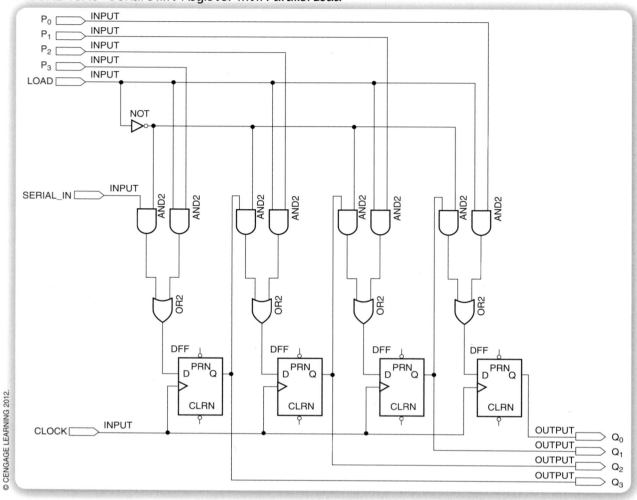

Your Turn

10.5 Can the D flip-flops in Figure 10.33 be replaced by JK flip-flops? If so, what modifications to the existing circuit are required?

10.6 SHIFT REGISTER COUNTERS

KEY TERMS

Ring counter A serial shift register with feedback from the output of the last flip-flop to the input of the first.

Johnson counter A serial shift register with complemented feedback from the output of the last flip-flop to the input of the first. Also called a twisted ring counter.

By introducing feedback into a serial shift register, we can create a class of synchronous counters based on continuous circulation, or rotation, of data.

If we feed back the output of a serial shift register to its input without inversion, we create a circuit called a **ring counter**. If we introduce inversion into the feedback loop, we have a circuit called a **Johnson counter**. These circuits can be decoded more easily than binary counters of similar size and are particularly useful for event sequencing.

Ring Counters

Figure 10.41 shows a 4-bit ring counter made from D flip-flops. This circuit could also be constructed from SR or JK flip-flops, as can any serial shift register.

FIGURE 10.41 **4-Bit Ring Counter**

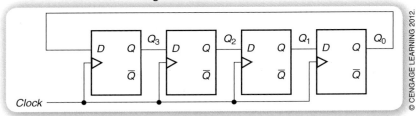

© CENGAGE LEARNING 2012.

A ring counter circulates the same data in a continuous loop. This assumes that the data have somehow been placed into the circuit upon initialization, usually by synchronous or asynchronous preset and clear inputs, which are not shown.

Figure 10.42 shows the circulation of a logic 1 through a 4-bit ring counter. If we assume that the circuit is initialized to the state $Q_3Q_2Q_1Q_0 = 1000$, it is easy to see that the 1 is shifted one place right with each clock pulse. The feedback connection from Q_0 to D_3 ensures that the input of flip-flop 3 will be filled by the contents of Q_0, thus recirculating the initial data. The final transition in the sequence shows the 1 recirculated to Q_3.

A ring counter is not restricted to circulating a logic 1. We can program the counter to circulate any data pattern we happen to find convenient.

Figure 10.43 shows a ring counter circulating a 0 by starting with an initial state of $Q_3Q_2Q_1Q_0 = 0111$. The circuit is the same as before; only the initial state has changed. *Figure 10.44* shows the timing diagrams for the circuit in Figure 10.42 and Figure 10.43.

Ring Counter Modulus and Decoding

The maximum modulus of a ring counter is the maximum number of unique states in its count sequence. In Figure 10.42 and Figure 10.43, the ring counters each had a maximum modulus of 4. We say that 4 is the *maximum* modulus of the ring counters shown, because we can change the modulus of a ring counter by loading different data at initialization.

For example, if we load a 4-bit ring counter with the data $Q_3Q_2Q_1Q_0 = 1000$, the following unique states are possible: 1000, 0100, 0010, and 0001. If we load the same circuit with the data $Q_3Q_2Q_1Q_0 = 1010$, there are only two unique states: 1010 and 0101. Depending on which data are loaded, the modulus is 4 or 2.

Most input data in this circuit will yield a modulus of 4. Try a few combinations.

Note . . .

The design of a ring counter simply ties the output from one flip-flop into the input of the next; however, this simply rotates the current values through the cycle. To load a specific value, additional circuitry to either SET or CLEAR the flip-flops is necessary.

Note . . .

The maximum modulus of a ring counter is the same as the number of bits in its output.

FIGURE 10.42 *Circulating a 1 in a Ring Counter*

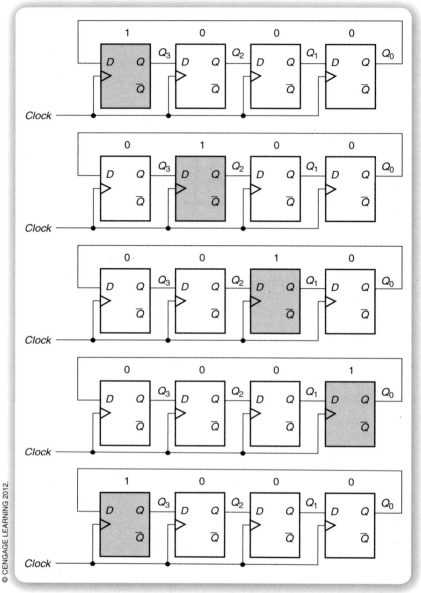

A ring counter requires more flip-flops than a binary counter to produce the same number of unique states. Specifically, for *n* flip-flops, a binary counter has 2^n unique states and a ring counter has *n*.

This is offset by the fact that a ring counter requires no decoding. A binary counter used to sequence eight events requires three flip-flops and eight 3-input decoding gates. To perform the same task, a ring counter requires eight flip-flops and no decoding gates.

As the number of output states of an event sequencer increases, the complexity of the decoder for the binary counter also increases. A circuit requiring sixteen output states can be implemented with a 4-bit binary counter and sixteen 4-input decoding gates. If you need eighteen output states, you must have a 5-bit counter ($2^4 < 18 < 2^5$) and eighteen 5-input decoding gates.

The only required modification to the ring counter is one more flip-flop for each additional state. A 16-state ring counter needs sixteen flip-flops and an 18-state ring counter must have eighteen flip-flops. No decoding is required for either circuit.

FIGURE 10.43 Circulating a 0 in a Ring Counter

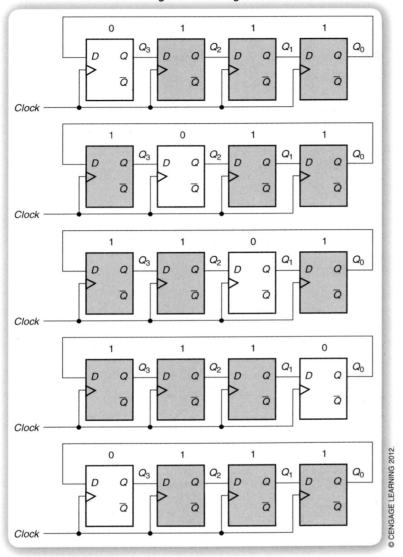

FIGURE 10.44 Timing Diagrams for Figure 10.42 and Figure 10.43

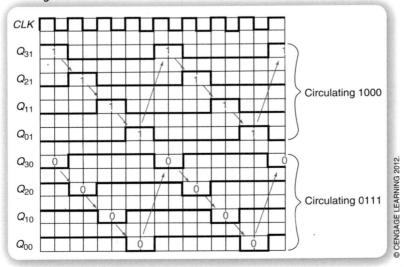

FIGURE 10.45 4-Bit Johnson Counter

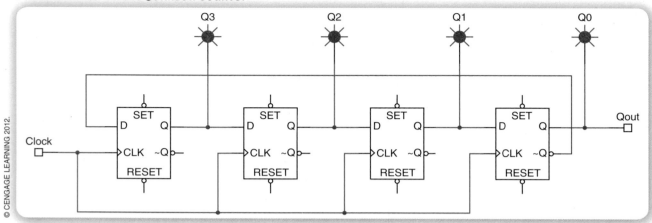

Johnson Counters

Figure 10.45 shows a 4-bit Johnson counter constructed from D flip-flops. It is the same as a ring counter except for the inversion in the feedback loop where $\overline{Q}_n$ is connected to D_3. The circuit output is taken from flip-flop outputs Q_3 through Q_0. Because the feedback introduces a "twist" into the recirculating data, a Johnson counter is also called a "twisted ring counter."

Figure 10.46 shows the progress of data through a Johnson counter that starts cleared ($Q_3Q_2Q_1Q_0 = 0000$). The shaded flip-flops represent 1's and the unshaded

FIGURE 10.46 Data Circulation in a 4-Bit Johnson Counter

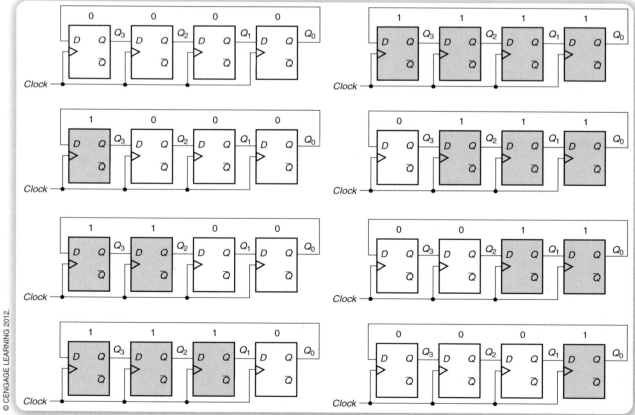

flip-flops are 0's. Every 0 at Q_0 is fed back to D_3 as a 1 and every 1 is fed back as a 0. The count sequence for this circuit is given in *Table 10.13*. There are eight unique states in the count sequence table. Open Multisim file **10.45 Johnson Counter. ms11** to see its operation.

Johnson Counter Modulus and Decoding

The Johnson counter represents a compromise between binary and ring counters, whose maximum moduli are, respectively, 2^n and n for an n-bit counter.

If it is used for event sequencing, a Johnson counter must be decoded, unlike a ring counter. Its output states are such that each state can be decoded uniquely by a 2-input AND or NAND gate, depending on whether you need active-HIGH or active-LOW indication. This yields a simpler decoder than is required for a binary counter.

Table 10.14 shows the count sequence for an 8-bit Johnson counter. *Table 10.15* shows the decoding of a 4-bit Johnson counter.

TABLE 10.13 *Count Sequence of a 4-Bit Johnson Counter*

Q_3	Q_2	Q_1	Q_0
0	0	0	0
1	0	0	0
1	1	0	0
1	1	1	0
1	1	1	1
0	1	1	1
0	0	1	1
0	0	0	1

© CENGAGE LEARNING 2012.

Note . . .
The maximum modulus of a Johnson counter is $2n$ for a circuit with n flip-flops.

TABLE 10.14 *Count Sequence of an 8-Bit Johnson Counter*

Q_7	Q_6	Q_5	Q_4	Q_3	Q_2	Q_1	Q_0
0	0	0	0	0	0	0	0
1	0	0	0	0	0	0	0
1	1	0	0	0	0	0	0
1	1	1	0	0	0	0	0
1	1	1	1	0	0	0	0
1	1	1	1	1	0	0	0
1	1	1	1	1	1	0	0
1	1	1	1	1	1	1	0
1	1	1	1	1	1	1	1
0	1	1	1	1	1	1	1
0	0	1	1	1	1	1	1
0	0	0	1	1	1	1	1
0	0	0	0	1	1	1	1
0	0	0	0	0	1	1	1
0	0	0	0	0	0	1	1
0	0	0	0	0	0	0	1

© CENGAGE LEARNING 2012.

Decoding a sequential circuit depends on the decoder responding uniquely to every possible state of the circuit outputs. If we want to use only 2-input gates in our decoder, it must recognize two variables for every state that are *both* active *only* in that state.

A Johnson counter decoder exploits what might be called the "1/0 interface" of the count sequence table. Careful examination of Table 10.13 and Table 10.14 reveals that for every state except where the outputs are all 1's or all 0's, there is a side-by-side 10 or 01 pair that exists only in that state.

Each of these pairs can be decoded to give unique indication of a particular state. For example, the pair $Q_3\overline{Q_2}$ uniquely indicates the second state because $Q_3 = 1$ AND $Q_2 = 0$ *only* in the second line of the count sequence table. (This is true for any size of Johnson counter; compare the second lines

TABLE 10.15   Decoding a 4-Bit Johnson Counter

Q_3	Q_2	Q_1	Q_0	Decoder Outputs	Comment
0	0	0	0	$\overline{Q}_3\overline{Q}_0$	MSB = LSB = 0
1	0	0	0	$Q_3\overline{Q}_2$	"1/0"
1	1	0	0	$Q_2\overline{Q}_1$	Pairs
1	1	1	0	$Q_1\overline{Q}_0$	
1	1	1	1	Q_3Q_0	MSB = LSB = 1
0	1	1	1	$\overline{Q}_3Q_2$	"0/1"
0	0	1	1	$\overline{Q}_2Q_1$	Pairs
0	0	0	1	$\overline{Q}_1Q_0$	

of Table 10.13 and Table 10.14. In the second line of both tables, the MSB is 1 and the second MSB is 0.)

For the states where the outputs are all 1's or all 0's, the most significant AND least significant bits can be decoded uniquely, these being the only states where MSB = LSB.

Figure 10.47 shows the decoder circuit for a 4-bit Johnson counter.

The output decoder of a Johnson counter does not increase in complexity as the modulus of the counter increases. The decoder will always consist of $2n$ 2-input AND or NAND gates for an n-bit counter. (For example, for an 8-bit Johnson counter, the decoder will consist of sixteen 2-input AND or NAND gates.)

FIGURE 10.47   4-Bit Johnson Counter with Output Decoding

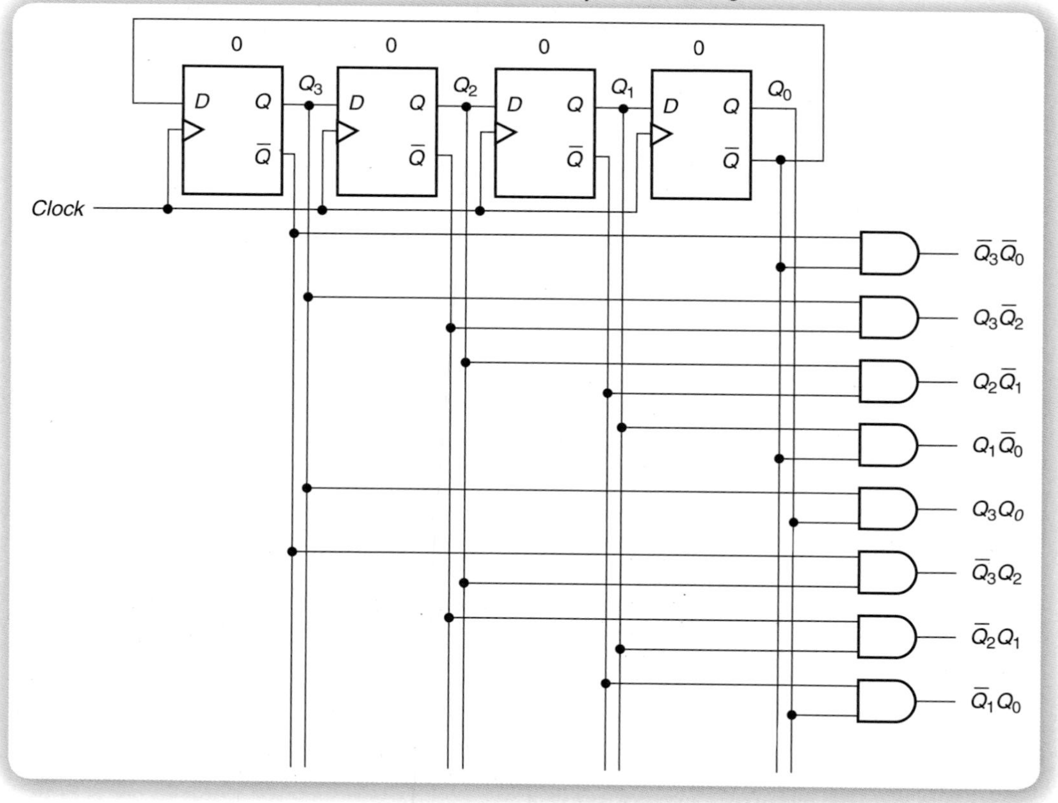

Example 10.11

Draw the timing diagram of the Johnson counter decoder of Figure 10.47, assuming that the counter is initially cleared.

Solution

Figure 10.48 shows the timing diagram of the Johnson counter and its decoder outputs.

FIGURE 10.48 **Example 10.11: Johnson Counter Decoder Outputs**

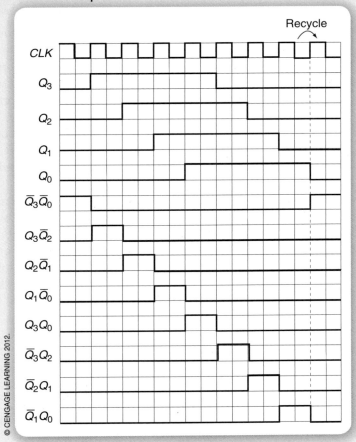

Your Turn

10.6 How many flip-flops are required to produce twenty-four unique states in each of the following types of counters: binary counter, ring counter, Johnson counter? How many and what type of decoding gates are required to produce an active-LOW decoder for each type of counter?

SUMMARY

1. A counter is a circuit that progresses in a defined sequence at the rate of one state per clock pulse.

2. The modulus of a counter is the number of states through which the counter output progresses before repeating.

3. A counter with an ascending sequence of states is called an UP counter. A counter with a descending sequence of states is called a DOWN counter.

4. In general, the maximum modulus of a counter is given by 2^n for an n-bit counter.

5. A counter whose modulus is 2^n is called a full-sequence counter. The count progresses from 0 to $2^n - 1$, which corresponds to a binary output of all 0's to all 1's.

6. A counter whose output is less than 2^n is called a truncated sequence counter.

7. The adjacent outputs of a full-sequence binary counter have a frequency ratio of 2:1. The less significant of the two bits has the higher frequency.

8. A synchronous counter consists of a series of flip-flops, all clocked from the same source, that stores the present state of the counter and a combinational circuit that monitors the counter's present state and determines its next state.

9. A synchronous counter can be analyzed by a formal procedure that includes the following steps:
   a. Write the Boolean equations for the synchronous inputs of the counter flip-flops in terms of the present state of the flip-flip outputs.
   b. Evaluate each Boolean equation for an initial state to find the states of the synchronous inputs.
   c. Use flip-flop function tables to determine each flip-flop next state.
   d. Set the next state to the new present state.
   e. Continue until the sequence repeats.

10. The analysis procedure just noted should be applied to any unused states of the counter to ensure that they will enter the count sequence properly.

11. A synchronous counter can be designed using a formal method that relies on the excitation tables of the flip-flops used in the counter. An excitation table indicates the required logic levels on the flip-flop inputs to effect a particular transition.

12. The synchronous counter design procedure is based on the following steps:
   a. Draw the state diagram of the counter and use it to list the relationship between the counter's present and next states. The table should list the counter's present states in binary order.
   b. For the initial design, unused states can be set to a known destination state, such as 0, or treated as don't care states.
   c. Use the flip-flop excitation table to determine the synchronous input levels for each present-to-next state transition.
   d. Use Karnaugh maps to find the simplest equations for the flip-flop inputs (JK, D, or T) in terms of Q.
   e. Unused states should be analyzed by substituting their values into the Boolean equations of the counter. This will verify whether or not an unused state will enter the count sequence properly.

13. If a counter must reset to 0 from an unused state, the flip-flops can be reset asynchronously to their initial states or the counter can be designed with the unused states always having 0 as their next state.

14. Some of the most common control features available in synchronous counters include:
   a. Synchronous or asynchronous parallel load, which allows the count to be set to any value whenever a *LOAD* input is asserted.

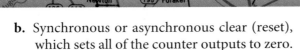

b. Synchronous or asynchronous clear (reset), which sets all of the counter outputs to zero.

c. Count enable, which allows the count sequence to progress when asserted and inhibits the count when deasserted.

d. Bidirectional control, which determines whether the counter counts up or down.

e. Output decoding, which activates one or more outputs when detecting particular states on the counter outputs.

f. Ripple carry out or ripple clock out (RCO), a special case of output decoding that produces a pulse upon detecting the terminal count, or last state, of a count sequence.

15. The parallel load function of a counter requires load *data* (the parallel input values) and a load *command* input, such as *LOAD*, that transfers the parallel data when asserted. If the load function is synchronous, a clock pulse is also required.

16. Synchronous load transfers data to the counter outputs on an active clock edge. Asynchronous load operates as soon as the load input activates, without waiting for the clock.

17. Synchronous load is implemented by a function select circuit that selects either the count logic or the direct parallel input to be applied to the synchronous input(s) of a flip-flop.

18. The count enable function enables or disables the count logic of a counter without affecting other functions, such as clock or clear. This can be done by ANDing the count logic with the count enable input signal.

19. An output decoder asserts one output for each counter state. A special case is a terminal count decoder that detects the last state of a count sequence.

20. RCO (ripple clock out) generates one clock pulse upon terminal count, with its positive edge at the end of the count cycle.

21. A shift register is a circuit for storing and moving data. Three basic movements in a shift register are serial (from one flip-flop to another), parallel (into all flip-flops at once), and rotation (serial shift with a connection from the last flip-flop output to the first flip-flop input).

22. Serial shifting can be left (toward the MSB) or right (away from the MSB). Some datasheets indicate the opposite relationship between right/left and LSB/MSB.

23. A ring counter is a serial shift register with the serial output fed back to the serial input so that the internal data is continuously circulated. The initial value is generally set by asynchronous preset and clear functions.

24. The maximum modulus of a ring counter is n for a circuit with n flip-flops, as compared to 2^n for a binary counter. A ring counter output is self-decoding, whereas a binary counter requires $m \pounds 2^n$ AND or NAND gates with n inputs each.

25. A Johnson counter is a ring counter where the feedback is complemented. A Johnson counter has $2n$ states for an n-bit counter, which can be uniquely decoded by $2n$ 2-input AND or NAND gates.

BRING IT HOME

10.1 Basic Concepts of Digital Counters

10.1 A parking lot at a football stadium is monitored before a game to determine whether or not there is available space for more cars. When a car enters the lot, the driver takes a ticket from a dispenser, which also produces a pulse for each ticket taken.

The parking lot has space for 4095 cars. Draw a block diagram that shows how you can use a digital counter to light a LOT FULL sign after 4095 cars have entered. (Assume that no cars leave the lot until after the game, so you don't need to keep track of cars leaving the lot.) How many bits should the counter have?

continued...

10.2 Draw the timing diagram for one complete cycle of a mod-8 counter, including waveforms for *CLK*, Q_0, Q_1, and Q_2, where Q_0 is the LSB.

10.3 How many bits are required to make a counter with a modulus of 64? Why? What is the maximum count of such a counter?

10.4 a. Draw the state diagram of a mod-10 UP counter.

    b. Use the state diagram drawn in part **a** to answer the following questions:

      i. The counter is at state 0111. What is the count after 7 clock pulses are applied?

      ii. After 5 clock pulses, the counter output is at 0001. What was the counter state prior to the clock pulses?

10.5 What is the maximum modulus of a 6-bit counter? A 7-bit? 8-bit?

10.6 Draw the count sequence table and timing diagram of a mod-10 UP counter.

10.2 Synchronous Counters

10.7 Draw the circuit for a synchronous mod-16 UP counter made from negative edge-triggered JK flip-flops.

10.8 Write the Boolean equations required to extend the counter drawn in Problem 10.7 to a mod-64 counter.

10.9 Analyze the operation of the synchronous counter in *Figure 10.49* drawing a state table showing all transitions, including unused states. Use this state table to draw a state diagram and a timing diagram. What is the counter's modulus?

FIGURE 10.49 Problem 10.9: Synchronous Counter

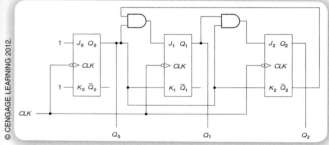

© CENGAGE LEARNING 2012.

10.3 Design of Synchronous Counters

10.10 Design a synchronous mod-10 counter, using positive edge-triggered JK flip-flops. Check that unused states properly enter the main sequence. Draw a state diagram showing the unused states.

10.11 Design a synchronous mod-10 counter, using positive edge-triggered D flip-flops. Check that unused states properly enter the main sequence. Draw a state diagram showing the unused states.

10.12 Multisim Problem

Design a synchronous mod-10 counter, using positive edge-triggered D flip-flops using Multisim.

10.4 Control Options for Synchronous Counters

10.13 Briefly explain the difference between asynchronous and synchronous parallel load in a synchronous counter.

10.14 Derive the Boolean equations for the synchronous DOWN-counter in Figure 10.25.

10.15 Write the Boolean equations for the count logic of the 4-bit bidirectional counter in Figure 10.27. Briefly explain how the logic works.

10.5 Shift Registers

10.16 The following bits are applied in sequence to the input of a 6-bit serial right-shift register: 0111111 (0 is applied first). Draw the timing diagram.

10.17 After the data in Problem 10.16 are applied to the 6-bit shift register, the serial input goes to 0 for the next 8 clock pulses and then returns to 1. Write the internal states, Q_5 through Q_0, of the shift register flip-flops after the first 2 clock pulses. Write the states after 6, 8, and 10 clock pulses.

10.18 Multisim Problem

Use Multisim to create the schematic for a 4-bit serial shift register based on JK flip-flops that shifts left, rather than right. Create a simulation to verify the operation of the shift register.

10.6 Shift Register Counters

10.19 Construct the count sequence table of a 5-bit Johnson counter, assuming the counter is initially cleared.

EXTRA MILE

10.1 Basic Concepts of Digital Counters

10.20 *Figure 10.50* shows a mod-16, which controls the operation of two digital sequential circuits, labeled Circuit 1 and Circuit 2. Circuit 1 is positive edge-triggered and clocked by counter output Q_1. Circuit 2 is negative edge-triggered and clocked by Q_3. (Q_3 is the MSB output of the counter.)

a. Draw the timing diagram for one complete cycle of the circuit operation. Draw arrows on the active edges of the waveforms that activate Circuit 1 and Circuit 2.

b. State how many times Circuit 1 is clocked for each time that Circuit 2 is clocked.

FIGURE 10.50 Problem 10.20: Mod-16 Counter Driving Two Sequential Circuits

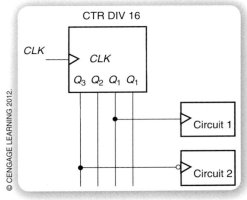

10.21 A mod-16 counter is clocked by a waveform having a frequency of 48 kHz. What is the frequency of each of the waveforms at Q_0, Q_1, Q_2, and Q_3?

10.22 A mod-10 counter is clocked by a waveform having a frequency of 48 kHz. What is the frequency of the Q_3 output waveform? The Q_0 waveform? Why is it difficult to determine the frequencies of Q_1 and Q_2?

10.2 Synchronous Counters

10.23 a. Write the equations for the J and K inputs of each flip-flop of the synchronous counter represented in *Figure 10.51*.

b. Assume that $Q_3Q_2Q_1Q_0 = 1010$ at some point in the count sequence. Use the equations from part **a** to predict the circuit outputs after each of three clock pulses.

10.3 Design of Synchronous Counters

10.24 Figure 10.17 shows the design of a synchronous mod-12 counter in which the unused states were set as don't cares. Redesign the mod-12 counter setting the unused states to all go to state 0000. Comment on whether the new design is much more complex than the original design using don't cares.

10.25 Figure 10.20 shows the design of a 4-bit counter. Create this circuit in Multisim and verify its operation.

FIGURE 10.51 Problem 10.23: Synchronous Counter

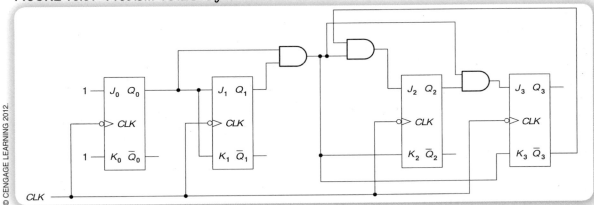

continues...

continued...

10.26 *Table 10.16* shows the count sequence for a *biquinary sequence* counter. The sequence has ten states, but does not progress in binary order. The advantage of the sequence is that its most significant bit has a divide-by-10 ratio, relative to a clock input, and a 50% duty cycle. Design the synchronous counter circuit for this sequence, using D flip-flops. *Hint:* When making the state table, list all *present states* in binary order. The next states *will not* be in binary order.

TABLE 10.16   Biquinary Sequence

Q_3	Q_2	Q_1	Q_0
0	0	0	0
0	0	0	1
0	0	1	0
0	0	1	1
0	1	0	0
1	0	0	0
1	0	0	1
1	0	1	0
1	0	1	1
1	1	0	0

© CENGAGE LEARNING 2012.

10.5   Shift Registers

10.27 Figure 10.40 shows the design of a serial shift register with parallel load. Create this circuit in Multisim and verify its operation.

10.6   Shift Register Counters

10.28 Figure 10.47 shows the design of a 4-bit Johnson counter with output decoding. Create this circuit in Multisim and verify its operation.

10.29 Figure 10.42 shows a basic 4-bit ring counter, which circulates a 4-bit number, but is not suited to load a 4-bit number into its initial state. Using Multisim, add the capability to this ring counter to load an initial value. Create this circuit in Multisim and verify its operation.

CHAPTER 11
State Machine Design

GPS DELUXE

Menu

START LOCATION   DISTANCE   END LOCATION

CHAPTER OBJECTIVES

Upon successful completion of this chapter, you will be able to:

1 Describe the components of a state machine.

2 Distinguish between Moore and Mealy implementations of state machines.

3 Draw the state diagram of a state machine from a verbal description.

4 Use the state table method of state machine design to determine the Boolean equations of the state machine.

5 Translate the Boolean equations of a state machine into a schematic.

6 Determine whether the output of a state machine is vulnerable to asynchronous changes of input.

7 Design state machine applications, such as a traffic light controller.

8 Troubleshoot state machines by examining next state tables.

PHOTO: ©iSTOCKPHOTO.COM/INOK.

A special type of counter circuit is a finite state machine, a circuit that counts through a specific sequence of numbers. These circuits don't have to count up or down, but just have to count in a predictable sequence. For example, imagine an elevator controller: if you are on the 5th floor, which floor is next? It depends on whether you plan to go up or down, but once we know that, we can figure out which is next. We can design a finite state machine to count through any sequence we would like.

A state machine is a digital circuit that goes through a predictable sequence of states depending on the current state and current inputs. Counters are special examples of state machines: a MOD-10 down counter that is currently on state 7 will go to state 6 on a clock edge. We can add inputs that will control the next state: for example, if we added an UP/DOWN input to a counter that told the circuit to count up, and the current state was 7, the next state would be 8. For this type of counter, its next state depends on its current state and the state of the UP/DOWN input.

Two types of state machines are Moore machines and Mealy machines. In a Moore machine, the next state and output of the circuit depend only on its present state. The output of a Mealy machine depends on the present state and the state of one or more control inputs. The dependence on input states can lead to the output accidentally changing asynchronously, that is, not on a clock edge. However, this problem is easy to fix and a Mealy machine can often be done more efficiently than a Moore machine.

11.1 STATE MACHINES

The synchronous counters and shift registers we examined in Chapter 10 are examples of a larger class of circuits known as **state machines**. As described for synchronous counters in Section 10.2, a state machine consists of a memory section that holds the present state of the machine and a control section that determines the machine's next state. These sections communicate via a series of command and status lines. Depending on the type of machine, the outputs will either be functions of the present state only or of the present and next states.

Figure 11.1 shows the block diagram of a **Moore machine**. The outputs of a Moore machine are determined solely by the present state of the machine's memory section. The output may be directly connected to the Q outputs of the internal flip-flops, or the Q outputs might pass through a decoder circuit. The output of a Moore machine is synchronous to the system clock, because the output can change *only* when the machine's internal **state variables** change.

The block diagram of a **Mealy machine** is shown in *Figure 11.2*. The outputs of the Mealy machine are derived from the combinational (control) and

FIGURE 11.1 *Moore-Type State Machine*

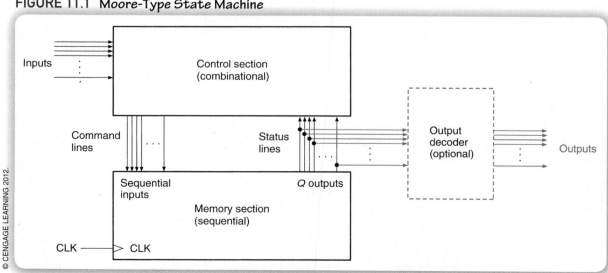

FIGURE 11.2 Mealy-Type State Machine

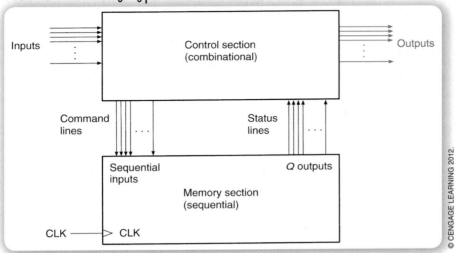

© CENGAGE LEARNING 2012.

the sequential (memory) parts of the machine. Therefore, the outputs can change asynchronously when the combinational circuit inputs change out of phase with the clock. (When we say that the outputs change asynchronously, we generally do not mean a change such as asynchronous reset directly on the machine's flip-flops. We simply mean a change that is not synchronized to the system clock.)

Your Turn

11.1 What is the main difference between a Moore-type state machine and a Mealy-type state machine?

11.2 STATE MACHINES WITH NO CONTROL INPUTS

KEY TERM

Bubble A circle in a state diagram containing the state name and values of the state variables.

A state machine can be designed using a classical technique, also called a state table design technique, similar to that used to design a synchronous counter. We will design several state machines using classical techniques. It is also possible to design a state machine with a specialized programming language, but we will not examine this method.

As an example of the classical design technique, we will design a state machine whose output depends only on the clock input: a 3-bit counter with a Gray code count sequence. A 3-bit Gray code, shown in *Table 11.1*, changes only one bit between adjacent codes and is therefore not a binary-weighted sequence.

TABLE 11.1   3-Bit Gray Code Sequence

Q_2	Q_1	Q_0
0	0	0
0	0	1
0	1	1
0	1	0
1	1	0
1	1	1
1	0	1
1	0	0

© CENGAGE LEARNING 2012.

FIGURE 11.3 *Gray Code on a Shaft Encoder*

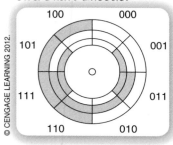

© CENGAGE LEARNING 2012.

Gray code is often used in situations where it is important to minimize the effect of single-bit errors. For example, suppose the angle of a motor shaft is measured by a detected code on a Gray-coded shaft encoder, shown in *Figure 11.3*. The encoder indicates a 3-bit number for each of eight angular positions by having three concentric circular segments for each code. A dark band indicates a 1 and a transparent band indicates a 0, with the most significant bit (MSB) as the outermost band. The dark or transparent bands are detected by three sensors that detect light shining through a transparent band. (A real shaft encoder has more bits to indicate an angle more precisely. For example, a shaft encoder that measures an angle of 1 degree would require nine bits, because there are 360 degrees in a circle and $2^8 \leq 360 \leq 2^9$.)

For most positions on the encoder, the error of a single bit results in a positional error of only one-eighth of the circle. This is not true with binary coding, where single bit errors can give larger positional errors. For example, if the positional decoder reads 100 instead of 000, this is a difference of 4 in binary, representing an error of one-half of the circle. These same codes differ by only one position in Gray code.

Classical Design Techniques

We can summarize the classical design technique for a state machine as follows:

1. Define the problem.
2. Draw a state diagram.
3. Make a state table that lists all possible present states and inputs, and the next state and output state for each present state/input combination. *List the present states and inputs in* **binary order**.
4. Use flip-flop excitation tables to determine at what states the flip-flop synchronous inputs must be to make the circuit go from each present state to its next state. *The next state variables are functions of the inputs and present state variables.*
5. Write the output value for each present state/input combination. *The output variables are functions of the inputs and present state variables.*
6. Simplify the Boolean expression for each output and synchronous input.
7. Use the Boolean expressions found in Step 6 to draw the required logic circuit.

Let us follow this procedure to design a 3-bit Gray code counter. We will modify the procedure to acknowledge there are no inputs other than the clock and no outputs that must be designed apart from the counter itself.

1. *Define the problem.* Design a counter whose outputs progress in the sequence defined in Table 11.1.

2. *Draw a state diagram.* The state diagram is shown in *Figure 11.4*. In addition to the values of state variables shown in each circle (or **bubble**), we also indicate a state name, such as s0, s1, s2, and so on. This name is independent of the value of state variables. We use numbered states (s0, s1, . . .) for convenience, but we could use any names we wanted to.

3. *Make a state table.* The state table, based on D flip-flops, is shown in *Table 11.2. Because there are eight unique states in the state diagram, we require three state variables* ($2^3 = 8$), *and hence three flip-flops.* Note that the present states are in binary-weighted order, even though the count does not progress in this order. In such a case, it is essential to have an accurate state diagram, from which we derive each next state. For example, if the present state is 010, the next state is not 011, as we would expect, but 110, which we derive by examining the state diagram.

FIGURE 11.4 *State Diagram for a 3-Bit Gray Code Counter*

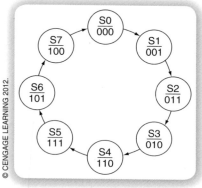

© CENGAGE LEARNING 2012.

TABLE 11.2   State Table for a 3-Bit Gray Code Counter

Present State			Next State			Synchronous Inputs		
Q_2	Q_1	Q_0	Q_2	Q_1	Q_0	D_2	D_1	D_0
0	0	0	0	0	1	0	0	1
0	0	1	0	1	1	0	1	1
0	1	0	1	1	0	1	1	0
0	1	1	0	1	0	0	1	0
1	0	0	0	0	0	0	0	0
1	0	1	1	0	0	1	0	0
1	1	0	1	1	1	1	1	1
1	1	1	1	0	1	1	0	1

Why list the present states in binary order, rather than the same order as the output sequence? By doing so, we can easily simplify the equations for the D inputs of the flip-flops by using a series of Karnaugh maps. This is still possible, but harder to do, if we list the present states in order of the output sequence.

4. *Use flip-flop excitation tables to determine at what states the flip-flop synchronous inputs must be to make the circuit go from each present state to its next state.* This is not necessary if we use D flip-flops, because Q follows D. The D inputs are the same as the next state outputs. This is easily verified in Table 11.2. For JK flip-flops, we would follow the same procedure as for the design of synchronous counters outlined in Chapter 10.

5. *Simplify the Boolean expression for each synchronous input.* Figure 11.5 shows three Karnaugh maps, one for each D input of the circuit. The K-maps yield three Boolean equations:

$$D_2 = Q_1\bar{Q}_0 + Q_2Q_0$$
$$D_1 = Q_1\bar{Q}_0 + \bar{Q}_2Q_0$$
$$D_0 = \bar{Q}_2\bar{Q}_1 + Q_2Q_1$$

6. *Draw the logic circuit for the state machine.* Figure 11.6 shows the circuit for a 3-bit Gray code counter drawn in Multisim. The *Set* inputs of the flip-flops are disabled by tying them HIGH. The flip-flop *Reset* inputs are all connected to a pushbutton that allows the counter to be manually reset to 000 at any time. A timing diagram for this circuit is shown in *Figure 11.7*, with the outputs shown as individual waveforms and as a group with a binary value.

FIGURE 11.5 Karnaugh Maps for 3-Bit Gray Code Counter

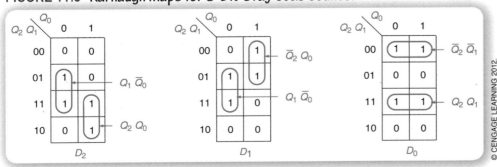

FIGURE 11.6 Logic Diagram of a 3-Bit Gray Code Counter

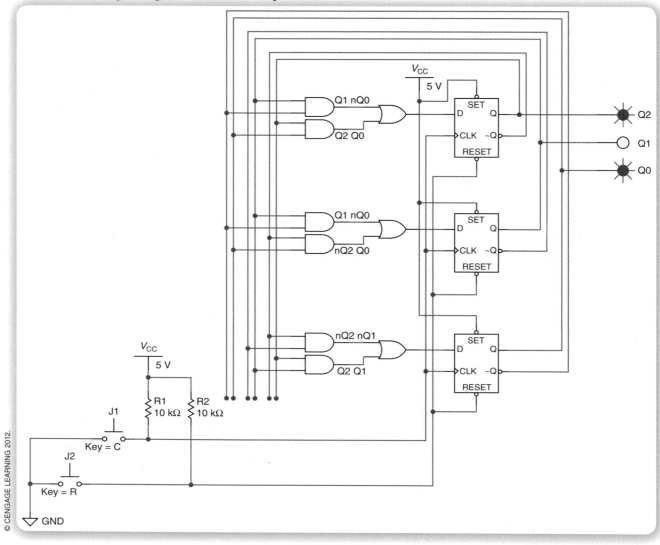

FIGURE 11.7 3-Bit Gray Code Waveforms

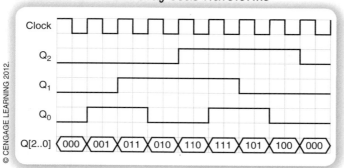

Your Turn

11.2 Write the Boolean equations for the *J* and *K* inputs of the flip-flops in a 3-bit Gray code counter based on JK flip-flops. The solution requires the use of the flip-flop excitation table shown in Table 10.7 of the previous chapter.

11.3 STATE MACHINES WITH CONTROL INPUTS

As an extension of the techniques used in the previous section, we will examine the design of state machines that use **control inputs**, and the clock, to direct their operation. Outputs of these state machines will not necessarily be the same as the states of the machine's flip-flops. As a result, this type of state machine requires a more detailed state diagram notation, such as that shown in *Figure 11.8*.

The state machine represented by the diagram in Figure 11.8 has two states, and thus requires only one state variable. Each state is represented by a bubble (circle) containing the state name and the value of the state variable. For example, the bubble containing the notation $\frac{\text{start}}{0}$ indicates that the state called **start** corresponds to a state variable with a value of 0. Each state must have a unique value for the state variable(s).

Transitions between states are marked with a combination of input and output values corresponding to the transition. The inputs and outputs are labeled **in1, in2,..., in$x$/ out1, out2,..., out$x$**. The inputs and outputs are sometimes simply indicated by the value of each variable for each transition. In this case, a legend indicates which variable corresponds to which position in the label.

For example, the legend in the state diagram of Figure 11.8 indicates that the inputs and outputs are labeled in the order **in1/out1, out2**. Thus, if the machine is in the **start** state and the input **in1** goes to 0, there is a transition to the state **continue**. During this transition, **out1** goes to 1 and **out2** goes to 0. This is indicated by the notation 0/10 beside the transitional arrow. This is called a **conditional transition** because the transition depends on the state of **in1**. The other possibility from the **start** state is a no-change transition, with both outputs at 0, if **in1** = 1. This is shown as **1/00**.

If the machine is in the state named **continue**, the notation **X/01** indicates that the machine makes a transition back to the **start** state, regardless of the value of **in1**, and that **out1** = 0 and **out2** = 1 upon this transition. Because the transition always happens, it is called an **unconditional transition**.

What does this state machine do? We can determine its function by analyzing the state diagram:

1. There are two states, called **start** and **continue**. The machine begins in the **start** state and waits for a LOW input on **in1**. As long as **in1** is HIGH, the machine waits and the outputs **out1** and **out2** are both LOW.
2. When **in1** goes LOW, the machine makes a transition to **continue** in one clock pulse. Output **out1** goes HIGH.

FIGURE 11.8 State Diagram Notation

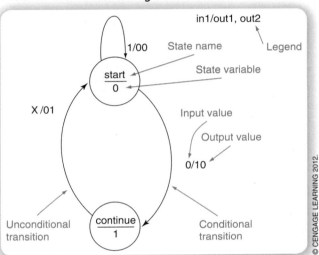

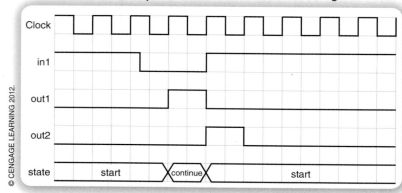

FIGURE 11.9 *Ideal Operation of State Machine in Figure 11.8*

3. On the next clock pulse, regardless of the values of the input, the machine goes back to **start**. The output **out2** goes HIGH and **out1** goes back LOW.

4. If **in1** is HIGH, the machine waits for a new LOW on **in1**, and **both** outputs are LOW again. If **in1** is LOW, the cycle repeats.

In summary, the machine waits for a LOW input on **in1**, then generates a pulse of one clock cycle duration on **out1**, then on **out2**. A timing diagram describing this operation is shown in *Figure 11.9*.

TABLE 11.3 *State Table for State Diagram in Figure 11.8*

Present State	Input	Next State	Sync. Inputs	Oututs	
Q	*in1*	*Q*	*JK*	*Out1*	*Out2*
0	0	1	1X	1	0
0	1	0	0X	0	0
1	0	0	X1	0	1
1	1	0	X1	0	1

TABLE 11.4 *JK Flip-Flop Excitation Table*

Transition	JK
0→0	0X
0→1	1X
1→0	X1
1→1	X0

Classical Design of State Machines with Control Inputs

We can use the classical design technique of the previous section to design a circuit that implements the state diagram of Figure 11.8.

1. *Define the problem.* Implement a digital circuit that generates a pulse on each of two outputs, as previously described. For this implementation, let us use JK flip-flops for the state logic. If we so chose, we could also use *D* flip-flops.

2. *Draw a state diagram.* The state diagram is shown in Figure 11.8.

3. *Make a state table.* The state table is shown in *Table 11.3*. The combinations of present state and input are listed in binary order, thus making Table 11.3 into a truth table for the next state and output functions. Because there are two states, we require one state variable, *Q*. The next state of *Q*, a function of the present state and the input **in1**, is determined by examining the state diagram. (Thus, if you are in state 0, the next state is 1 if **in1** = 0 and 0 if **in1** = 1. If you are in state 1, the next state is always 0.)

4. *Use flip-flop excitation tables to determine at what states the flip-flop synchronous inputs must be to make the circuit go from each present state to its next state.* Table 11.4 shows the flip-flop excitation table for a *JK* flip-flop, specifying the necessary values of *J* and *K* to give the transition shown. The synchronous inputs are derived from the present-to-next state transitions in Table 11.4 and entered into Table 11.3. (Refer to Table 10.7 and the synchronous counter design process in Chapter 10 for more detail about using flip-flop excitation tables.)

5. *Write the output values for each present state/input combination.* These can be determined from the state diagram and are entered in the last two columns of Table 11.3.

6. *Simplify the Boolean expression for each output and synchronous input.* The following equations represent the next state and output logic of the state machine:

$$J = \overline{Q} \cdot \overline{in1} + Q \cdot \overline{in1} = (\overline{Q} + Q)\overline{in1} = \overline{in1}$$
$$K = 1$$
$$out1 = \overline{Q} \cdot \overline{in1}$$
$$out2 = Q \cdot \overline{in1} + Q \cdot in1 = Q(\overline{in1} + in1) = Q$$

7. *Use the Boolean expressions found in Step 6 to draw the required logic circuit.* Figure 11.10 shows the circuit of the state machine. Because **out1** is a function

FIGURE 11.10 Logic Circuit of the State Machine Described in Figure 11.8

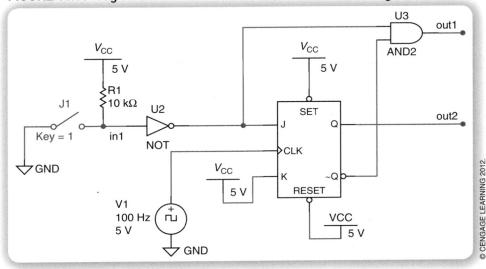

of the control section and the memory section of the machine, we can categorize the circuit as a Mealy machine. (All counter circuits that we have previously examined have been Moore machines because their outputs are derived solely from the flip-flop outputs of the circuit.)

The circuit is a Mealy machine, so it is vulnerable to asynchronous changes of output due to asynchronous input changes. This is shown in the simulation waveforms of *Figure 11.11*.

Ideally, **out1** should not change until the first positive clock edge after **in1** goes LOW. However, **out1** is derived from a combinational output, so it will change

FIGURE 11.11 Simulation of State Machine Circuit of Figure 11.10 Showing an Asynchronous Pulse Error

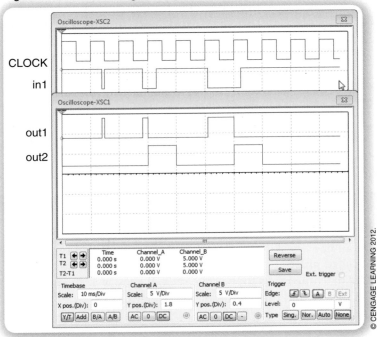

FIGURE 11.12 State Machine with Synchronizing Flip-Flops on the Outputs

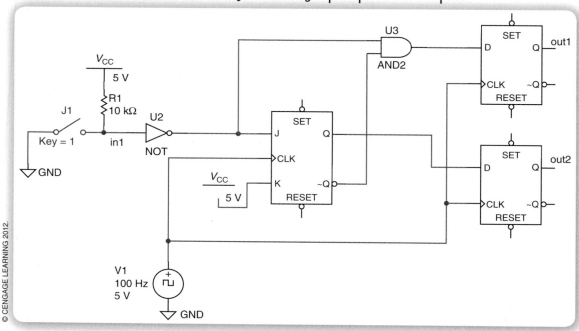

FIGURE 11.13 Properly Synchronized Outputs of the State Machine in Figure 11.12

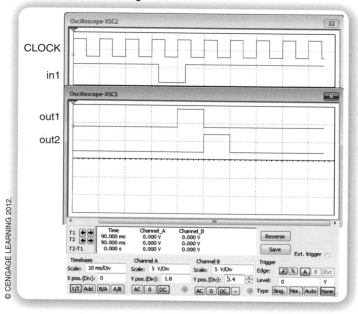

as soon as **in1** goes LOW, after allowing for a short propagation delay. If **in1** is not held LOW past the first positive clock edge, **out1** will still pulse asynchronously, but the machine will not move on to the next state.

If output synchronization is a problem (and it may not be), it can be fixed by adding a synchronizing D flip-flop to each output, as shown in *Figure 11.12*.

The state variable is stored as the state of the JK flip-flop. This state is clocked through a D flip-flop to generate **out2** and combined with **in1** to generate **out1** via another flip-flop. The simulation for this circuit, shown in *Figure 11.13*, indicates that the two outputs are synchronous with the clock, with each output pulse lasting for the full time of one clock cycle.

Example 11.1

A state machine called a single-pulse generator operates as follows:

1. The circuit has two states: **seek** and **find**, an input called **sync**, and an output called **pulse**.
2. The state machine resets to the state **seek**. If **sync** = 1, the machine remains in **seek** and the output, **pulse**, remains LOW.

continued...

FIGURE 11.14 Example 11.1:
State Diagram for a Single-
Pulse Generator

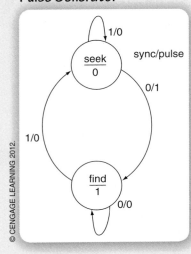

FIGURE 11.15 Example 11.1: Single-Pulse Generator

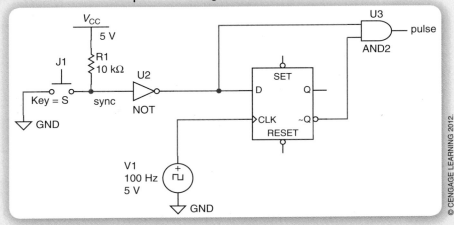

3. When **sync** = 0, the machine makes a transition to **find**. In this transition, **pulse** goes HIGH.
4. When the machine is in state **find** and **sync** = 0, the machine remains in **find** and **pulse** goes LOW.
5. When the machine is in **find** and **sync** = 1, the machine goes back to **seek** and **pulse** remains LOW.

Design the circuit for the single-pulse generator, using D flip-flops for the state logic. Use Multisim to draw the state machine circuit. Create a simulation to verify the design operation. Briefly describe what this state machine does.

TABLE 11.5 State Table for Single-Pulse Generator

Present State	Input	Next State	Sync. Input	Output
Q	*sync.*	*Q*	*D*	*pulse*
0	0	1	1	1
0	1	0	0	0
1	0	1	1	0
1	1	0	0	0

Solution

Figure 11.14 shows the state diagram derived from the description of the state machine. The state table is shown in *Table 11.5*. Because Q follows D, the D input is the same as the next state of Q.

The next-state and output equations are:

$$D = \overline{Q} \cdot \overline{sync} + Q \cdot \overline{sync} = \overline{sync}$$

$$pulse = \overline{Q} \cdot \overline{sync}$$

Figure 11.15 shows the state machine circuit derived from these Boolean equations. The simulation for this circuit is shown in *Figure 11.16*. The simulation shows that the circuit generates one pulse when the input **sync** goes LOW, regardless of the length of time that **sync** is LOW. The circuit could be used in conjunction with a debounced pushbutton to produce exactly one pulse, regardless of how long the pushbutton was held down. *Figure 11.17* shows such a circuit.

FIGURE 11.16 Example 11.1: Simulation Waveforms for a Single-Pulse Generator

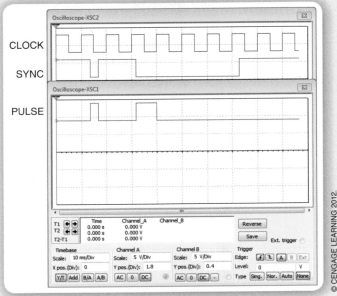

continued...

FIGURE 11.17 Example 11.1: Single-Pulse Generator Used with a Debounced Pushbutton

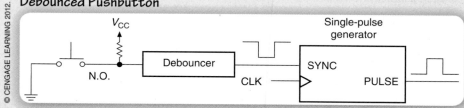

© CENGAGE LEARNING 2012.

Example 11.2

The state machine of Example 11.1 is vulnerable to asynchronous input changes. How do we know this from the circuit schematic and from the simulation waveform? Modify the circuit to eliminate the asynchronous behavior and show the effect of the change on a simulation of the design. How does this change improve the design?

Solution

The output, **pulse**, in the state machine of Figure 11.15 is derived from the state flip-flop and the combinational logic of the circuit. The output can be affected by a change that is purely combinational, thus making the output asynchronous. This is demonstrated on the first pulse of the simulation in Figure 11.16, where **pulse** momentarily goes HIGH between clock edges. Because no clock edge was present when either the input, **sync**, changed or when **pulse** changed, the output pulse must be due entirely to changes in the combinational part of the circuit.

The circuit output can be synchronized to the clock by adding an output flip-flop, as shown in *Figure 11.18*. A simulation of this circuit is shown in *Figure 11.19*. With the synchronized output, the output pulse is always the same width: one clock period. This gives a more predictable operation of the circuit.

FIGURE 11.18 Example 11.2: Single-Pulse Generator with a Synchronizing Flip-Flop

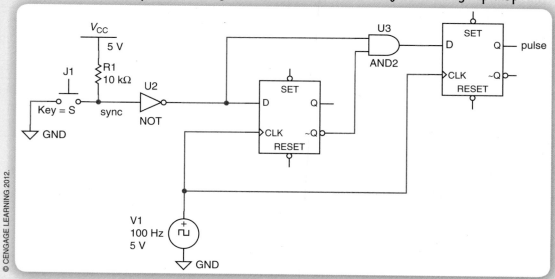

© CENGAGE LEARNING 2012.

continued...

FIGURE 11.19 Example 11.2: Simulation Waveforms for a
Single-Pulse Generator with a Synchronizing Flip-Flop

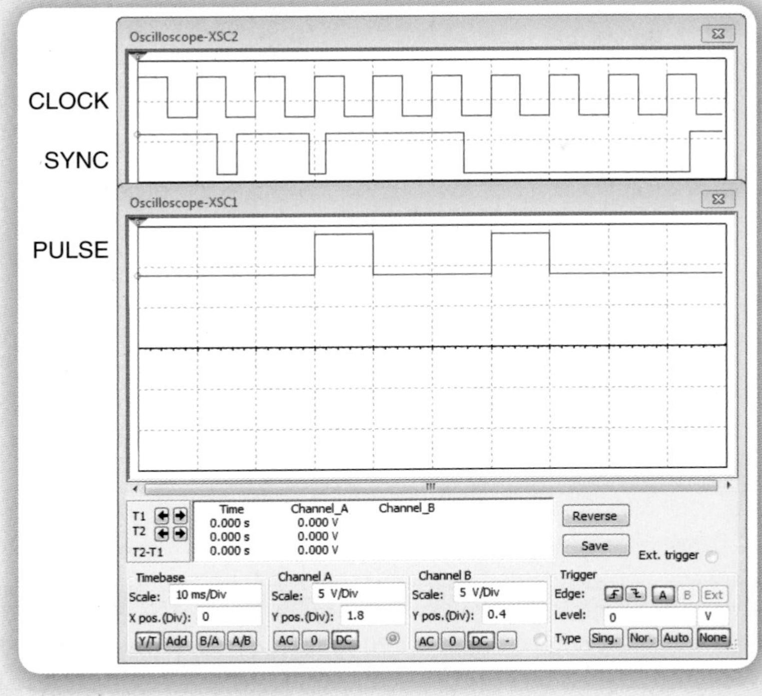

© CENGAGE LEARNING 2012.

Your Turn

11.3 Briefly explain why the single-pulse circuit in Figure 11.18 has a flip-flop on its output.

11.4 UNUSED STATES IN STATE MACHINES

In our study of counter circuits in Chapter 10, we found that when a counter modulus is not equal to a power of 2 there are unused states in the counter's sequence. For example, a mod-10 counter has six unused states, as the counter requires four bits to express ten states and the maximum number of 4-bit states is sixteen. The unused states (1010, 1011, 1100, 1101, 1110, and 1111) have to be accounted for in the design of a mod-10 counter.

The same is true of state machines whose number of states does not equal a power of 2. For instance, a machine with five states requires three state variables. There are up to eight states available in a machine with three state variables, leaving three unused states. *Figure 11.20* shows the state diagram of such a machine.

Unused states can be dealt with in two ways: they can be treated as don't care states, or they can be assigned specific destinations in the state diagram. In the latter case, the safest destination is the first state, in this case the state called **start**.

FIGURE 11.20 State Diagram for a Two-Pulse Generator

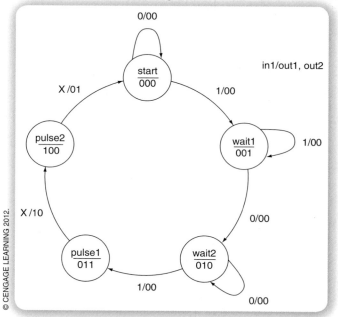

Example 11.3

Redraw the state diagram of Figure 11.20 to include the unused states of the machine's state variables. Set the unused states to have a destination state of **start**. Briefly describe the intended operation of the state machine.

Solution

Figure 11.21 shows the revised state diagram.

The machine begins in state **start** and waits for a HIGH on **in1**. The machine then makes a transition to **wait1** and stays there until **in1** goes LOW

FIGURE 11.21 Example 11.3: State Diagram for a Two-Pulse Generator Showing Unused States

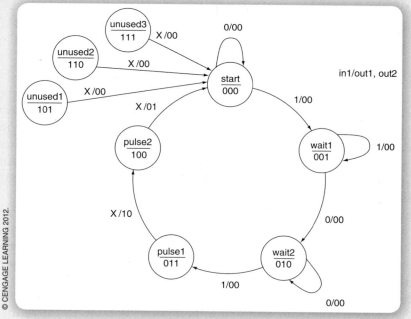

continued...

again. The machine goes to **wait2** and stays there until **in1** goes HIGH and then makes an unconditional transition to **pulse1** on the next clock pulse. Until this point, there is no change in either output.

The machine makes an unconditional transition to **pulse2** and makes **out1** go HIGH. The next transition, also unconditional, is to **start**, when **out1** goes LOW and **out2** goes HIGH. If **in1** is LOW, the machine stays in **start**. Otherwise, the cycle continues as outlined. In either case, **out2** goes LOW again.

Thus, the machine waits for a HIGH-LOW-HIGH input sequence and generates a pulse sequence on two outputs.

Example 11.4

Design the state machine described in the modified state diagram of Figure 11.21. Draw the state machine in Multisim and verify its function with a simulation.

Solution

Table 11.6 shows the state table of the state machine represented by Figure 11.21.

Figure 11.22 shows the Karnaugh maps used to simplify the next-state equations for the state variable flip-flops. The output equations can be simplified by inspection.

The next-state and output equations for the state machine are:

$$D_2 = \bar{Q}_2 Q_1 Q_0$$
$$D_1 = \bar{Q}_2 Q_1 \bar{Q}_0 + \bar{Q}_2 \bar{Q}_1 Q_0 \overline{in1}$$
$$D_0 = \bar{Q}_2 \bar{Q}_0 in1 + \bar{Q}_2 \bar{Q}_1 in1$$
$$out1 = \bar{Q}_2 Q_1 Q_0$$
$$out2 = Q_2 \bar{Q}_1 \bar{Q}_0$$

TABLE 11.6   State Table for State Machine of Figure 11.21

Present State			Input	Next State			Outputs	
Q_2	Q_1	Q_0	$in1$	Q_2	Q_1	Q_0	Out1	Out2
0	0	0	0	0	0	0	0	0
0	0	0	1	0	0	1	0	0
0	0	1	0	0	1	0	0	0
0	0	1	1	0	0	1	0	0
0	1	0	0	0	1	0	0	0
0	1	0	1	0	1	1	0	0
0	1	1	0	1	0	0	1	0
0	1	1	1	1	0	0	1	0
1	0	0	0	0	0	0	0	1
1	0	0	1	0	0	0	0	1
1	0	1	0	0	0	0	0	0
1	0	1	1	0	0	0	0	0
1	1	0	0	0	0	0	0	0
1	1	0	1	0	0	0	0	0
1	1	1	0	0	0	0	0	0
1	1	1	1	0	0	0	0	0

continued...

FIGURE 11.22 Example 11.4: Karnaugh Maps for Two-Pulse Generator Showing Unused States

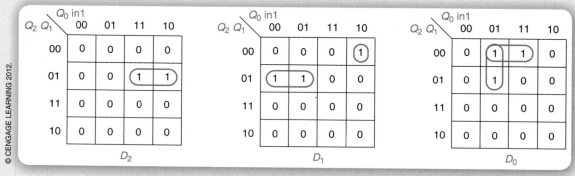

Figure 11.23 shows the Multisim schematic for the state machine. *Figure 11.24* shows the simulation waveforms.

FIGURE 11.23 Example 11.4: Two-Pulse Generator

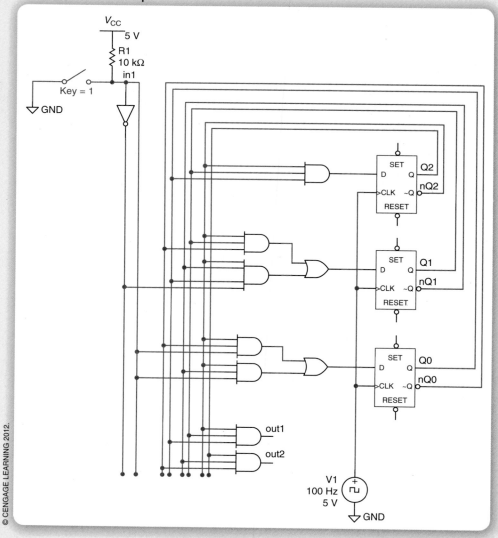

continued...

FIGURE 11.24 Example 11.4: Simulation of a Two-Pulse Generator

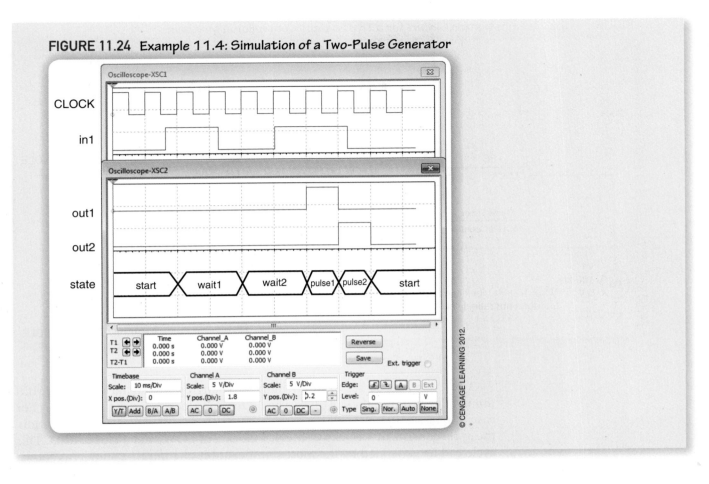

 Your Turn

11.4 Is the state machine designed in Example 11.4 a Moore machine or a Mealy machine? Why?

11.5 TRAFFIC LIGHT CONTROLLER

A simple traffic light controller can be implemented by a state machine with a state diagram such as the one shown in *Figure 11.25*.

The control scheme assumes control over a north-south road and an east-west road. The north-south lights are controlled by outputs called **nsr**, **nsy**, and **nsg** (north-south red, yellow, green). The east-west road is controlled by similar outputs called **ewr**, **ewy**, and **ewg**. A HIGH controller output turns on a light. Thus, an output 100001 corresponds to the north-south red and east-west green lights.

An input called *TIMER* controls the length of the two green-light cycles. When *TIMER* = 1, there is a transition from s0 to s1 or from s2 to s3 on the next positive clock edge (s0 represents the EW green; s2 the NS green). This transition accompanies a change from green to yellow on the active road. The light on the other road stays red. An unconditional transition follows, changing the yellow light to red on one road and the red light to green on the other.

FIGURE 11.25 *State Diagram for a Traffic Light Controller*

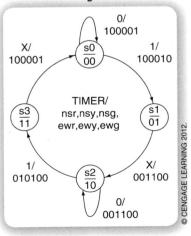

FIGURE 11.26 Demonstration Circuit for a Traffic Light Controller in Multisim

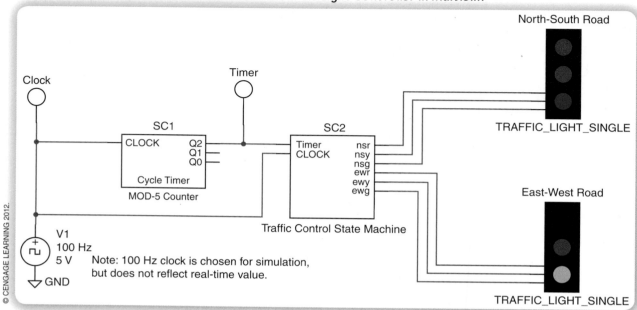

The cycle can be set to any length by changing the signal on the *TIMER* input. (The yellow light will always be on for one clock pulse in this design.) For ease of observation, we will use a cycle of ten clock pulses. For either direction, the cycle consists of 4 clocks GREEN, 1 clock YELLOW, and 5 clocks RED. This cycle can be generated by the most significant bit of a mod-5 counter, as shown in *Figure 11.26*.

Figure 11.27 shows a simulation of the mod-5 counter and output controller. The MSB of the counter goes HIGH for one clock period, then LOW for four. When applied to the *TIMER* input of the output controller, this signal directs the controller from state to state. The north-south lights are red for five clock pulses (shown by 100 in the **north_south** waveform). At the same time, the east-west lights are green for four clock pulses (**east_west** = 001), followed by yellow for one clock pulse (**east_west** = 010). The cycle continues with an **east-west** red and **north-south** green and then yellow.

FIGURE 11.27 Timing Diagram for a Traffic Light Controller

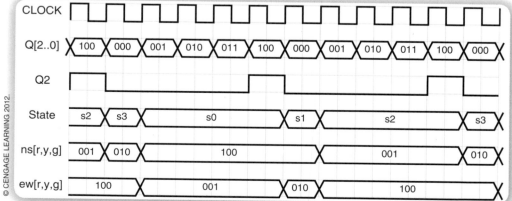

Example 11.5

Multisim Example

Multisim File: 11.08 Traffic Light Controller.ms11

Open the Multisim file for this example and run it as a simulation.

a. How many times does the Clock light pulse when the Timer light is off? How many times does the Clock light pulse when the Timer light is on?
b. When does one of the yellow lights turn on relative to the Timer light?
c. How can you make the overall traffic cycle longer with the components shown? Adjust the value of a component to make the traffic cycle twice as long.
d. What change would need to be made to the circuit to change the traffic cycle to: GREEN for eight pulses, YELLOW for one pulse, and RED for nine pulses?

FIGURE 11.28 Example 11.5: Adjusting the Frequency of the Clock Voltage Generator in the Traffic Light Controller of Figure 11.26

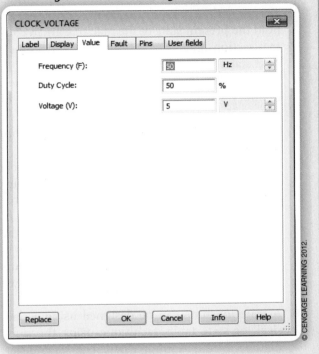

Solution

a. The Clock light pulses four times when the Timer light is off and once when it is on.
b. Either the north-south yellow or the east-west yellow light turns on when the Timer light is on. Which light turns on depends on which green light (north-south or east-west) was last on.
c. The overall cycle can be made longer or shorter by adjusting the frequency of the clock signal. If you double-click on the clock voltage generator, you will see a dialog box like the one shown in *Figure 11.28*. To make the overall cycle twice as long, change the clock frequency to 50 Hz.
d. The mod-5 counter could be replaced by a mod-9 counter. This counter has a most significant bit that is LOW for eight pulses and HIGH for one pulse, which meets the requirements of the circuit.

SUMMARY

1. A state machine is a synchronous sequential circuit with a memory section (flip-flops) to hold the present state of the machine and a control section (gates) to determine the machine's next state.

2. The number of flip-flops in a state machine's memory section is the same as the number of state variables.

3. Two main types of state machine are the Moore machine and the Mealy machine.

4. The outputs of a Moore machine are entirely dependent on the states of the machine's flip-flops. Output changes will always be synchronous with the system clock.

5. The outputs of a Mealy machine depend on the states of the machine's flip-flops and the gates in the control section. A Mealy machine's outputs can change asynchronously, relative to the system clock.

6. A state machine can be designed using a classical, or state table, technique using the same method as in designing a synchronous counter, as follows:

   a. Define the problem and draw a state diagram.
   b. Construct a table of present and next states.
   c. Use flip-flop excitation tables to determine the flip-flop inputs for each state transition.
   d. Use Boolean algebra or K-maps to find the simplest Boolean expression for flip-flop inputs (D or JK) in terms of outputs (Q).
   e. Draw the logic diagram of the state machine.

7. The state names in a state machine can be named numerically (**s0, s1, s2, . . .**) or literally (**start, idle, read, write**), depending on the machine function. State names are independent of the values of the state variables.

8. Notation for a state diagram includes a series of bubbles (circles) containing state names and values of state variables in the form $\frac{\text{state name}}{\text{state variables}}$.

9. The inputs and outputs of a state machine are labeled **in1, in2, . . . , in$x$/out1, out2, . . . , out$x$**.

10. Transitions between states can be conditional or unconditional. A conditional transition happens only under certain conditions of a control input and is labeled with the relevant input condition. An unconditional transition happens under all conditions of input and is labeled with an X for each input variable.

11. Mealy machine outputs are susceptible to asynchronous output changes if a combinational input changes out of synchronization with the clock. This can be remedied by clocking each output through a separate synchronizing flip-flop.

12. A maximum of 2^n states can be assigned to a state machine that has n state variables. If the number of states is less than 2^n, the unused states must be accounted for. Either they can be treated as don't care states, or they can be assigned a specific destination state, usually the reset state.

11.1 State Machines

11.1 Is the state machine in *Figure 11.29* a Moore machine or a Mealy machine? Explain your answer.

11.2 Is the state machine in *Figure 11.30* a Moore machine or a Mealy machine? Explain your answer.

FIGURE 11.29 Problem 11.1: State Machine Circuit

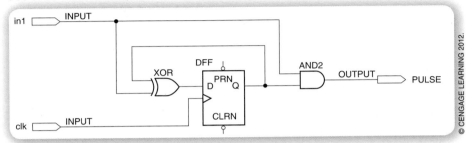

FIGURE 11.30 Problem 11.2: State Machine Circuit

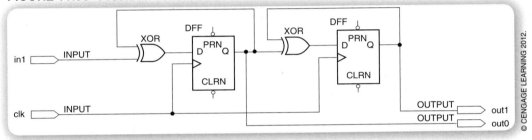

11.2 State Machines with No Control Inputs

11.3 A 4-bit Gray code sequence is shown in *Table 11.7*. Use classical design methods to design a counter with this sequence, using D flip-flops. Draw the resulting circuit diagram in Multisim and verify the circuit operation.

11.4 Use classical state machine design techniques to design a counter whose output sequence is shown in *Table 11.8*. (This is a divide-by-twelve counter in which the MSB output has a duty cycle of 50%.) Draw the state diagram, derive synchronous equations of the flip-flops, draw the circuit implementation in Multisim, and verify the circuit's function.

11.3 State Machines with Control Inputs

11.5 Use classical state machine design techniques to find the Boolean next state and output equations for the state machine represented

TABLE 11.7 4-Bit Gray Code Sequence for Problem 11.3

Q_3	Q_2	Q_1	Q_0
0	0	0	0
0	0	0	1
0	0	1	1
0	0	1	0
0	1	1	0
0	1	1	1
0	1	0	1
0	1	0	0
1	1	0	0
1	1	0	1
1	1	1	1
1	1	1	0
1	0	1	0
1	0	1	1
1	0	0	1
1	0	0	0

continues...

continued...

TABLE 11.8 Counter Sequence for Problem 11.4

Q_3	Q_2	Q_1	Q_0
0	0	0	0
0	0	0	1
0	0	1	0
0	0	1	1
0	1	0	0
0	1	0	1
1	0	0	0
1	0	0	1
1	0	1	0
1	0	1	1
1	1	0	0
1	1	0	1

© CENGAGE LEARNING 2012.

by the state diagram in *Figure 11.31*. Draw the state machine circuit in Multisim and verify the operation of the circuit. Briefly explain the intended function of the state machine.

FIGURE 11.31 Problem 11.5: State Diagram

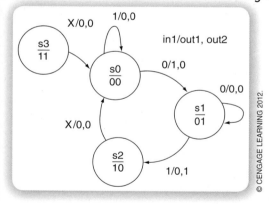

© CENGAGE LEARNING 2012.

11.6 Using classical design techniques, find the Boolean expressions for the next state and output for the state diagram shown in Figure 11.14. Use JK flip-flops.

11.4 Unused States in State Machines

11.7 Refer to the state diagram in *Figure 11.32*.

    a. How many state variables are required to implement this state machine? Why?

    b. How many unused states are there for this state machine? List the unused states.

    c. Complete the partial timing diagram shown in *Figure 11.33* to illustrate one complete cycle of the state machine represented by the state diagram of Figure 11.32.

FIGURE 11.32 Problem 11.7: State Diagram

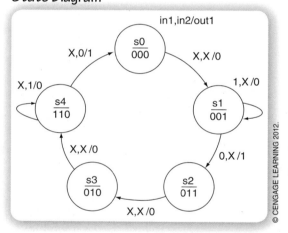

© CENGAGE LEARNING 2012.

11.8 Use classical state machine design techniques to implement the state machine described by the state diagram of Figure 11.32 using Multisim. Verify the operation of the circuit by running the Multisim files as a simulation.

FIGURE 11.33 Problem 11.7: Partial Timing Diagram

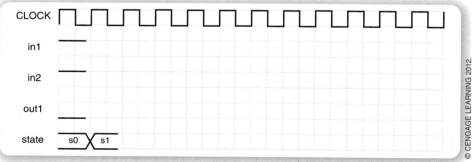

© CENGAGE LEARNING 2012.

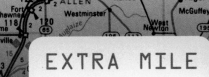

11.3 State Machines with Control Inputs

11.9 Referring to the simulation for the state machine in Problem 11.5, briefly explain why it is susceptible to asynchronous input changes. Modify the state machine circuit to eliminate the asynchronous behavior of the outputs. Verify the function of the modified state machine.

11.10 A state machine is used to control an analog-to-digital converter, as shown in the block diagram of *Figure 11.34*.

FIGURE 11.34 Problem 11.10: Analog-to-Digital Converter and Controller

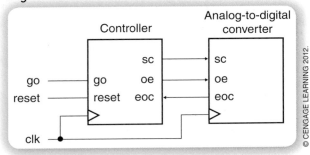

The controller has four states, defined by state variables Q_1 and Q_0 as follows: **idle** (00), **start** (01), **waiting** (11), and **read** (10). There are two outputs: **sc** (Start Conversion; active-HIGH) and **oe** (Output Enable; active-LOW). There are four inputs: **clock**, **go** (active-LOW), **eoc** (End of Conversion), and asynchronous reset (active-LOW). The machine operates as follows:

a. In the **idle** state, the outputs are: **sc** = 0, **oe** = 1. The machine defaults to the **idle** state when the machine is reset.

b. Upon detecting a 0 at the **go** input, the machine makes a transition to the **start** state. In this transition, **sc** = 1, **oe** = 1.

c. The machine makes an unconditional transition to the **waiting** state; **sc** = 0, **oe** = 1. It remains in this state, with no output change, until input **eoc** = 1.

d. When **eoc** = 1, the machine goes to the **read** state; **sc** = 0, **oe** = 0.

e. The machine makes an unconditional transition to the **idle** state; **sc** = 0, **oe** = 1.

Use classical state machine design techniques to design the controller. Draw the required circuit in Multisim and verify its operation. Is this machine vulnerable to asynchronous input change?

11.4 Unused States in State Machines

11.11 Use classical state machine design techniques to design a state machine described by the state diagram of *Figure 11.35*. Briefly describe the intended operation of the circuit. Verify the operation of the state machine design. Unused states may be treated as don't care states, but unspecified outputs should always be assigned to 0.

FIGURE 11.35 Problem 11.11: State Diagram

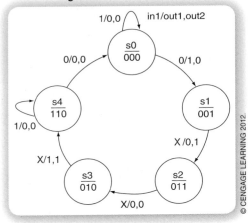

11.12 Determine the next state for each of the unused states of the state machine designed in Problem 11.11. Use this analysis to redraw the state diagram of Figure 11.35 so that it properly includes the unused states. (There is more than one right answer, depending on the result of the Boolean simplification process used in Problem 11.11 to simplify the equation for $D1$.)

11.5 Traffic Light Controller

11.13 Use classical state machine design techniques to design the state machines for the mod-5 counter and traffic light controller shown in Figure 11.26.

CHAPTER 12
Memory

Menu

START LOCATION	DISTANCE	END LOCATION

Chapter Objectives

Upon successful completion of this chapter, you will be able to:

1 Describe basic memory concepts of address and data.

2 Understand how latches and flip-flops act as simple memory devices and sketch simple memory systems based on these devices.

3 Distinguish between random access read/write memory (RAM) and read-only memory (ROM).

4 Describe the uses of tristate logic in data busing.

5 Sketch a block diagram of a static or dynamic RAM chip.

6 Describe the basic configuration of flash memory.

7 Describe the basic configuration and operation of two types of sequential memory: First In, First Out (FIFO) and Last In, First Out (LIFO).

8 Describe how dynamic RAM is configured into high-capacity memory modules.

9 Sketch a basic memory system, consisting of several memory devices, an address and a data bus, and address decoding circuitry.

In recent years, memory has become one of the most important topics in digital electronics. This is tied closely to the increasing prominence of cheap and readily available microprocessor chips. The simplest memory is a device we are already familiar with: the D flip-flop. This device stores a single bit of information as long as necessary. This simple concept is at the heart of all memory devices.

The other basic concept of memory is the organization of stored data. Bits are stored in locations specified by an "address," a unique number that tells a digital system how to find data that have been previously stored. (By analogy, think of your street address: a unique way to find you and anyone you live with.)

Some memory can be written to and read from in random order; this is called random access read/write memory (RAM). Other memory can be read only: read-only memory (ROM). Yet another type of memory, sequential memory, can be read or written only in a specific sequence. There are several variations on all these basic classes.

Memory devices are usually part of a larger system, including a microprocessor, peripheral devices, and a system of tristate buses. If dynamic RAM is used in such a system, it is often in a memory module of some type. The capacity of a single memory chip is usually less than the memory capacity of the microprocessor system in which it is used. To use the full system capacity, a method of memory address decoding is necessary to select a particular RAM device for a specified portion of system memory.

405

12.1 BASIC MEMORY CONCEPTS

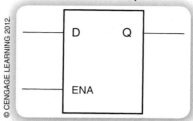

FIGURE 12.1 D-Type Latch

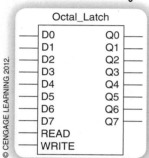

FIGURE 12.3 Octal Latch as an 8-Bit Memory

KEY TERMS

Memory A device for storing digital data so that they can be recalled for later use in a digital system.

Bit *B*inary dig*it*. A 0 or a 1.

Data Binary digits (0's and 1's) that contain some kind of information. The digital contents of a memory device.

Byte A group of 8 bits.

Address A number, represented by the binary states of a group of inputs or outputs, uniquely defining the location of data stored in a memory device.

Write Store data in a memory device.

Read Retrieve data from a memory device.

Address and Data

A **memory** is a digital device or circuit that can store one or more **bits** of **data**. The simplest memory device, a D-type latch, shown in *Figure 12.1*, can store 1 bit. A 0 or 1 is stored in the latch and remains there until changed.

A simple extension of the single D-type latch is an array of latches, shown in *Figure 12.2*, that can store 8 bits (1 **byte**) of data. *Figure 12.3* shows this octal latch built as a hierarchical block in Multisim and configured as an 8-bit memory.

FIGURE 12.2 Octal Latch

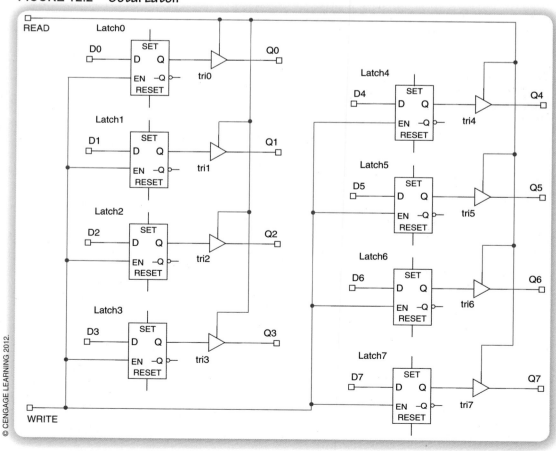

FIGURE 12.4   4 × 8-Bit Memory from Octal Latches

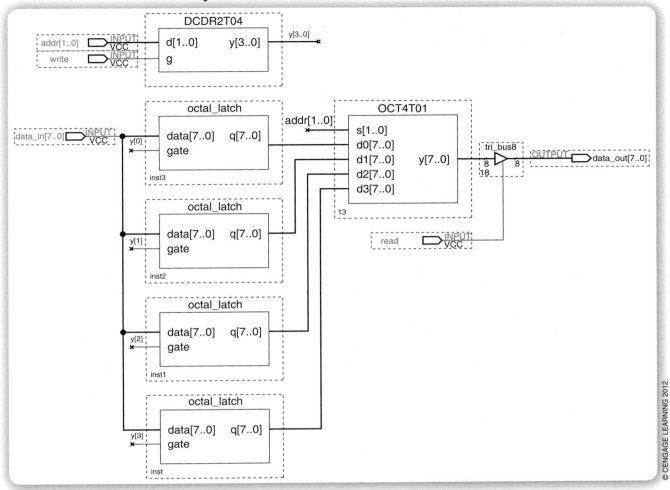

When the **WRITE** line goes HIGH, then LOW, data at the *DATA_IN* pins are stored in the eight latches. Data are available at the *DATA_OUT* pins when **READ** is HIGH. Note that although the *READ* and *WRITE* inputs are separate in this design, their functions can be implemented as opposite logic levels of the same pin. When *READ* is inactive, the outputs are in a high-impedance state.

Figure 12.4 shows a block diagram of an expanded version of the octal latch memory circuit. Four octal latches are configured to make a 4 × 8-bit memory that can store and recall four separate 8-bit words.

Each of the four octal latches can store one byte of data. The **address**, a binary number indicating which latch to use, is decoded to activate the WRITE line of only one latch. The contents of the other latches remain unchanged.

The 8-bit input data are applied to the inputs of all four octal latches simultaneously. Data are written to a particular latch when a 2-bit address and a HIGH on *WRITE* cause an output of a 2-line-to-4-line decoder (DCDR2TO4) to enable the selected latch. For example, when *ADDR[1..0] = 01* AND *WRITE = 1*, decoder output y[1] goes HIGH, activating the *ENABLE* input on latch 1. The values at *DATA_IN[7..0]* are transferred to latch 1 and stored there when *WRITE* goes LOW.

The latch outputs are applied to the data inputs of an octal 4-to-1 multiplexer (OCT4TO1). Recall that this circuit will direct one of four 8-bit inputs to an 8-bit output. The selected set of inputs correspond to the binary value at the MUX select inputs, which is the same as the address applied to the decoder in the write phase. The MUX output is directed to the *DATA_OUT* lines by an octal tristate bus driver, which is enabled by the *READ* line. To read the contents of latch 1, we set the address to 01, as before, and make the *READ* line HIGH. If *READ* is LOW, the *DATA_OUT* lines are in the high-impedance state.

RAM and ROM

> **KEY TERMS**
>
> **Random access memory (RAM)** A type of memory device where data can be accessed in any order, that is, randomly. The term usually refers to random access read/write memory.
>
> **Read-only memory (ROM)** A type of memory where data are permanently stored and can only be read, not written.

The memory circuit in Figure 12.4 is one type of **random access memory**, or **RAM**. Data can be stored in or retrieved from any address at any time. The data can be accessed randomly, without the need to follow a sequence of addresses, as would be necessary in a sequential storage device such as magnetic tape, sometimes used to back up large computer storage units or servers.

RAM has come to mean random access read/write memory, memory that can have its data changed by a write operation, and have its data read. The data in another type of memory, called **read-only memory**, or **ROM**, can also be accessed randomly, although it cannot be changed, or at least not changed as easily as RAM; there is no write function; hence the name "read only." Even though both types of memory are random access, we generally do not include ROM in this category.

Memory Capacity

> **KEY TERMS**
>
> **b** Bit.
>
> **B** Byte.
>
> **K** 1024 ($= 2^{10}$). Analogous to the metric prefix "k" (kilo-).
>
> **M** 1,048,576 ($= 2^{20}$). Analogous to the metric prefix "M" (mega-).
>
> **T** 1,099,511,627,776 ($= 2^{40}$). Analogous to the metric prefix "T" (terra-).
>
> **G** 1,073,741,824 ($= 2^{30}$). Analogous to the metric prefix "G" (giga-).

The capacity of a memory device is specified by the address and data sizes. The circuit shown in Figure 12.4 has a capacity of 4×8 bits ("four-by-eight"). This tells us that the memory can store 32 bits (32**b**), organized in groups of 8 bits at 4 different locations. This memory can also be described as having a capacity of 4 bytes (4**B**), since there are 8 bits per byte.

For large memories, with capacities of thousands to trillions of bits, we use the shorthand designations **K**, **M**, or **T** as prefixes for large binary numbers. The prefix K is analogous to, but not the same as, the metric prefix k (kilo). The metric kilo (lowercase k) indicates a multiplier of $10^3 = 1000$; the binary prefix K (uppercase) indicates a multiplier of $2^{10} = 1024$. Thus, one kilobit (Kb) is 1024 bits.

Similarly, the binary prefix M is analogous to the metric prefix M (mega). Both, unfortunately, are represented by uppercase M. The metric prefix represents a multiplier of $10^6 = 1,000,000$; the binary prefix M represents a value of $2^{20} = 1,048,576$. One megabit (Mb) is 1,048,576 bits. The next extension of this system is the multiplier G ($= 2^{30}$), which is analogous to the metric prefix **G** (giga; 10^9), then is the multiplier T ($= 2^{40}$), which is analogous to the metric prefix T (tera; 10^{12}).

Example 12.1

A small microcontroller system (i.e., a stand-alone microcomputer system designed for a particular control application) has a memory with a capacity of 64 Kb, organized as 8K $\times$ 8. What is the total memory capacity of the system in bits? What is the memory capacity in bytes?

Solution

The total number of bits in the system memory is:

$$8K \times 8 = 8 \times 8 \times 1K = 64 \text{ Kb} = 64 \times 1024 \text{ bits} = 65,536 \text{ bits}$$

The number of bytes in system memory is:

$$\frac{64 \text{ Kb}}{8 \text{b/B}} = 8 \text{ KB}$$

Usually, the range of numbers spanning 1K is expressed as the 1024 numbers from 0_{10} to 1023_{10} (0000000000_2 to 1111111111_2). This is the full range of numbers that can be expressed by 10 bits. In hexadecimal, the range of numbers spanning 1K is from 000H to 3FFH. The range of numbers in 1M is given as the full hexadecimal range of 20-bit numbers: 00000H to FFFFFH.

The range of numbers spanning 8K can be written in 13 bits ($8 \times 1K = 2^3 \times 2^{10} = 2^{13}$). The binary addresses in an 8K $\times$ 8 memory range from 0000000000000 to 1111111111111, or 0000 to 1FFF in hexadecimal. Thus, a memory device that is organized as 8K $\times$ 8 has 13 address lines and 8 data lines.

Figure 12.5 shows the address and data lines of an 8K $\times$ 8 memory and a map of its contents. The addresses progress in binary order, but the contents of any location are the last data stored there. Because there is no way to predict what

FIGURE 12.5   Address and Data in an 8K $\times$ 8 Memory

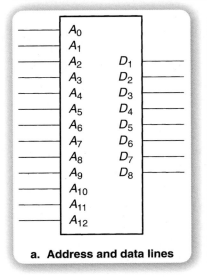

a. **Address and data lines**

b. **Contents (data) and location (address)**

D_8 ... D_1	Binary	Hexadecimal
1 0 1 1 0 1 0 1	0 0000 0000 0000	0000
0 0 0 1 1 0 1 1	0 0000 0000 0001	0001
1 1 0 1 0 0 1 1	0 0000 0000 0010	0002
0 0 0 0 0 1 1 1	0 0000 0000 0011	0003
0 1 1 1 0 1 1 1	0 0000 0000 0100	0004
1 0 0 0 1 0 1 0	0 0000 0000 0101	0005
0 1 0 1 1 1 1 1	0 0000 0000 0110	0006
⋮	⋮	⋮
1 0 1 0 1 0 1 0	1 1111 1111 1101	1FFD
0 0 0 1 1 1 1 1	1 1111 1111 1110	1FFE
1 1 0 0 1 0 1 1	1 1111 1111 1111	1FFF

those data are, they are essentially random. For example, in Figure 12.5, the byte at address 0000000000100_2 (0004H) is 01110111_2 (77H). (One can readily see the advantage of using hexadecimal notation.)

Example 12.2

How many address lines are needed to access all addressable locations in a memory that is organized as 64K × 4? How many data lines are required?

Solution

Addressable locations: $2^n = 64K$

$$64K = 64 \times 1K = 2^6 \times 2^{10} = 2^{16}$$
$$n = 16 \text{ address lines}$$

Data lines: There are 4 data bits for each addressable location. Thus, the memory requires 4 data lines.

Control Signals

Two memory devices are shown in *Figure 12.6*. The device in Figure 12.6a is a 1K × 4 random access read/write memory (RAM). Figure 12.6b shows 8K × 8 erasable programmable read-only memory (EPROM). The address lines are designated by A and the data lines by DQ. The dual notation DQ indicates that these lines are used for both input (D) and output (Q) data, using the conventional designations of D-type latches. (Note that the data inputs on the ROM can be used only under special conditions. They are used to load permanent or semi-permanent data into the device, a process known as **programming** or **burning**.) The input and output data are prevented from interfering with one another by a pair of opposite-direction tristate buffers on each input/output pin. One buffer goes to a memory cell input; the other comes from the memory cell output. The tristate outputs on the devices in Figure 12.6 allow the outputs to be electrically isolated from a system data bus that would connect several such devices to a microprocessor.

In addition to the address and data lines, most memory devices, including those in Figure 12.6, have one or more of the following control signal inputs. (Different manufacturers use different notation, so several alternate designations for each function are listed.)

$\overline{E}$(or $\overline{CE}$ or $\overline{CS}$). Enable (or $\overline{\text{Chip Enable}}$ or $\overline{\text{Chip Select}}$). The memory is enabled when this line is pulled LOW. If this line is HIGH, the memory cannot be written to or read from.

FIGURE 12.6 Address, Data, and Control Signals

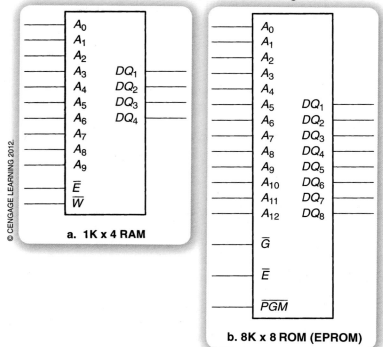

a. 1K x 4 RAM

b. 8K x 8 ROM (EPROM)

© CENGAGE LEARNING 2012.

$\overline{W}$ (or $\overline{WE}$ or $R/\overline{W}$). Write (or $\overline{\text{Write Enable}}$ or Read/$\overline{\text{Write}}$). This input is used to select the read or write function when data input and output are on the same lines. When HIGH, this line selects the read (output) function if the chip is selected. When LOW, the write (input) function is selected.

$\overline{G}$ (or $\overline{OE}$). **Gate** (or **Output Enable**). Some memory chips have a separate control to enable their tristate output buffers. When this line is LOW, the output buffers are enabled and the memory can be read. If this line is HIGH, the output buffers are in the high-state. The chip select performs this function in devices without output enable pins.

The electrical functions of these control signals are illustrated in *Figure 12.7*.

FIGURE 12.7 *Memory Control Signals*

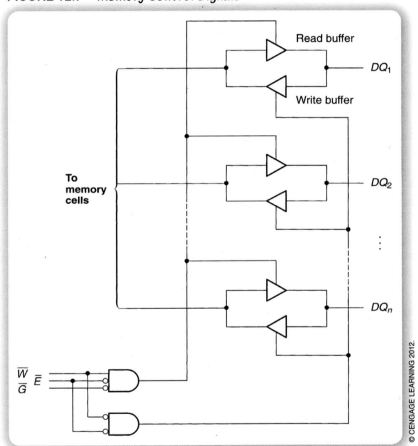

© CENGAGE LEARNING 2012.

12.2   RANDOM ACCESS READ/WRITE MEMORY (RAM)

KEY TERMS

Volatile A memory is volatile if its stored data are lost when electrical power is lost.

Static RAM A random access memory that can retain data indefinitely as long as electrical power is available to the chip.

RAM cell The smallest storage unit of a RAM, capable of storing 1 bit.

Dynamic RAM A random access memory that cannot retain data for more than a few milliseconds without being "refreshed."

Random access read/write memory (RAM) is used for temporary storage of large blocks of data. An important characteristic of RAM is that it is **volatile**. It can retain its stored data only as long as power is applied to the memory. When power is lost, so are the data. There are two main RAM configurations: static (SRAM) and dynamic (DRAM).

Static RAM (SRAM) consists of arrays of memory cells that are essentially flip-flops. Data can be stored in a static **RAM cell** and left there indefinitely, as long as power is available to the RAM.

A **dynamic RAM** cell stores a bit as the charged or discharged state of a small capacitor. Because the capacitor can hold its charge for only a few milliseconds, the charge must be restored ("refreshed") regularly. This makes a dynamic RAM (DRAM) system more complicated than SRAM, as it introduces a requirement for memory refresh circuitry.

DRAMs have the advantage of large memory capacity over SRAMs. Memory capacity figures are constantly increasing and are never up to date for very long. The most famous estimate of the growth rate of semiconductor memory capacity, Moore's law, estimates that it doubles every two years.

Static RAM Cell Arrays

> **KEY TERMS**
>
> **Word** Data accessed at one addressable location.
>
> **Word-organized** A memory is word-organized if one address accesses one word of data.
>
> **Word length** Number of bits in a word.

Static RAM cell arrays are arranged in a square or rectangular format, accessible by groups in rows and columns. Individual cells are selected by activating the appropriate ROW and COL lines, as shown in *Figure 12.8.*

FIGURE 12.8 SRAM Cell Array

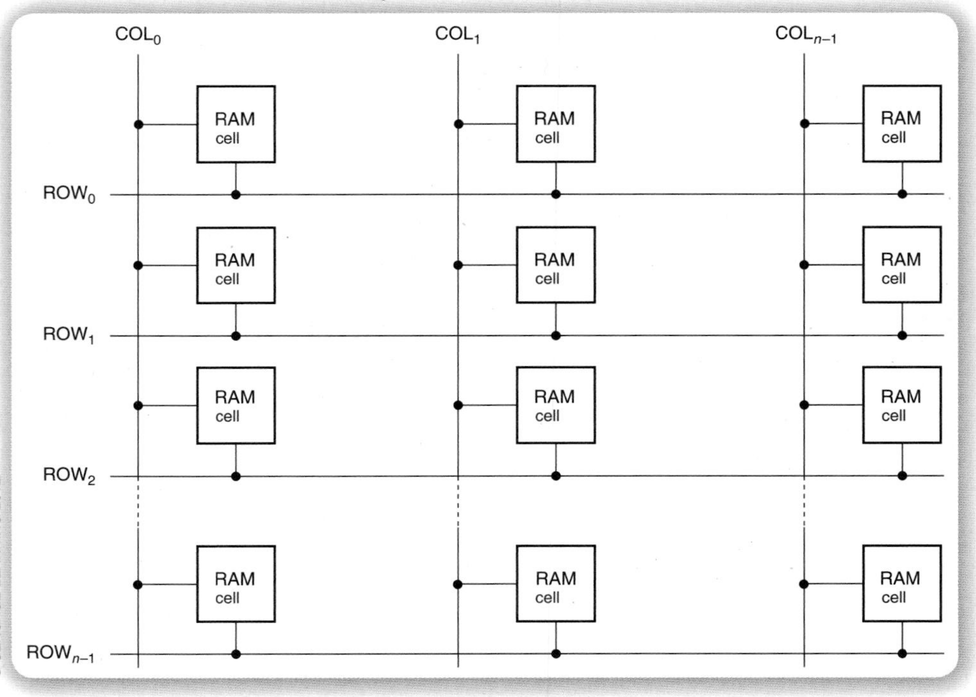

© CENGAGE LEARNING 2012.

FIGURE 12.9 *SRAM Cell Array*

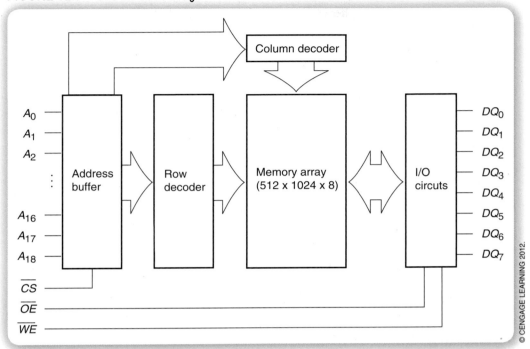

Figure 12.9 shows the block diagram of a 4-megabit (Mb) SRAM array, including blocks for address decoding and output circuitry. The RAM cells are arrayed in a pattern of 512 rows and 8192 columns for efficient packaging. When a particular address is applied to address lines $A_{18} \ldots A_0$, the row and column decoders select an SRAM cell in the memory array for a read or write by activating the associated sense amps for the column and the row select line for the cell. (A sense amp is a circuit, shared by a column of RAM cells, that amplifies the charge on the selected RAM cell's bit output line.)

The columns are further subdivided into groups of eight, so that one column address selects eight bits (one byte) for a read or write operation. Thus, there are 512 separate row addresses (9 bits) and 1024 separate column addresses (10 bits) for every unique group of 8 data bits, requiring a total of 19 address lines and 8 data lines. The capacity of the SRAM can be written as $512 \times 1024 \times 8$.

Because one address reads or writes 8 cells (an 8-bit **word**), we say that the SRAM in Figure 12.9 is **word-organized** and that the **word length** of the SRAM is 8 bits. Other popular word lengths for various memory arrays are 4, 16, 32, and 64 bits.

Your Turn

12.1 If an SRAM array is organized as $512 \times 512 \times 16$, how many address and data lines are required? How does the bit capacity of this SRAM compare to that of Figure 12.9?

Dynamic RAM Cells

KEY TERM

Refresh cycle The process that periodically recharges the storage capacitors in a dynamic RAM.

A dynamic RAM (DRAM) cell consists of a capacitor and a pass transistor (a MOSFET), as shown in *Figure 12.10*. A bit is stored in the cell as the charged or discharged state of the capacitor. The bit location is read from or written to by activating the cell MOSFET via the Word Select line, thus connecting the capacitor to the *BIT* line.

The major disadvantage of dynamic RAM is that the capacitor will eventually discharge by internal leakage current and must be recharged periodically to maintain integrity of the stored data. The recharging of the DRAM cell capacitors, known as refreshing the memory, must be done every 8 ms to 64 ms, depending on the device.

The **refresh cycle** adds an extra level of complication to the DRAM hardware and also to the timing of the read and write cycles, because the memory might have to be refreshed between read and write tasks. DRAM timing cycles are much more complicated than the equivalent SRAM cycles.

This inconvenience is offset by the high bit densities of DRAM, which are possible due to the simplicity of the DRAM cell.

FIGURE 12.10 *Dynamic RAM Cell*

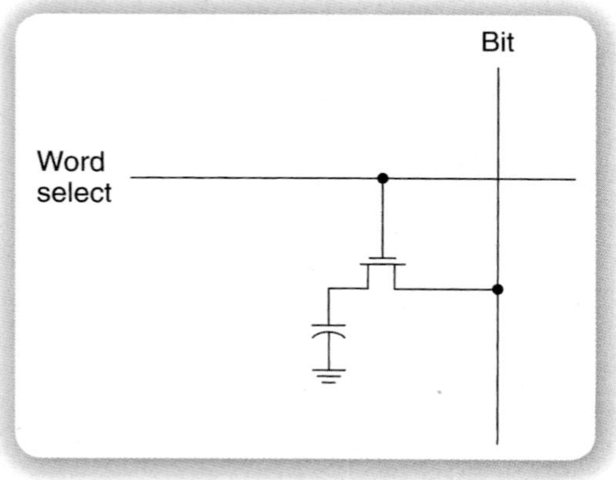

DRAM Cell Arrays

KEY TERMS

Bit-organized A memory is bit-organized if one address accesses one bit of data.

Address multiplexing A technique of addressing storage cells in a dynamic RAM that sequentially uses the same inputs for the row address and column address of the cell.

RAS **(row address strobe)** A signal used to latch the row address into the decoding circuitry of a dynamic RAM with multiplexed addressing.

CAS **(column address strobe)** A signal used to latch the column address into the decoding circuitry of a dynamic RAM with multiplexed addressing.

Dynamic RAM is sometimes **bit-organized** rather than word-organized. That is, one address will access one bit rather than one word of data. A bit-organized DRAM with a large capacity requires more address lines than a static RAM (e.g., 4 Mb $\times$ 1 DRAM requires 22 address lines [2^{22} = 4,194,304 = 4M] and 1 data line to access all cells).

To save pins on the IC package, a system of **address multiplexing** is used to specify the address of each cell. Each cell has a row address and a column address, which use the same input pins. Two negative-edge signals called **row address strobe** (*RAS*) and **column address strobe** (*CAS*) latch the row and column addresses into the DRAM's decoding circuitry. *Figure 12.11* shows a simplified block diagram of the row and column addressing circuitry of a 1M $\times$ 1 dynamic RAM.

Figure 12.12 shows the relative timing of the address inputs of a dynamic RAM. The first part of the address is applied to the address pins and latched into the row address buffers when *RAS* goes LOW. The second part of the address is then applied to the address pins and latched into the column address buffers by

FIGURE 12.11   Row and Column Decoding in a 1M × 1 Dynamic RAM

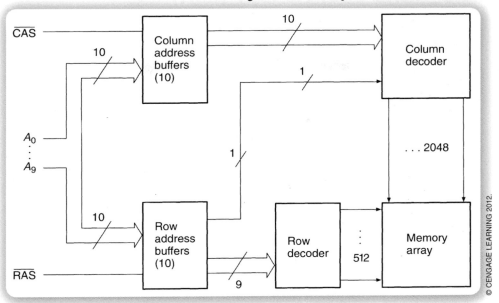

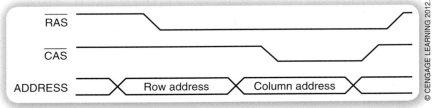

FIGURE 12.12   DRAM Address Latch Signals

the *CAS* signal. This allows a 20-bit address to be implemented with 12 pins: 10 address and 2 control lines. Adding another address line effectively adds 2 bits to the address, allowing access to 4 times the number of cells.

The memory cell array in Figure 12.11 is rectangular, not square. One of the row address lines is connected internally to the column address decoder, resulting in a 512-row-by-2048-column memory array.

One advantage to the rectangular format shown is that it cuts the memory refresh time in half, because all the cells are refreshed by accessing the rows in sequence. Fewer rows means a faster refresh cycle. All cells in a row are also refreshed by normal read and write operations.

Your Turn

12.2 How many address and data lines are required for the following sizes of dynamic RAM, assuming that each memory cell array is organized in a square format, with common row and column address pins?

a. 1M × 1

b. 1M × 4

c. 4M × 1

12.3 READ-ONLY MEMORY (ROM)

KEY TERMS

Software Programming instructions required to make hardware perform specified tasks.

Hardware The electronic circuit of a digital or computer system.

Firmware Software instructions permanently stored in ROM.

The main advantage of read-only memory (ROM) over random access read/write memory (RAM) is that ROM is nonvolatile. It will retain data even when electrical power is lost to the ROM chip. The disadvantage is that stored data are difficult or impossible to change.

ROM is used for storing data required for tasks that never or rarely change, such as software instructions for a bootstrap loader in a personal computer or microcontroller (the hardware).

Examples of applications suitable for ROM include:

▶ Bootstrap loaders and BIOS (basic input/output system) for PCs.
▶ Character generators (decoders that convert ASCII codes into alphanumeric characters on a CRT or LCD display).
▶ Function lookup tables (tables corresponding to binary values of trigonometric, exponential, or other functions).
▶ Special software instructions that must be permanently stored and never changed. Software instructions stored in ROM are called firmware.

Types of ROM

KEY TERM

Mask-programmed ROM A type of read only memory (ROM) where the stored data are permanently encoded into the memory device during the manufacturing process.

EPROM Erasable programmable read-only memory. A type of ROM that can be programmed ("burned") by the user and erased later, if necessary, by exposing the chip to ultraviolet radiation.

EEPROM (or E^2PROM) Electrically erasable programmable read-only memory. A type of read-only memory that can be field-programmed and selectively erased while still in a circuit.

The most permanent form of read-only memory is the mask-programmed ROM, where the stored data are manufactured into the memory chip. Due to the inflexibility of this type of ROM and the relatively high cost of development, it is used only for well-developed high-volume applications. However, even though development cost of a mask-programmed ROM is high, volume production is cheaper than for some other types of ROM.

Erasable programmable read-only memory (EPROM) combines the nonvolatility of ROM with the ability to change the internal data if necessary. This erasability is particularly useful in the development of a ROM-based system. To erase an EPROM, the die (i.e., the silicon chip itself) must be exposed for about 20 to 45 minutes to high-intensity ultraviolet (UV) light of a specified wavelength (2537 angstroms) at a distance of 2.5 cm (1 inch).

Note . . .

The bootstrap loader—a term derived from the whimsical idea of "pulling oneself up by one's bootstraps," that is, starting from nothing—is the software that gives the personal computer its minimum startup information. Generally, it contains the instructions needed to read a magnetic disk containing further operating instructions. This task is always the same for any given machine and is needed every time the machine is turned on, thus making it the ideal candidate for ROM storage.

EPROMs are manufactured with a quartz window over the die to allow the UV radiation in. Because both sunlight and fluorescent light contain UV light of the right wavelength to erase the EPROM over time (several days to several years, depending on the intensity of the source), the quartz window is usually covered by an opaque label after the EPROM has been programmed.

Electrically erasable programmable read-only memory (EEPROM or E^{2}PROM) provides the advantage of allowing erasure of selected bits while the chip is in the circuit; it combines the read/write properties of RAM with the nonvolatility of ROM. EEPROM is useful for storage of data that need to be changed occasionally, but that must be retained when power is lost to the EEPROM chip. One example is the memory circuit in an electronically tuned car radio that stores the channel numbers of local stations.

Flash Memory

KEY TERMS

Flash memory A nonvolatile type of memory that can be programmed and erased in sectors, rather than byte-at-a-time.

Sector A segment of flash memory that forms the smallest amount that can be erased and reprogrammed at one time.

Boot block A sector in a flash memory reserved for primary firmware.

Top boot block A boot block sector in a flash memory placed at the highest address in the memory.

Bottom boot block A boot block sector in flash memory placed at the lowest address in the memory.

Flash memory is a type of electrically erasable programmable memory which can be programmed and reprogrammed while in a circuit. This type of nonvolatile memory generally has a large byte capacity (many GB) and thus can be used to store large amounts of firmware, such as the BIOS (basic input/output system) of a PC. Flash is commonly used as a storage medium for personal data, such as documents, digital photos, videos, or music.

A flash memory is divided into sectors, groups of bytes that are programmed and erased at one time. One sector is designated as the boot block, which is either the sector with the highest (top boot block) or lowest (bottom boot block) address. The primary firmware is usually stored in the boot block, with the idea that the system using the flash memory is configured to look there first for firmware instructions. The boot block can also be protected from unauthorized erasure or modification (e.g., by a virus), thus adding a security feature to the device.

Figure 12.13 shows the arrangement of sectors of a 512K × 8-bit (4 Mb) flash memory with a bottom boot block architecture. The range of addresses is shown alongside the blocks. For example, sector S0 (the boot block) has a 16 KB address range of 00000H to 03FFFH. Sector S1 has an 8 KB address range from 04000H to 05FFFH. The first 64 KB of the memory are divided into one 16 KB, two 8 KB, and one 32 KB sector. The remainder of the memory is divided into equal 64 KB sectors. Note that even though the boot block is drawn at the top of Figure 12.13, it is a bottom boot block because it is the sector with the lowest address.

FIGURE 12.13 *Sectors in a 512k × 8b Flash Memory (Bottom Boot Block)*

Address	Sector	Size
00000H	S0 (Boot block)	16 KB
04000H	S1	8 KB
06000H	S2	8 KB
08000H	S3	32 KB
10000H	S4	64 KB
20000H	S5	64 KB
30000H	S6	64 KB
40000H	S7	64 KB
50000H	S8	64 KB
60000H	S9	64 KB
70000H	S10	64 KB
7FFFFH		

© CENGAGE LEARNING 2012.

A flash memory with a top boot block would have the same proportions given over to its sectors, but mirror-image to the diagram in Figure 12.13. That is, S10 (boot block) would be a 16 KB sector from 7C000H to 7FFFFH. The other sectors would be identical to the bottom boot block architecture, but in reverse order.

Your Turn

12.3 A flash memory has a capacity of 8 Mb, organized as 1M × 8 bit. List the address range for the 32-KB boot block sector of the memory if the device has a bottom boot block architecture and if it has a top boot block architecture.

12.4 SEQUENTIAL MEMORY: FIFO AND LIFO

KEY TERMS

Sequential memory Memory in which the stored data cannot be read or written in random order, but must be addressed in a specific sequence.

Queue A FIFO memory.

Stack A LIFO memory.

FIFO (First In, First Out) A sequential memory in which the stored data can be read only in the order in which it was written.

LIFO (Last In, First Out) A sequential memory in which the last data written are the first data read.

Stack pointer The address in memory of the top of the stack: the stack pointer itself is also stored in memory.

The RAM and ROM devices we have examined up until now have all been random access devices. That is, any data could be read from or written to any sequence of addresses in any order. There is another class of memory in which the data must be accessed in a particular order. Such devices are called **sequential memory**.

There are two main ways of organizing a sequential memory—as a **queue** or as a **stack**. *Figure 12.14* shows the arrangement of data in each of these types of memory.

A queue is a **First In, First Out** (**FIFO**) memory, meaning that the data can be read only in the same order they are written, much as railway cars always come out of a tunnel in the same order they go in.

One common use for FIFO memory is to connect two devices that have different data rates. For instance, a computer can send data to a printer much faster than the printer can use it. To keep the computer from either waiting for the printer to print everything or periodically interrupting the computer's operation to continue the print task, data can be sent in a burst to a FIFO, where the printer can read them as needed. The only provision is that there must be some logic signal to the computer telling it when the queue is full and not to send more data and another signal to the printer letting it know that there are some data to read from the queue.

The **Last In, First Out** (**LIFO**), or stack, memory configuration, also shown in Figure 12.14, is not available as a special chip, but rather is a way of organizing RAM in a memory system.

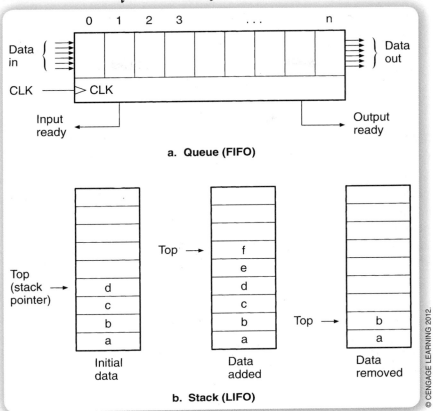

FIGURE 12.14 *Sequential Memory*

0 1 2 3 ... n

Data in { — CLK — > CLK — } Data out

Input ready — Output ready

a. Queue (FIFO)

Top (stack pointer) → | d | c | b | a |

Initial data

Top → | f | e | d | c | b | a |

Data added

Top → | b | a |

Data removed

b. Stack (LIFO)

© CENGAGE LEARNING 2012.

The term "stack" is analogous to the idea of a spring-loaded stack of plates in a cafeteria line. When you put a bunch of plates on the stack, they settle into the recessed storage area. When a plate is removed, the stack springs back slightly and brings the second plate to the top level. (The other plates, of course, all move up a notch.) The top plate is the only one available for removal from the stack, and plates are always removed in reverse order from that in which they were loaded.

Figure 12.14b shows how data are transferred to and from a LIFO memory. A block of addresses in a RAM is designated as a stack, and several bytes of data in the RAM store a number called the **stack pointer**, which is the current address of the top of the stack. (The size of the stack pointer depends on the width of the system address bus.)

In Figure 12.14, the value of the stack pointer changes with every change of data in the stack, pointing to the last-in data in every case. When data are removed from the stack, the stack pointer is used to locate the data that must be read first. After the read, the stack pointer is modified to point to the next-out data. Some stack configurations have the stack pointer pointing to the next empty location on the stack.

A common application for LIFO memory is in a computer system. If a program is interrupted during its execution by a demand from the program or some piece of hardware that needs attention, the status of various registers within the computer are stored on a stack and the computer can pay attention to the new demand, which will certainly change its operating state. After the interrupting task is finished, the original operating state of the computer can be taken from the top of the stack and reloaded into the appropriate registers, and the program can resume where it left off.

Your Turn

12.4 State the main difference between a stack and a queue.

12.5 DYNAMIC RAM MODULES

KEY TERMS

Memory modules A small circuit board containing several dynamic RAM chips.

Single in-line memory module (SIMM) A memory module with DRAMs and connector pins on one side of the board only.

Dual in-line memory module (DIMM) A memory module with DRAMs and connector pins on both sides of the board.

Small-outline dual in-line memory module (SoDIMM) A smaller version of a DIMM often used in laptop or tablet computers.

Dynamic RAM chips are often combined on a small circuit board to make a **memory module**. This is because the data bus widths of systems requiring the DRAMs are not always the same as the DRAMs themselves. For example, *Figure 12.15* shows how four 64M × 8 DRAMs are combined to make a 64M × 32 memory module. The block diagram of the module is shown in Figure 12.15, and the mechanical outline is shown in *Figure 12.16*. The data input/output lines are separate from one another so that there are 32 data I/Os (*DQ*). The address lines (*ADDR*[12..0]) for the module are parallel on all chips. With address multiplexing, this 13-bit address bus yields a 26-bit address, giving a 64M address range. Chip selects (*CS*) for all devices are connected together so that selecting the module selects all chips on the module.

This particular memory module is configured as a **single in-line memory module (SIMM)**, which has the DRAM chips and pin connections on one side of the board only. A **dual in-line memory module (DIMM)** or **small-outline dual in-line memory module (SoDIMM)** has the DRAMs mounted on both sides of the circuit board and pin connections on both sides of the board as well.

Figure 12.17 shows the layout of a 168-pin DRAM. There are 84 pins on each side of the board. So DIMMs with a higher memory capacity can have 200 pins (100 on each side), and DIMMs can have 240 pins (120 on each side).

FIGURE 12.15 SIMM Block Diagram

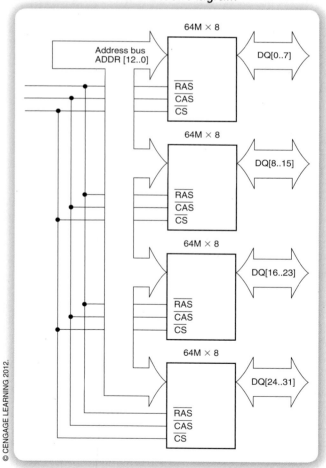

FIGURE 12.16 144-Pin SIMM Layout

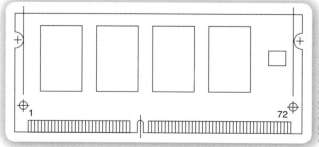

FIGURE 12.17   168-Pin DIMM Layout

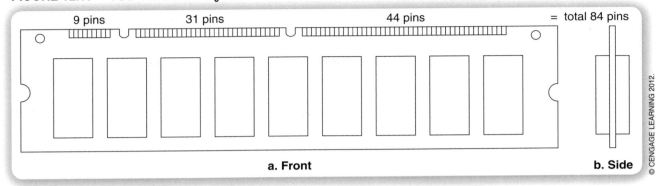

9 pins   31 pins   44 pins   = total 84 pins

a. Front   b. Side

© CENGAGE LEARNING 2012.

DRAM Types

KEY TERMS

SDRAM (synchronous DRAM) Dynamic RAM whose data are synchronously transferred to and from the data bus.

DDR (Double Data Rate) SDRAM that uses both edges of the clock to transfer data.

DDR2 (also called Double Data Rate) SDRAM that uses both edges of the clock and a different internal clock speed to transfer data twice as fast as DDR RAM.

A limitation of standard dynamic RAM is the speed of data transfer between the RAM and the system using it. Historically, data transfer in DRAM has been asynchronous; after various control inputs ($\overline{WE}$, $\overline{RAS}$, $\overline{CAS}$) have been applied, data are available on the bus as soon as they can get there. This asynchronous behavior has made data transfer to and from these DRAMs difficult to interface with other system components, such as a microprocessor, often requiring wait states that add up over time, slowing down general system performance.

One solution has been the development of **SDRAM** (synchronous DRAM), which uses a clock signal to synchronously transfer data to the system bus. The synchronous nature of this type of DRAM makes the availability of data more predictable, thus speeding the transfer of data to and from other system components.

A further development along this line is **DDR (Double Data Rate)** SDRAM, which uses both the rising and falling edges of the clock signal to transfer data. This allows a doubling of the data I/O rate, while using the same clock frequency as for the original SDRAM system. Systems that incorporate **DDR2** (also called **Double Data Rate**) SDRAM use a clock signal that is half of the system clock, effectively transferring data at 4 times the normal rate.

Your Turn

12.5 A SIMM has a capacity of 16M × 32. How many 16M × 8 DRAMs are required to make this SIMM? How many address lines does the SIMM require? How should the DRAMs be connected?

12.6 MEMORY SYSTEMS

In the section on memory modules, we saw how multiple memory devices can be combined to make a system that has the same number of addressable locations as the individual devices making up the system, but with a wider data bus. We can also create memory systems where the data I/O width of the system is the same as the individual chips, but where the system has more addressable locations than any chip within the system.

In such a system, the data I/O and control lines from the individual memory chips are connected in parallel, as are the lower bits of an address bus connecting the chips. However, it is important that only one memory device be enabled at any given time, to avoid **bus contention**, the condition that results when more than one output attempts to drive a common bus line. To avoid bus contention, one or more additional address lines must be decoded by an **address decoder** that allows only one chip to be selected at a time.

Figure 12.18 shows two 32K $\times$ 8 SRAMs connected to make a 64K $\times$ 8 memory system. A single 32K $\times$ 8 SRAM, as shown in Figure 12.18a, requires 15 address lines, 8 data lines, a write enable ($\overline{WE}$), and chip select ($\overline{CS}$) line. To make a 64K $\times$ 8 SRAM system, all of these lines are connected in parallel, except the ($\overline{CS}$) lines. To enable only one at a time, we use one more address line, A_{15}, and enable the top SRAM when $A_{15} = 0$ and the bottom SRAM when $A_{15} = 1$.

The address range of one 32K $\times$ 8 SRAM is given by the range of states of the address lines $A[14..0]$:

Lowest single-chip address:	000 0000 0000 0000 = 0000H
Highest single-chip address:	111 1111 1111 1111 = 7FFFH

The address range of the whole system must also account for the A_{15} bit:

Lowest system address:	0000 0000 0000 0000 = 0000H
Highest system address:	1111 1111 1111 1111 = FFFFH

Within the context of the system, each individual SRAM chip has a range of addresses, depending on the state of A_{15}. Assume that $SRAM_0$ is selected when $A_{15} = 0$ and $SRAM_1$ is selected when $A_{15} = 1$.

Lowest $SRAM_0$ address:	0000 0000 0000 0000 = 0000H
Highest $SRAM_0$ address:	0111 1111 1111 1111 = 7FFFH
Lowest $SRAM_1$ address:	1000 0000 0000 0000 = 8000H
Highest $SRAM_1$ address:	1111 1111 1111 1111 = FFFFH

Figure 12.19 shows a **memory map** of the 64K $\times$ 8 SRAM system, indicating the range of addresses for each device in the system. The total range of addresses in the system is called the **address space**.

FIGURE 12.18   Expanding Memory Space

FIGURE 12.19   Memory Map

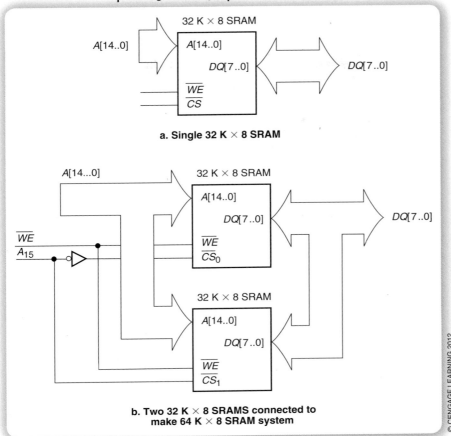

a. Single 32 K × 8 SRAM

b. Two 32 K × 8 SRAMS connected to make 64 K × 8 SRAM system

© CENGAGE LEARNING 2012.

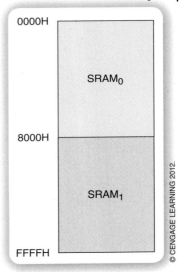

© CENGAGE LEARNING 2012.

Example 12.3

Figure 12.20 shows a memory map for a system with an address space of 64K (16 address lines). Two 16K × 8 blocks of SRAM are located at start addresses of 0000H and 8000H, respectively. Sketch a memory system that implements the memory map of Figure 12.20.

Solution

A 16K address block requires 14 address lines, because

$$16K = 16 \times 1024 = 2^4 \times 2^{10} = 2^{14}$$

The entire 64K address space requires 16 address lines, because

$$64K = 64 \times 1024 = 2^6 \times 2^{10} = 2^{16}$$

The highest address in a block is the start address plus the block size.

	16K block size:	11 1111 1111 1111 = 3FFFH
$SRAM_0$:	Lowest address:	0000 0000 0000 0000 = 0000H
	Highest address:	0011 1111 1111 1111 = 3FFFH
$SRAM_2$:	Lowest Address:	1000 0000 0000 0000 = 8000H
	Highest Address:	1011 1111 1111 1111 = BFFFH

FIGURE 12.20   Example 12.3: Memory Map Showing Noncontiguous Decoded Blocks

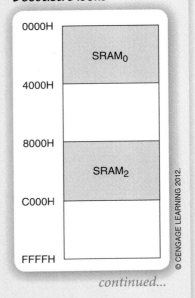

© CENGAGE LEARNING 2012.

continued...

$A_{15}A_{14} = 00$ for the entire range of the SRAM$_0$ block. $A_{15}A_{14} = 10$ for the entire SRAM$_2$ range. These can be decoded by the gates shown in *Figure 12.21*.

FIGURE 12.21 Example 12.3: 32K × 8 SRAM with Noncontiguous Blocks

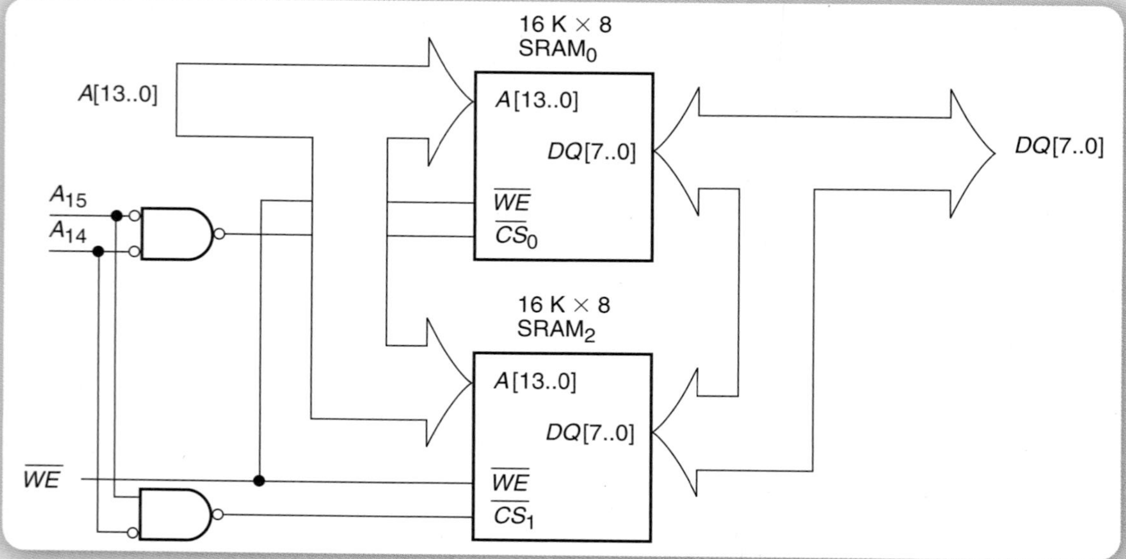

Address Decoding with *n*-Line-to-*m*-Line Decoders

Figure 12.22 shows a 64K memory system with four 16K chips: one EPROM at 0000H and three SRAMs at 4000H, 8000H, and C000H, respectively. In this circuit, the address decoding is done by a 2-line-to-4-line decoder, which can be an off-the-shelf MSI decoder, such as a 74HC139 decoder or can be programmed directly into the design on a programmable device.

Table 12.1 shows the address ranges decoded by each decoder output. The first two address bits are the same throughout any given address range. *Figure 12.23* shows the memory map for the system.

TABLE 12.1 Address Decoding for Figure 12.22

A_{15}	A_{14}	Active Decoder Output	Device	Address Range
0	0	Y_0	EPROM	0000 0000 0000 0000 = 0000H 0011 1111 1111 1111 = 3FFFH
0	1	Y_1	SRAM$_1$	0100 0000 0000 0000 = 4000H 0111 1111 1111 1111 = 7FFFH
1	0	Y_2	SRAM$_2$	1000 0000 0000 0000 = 8000H 1011 1111 1111 1111 = BFFFH
1	1	Y_3	SRAM$_3$	1100 0000 0000 0000 = C000H 1111 1111 1111 1111 = FFFFH

FIGURE 12.22 *64K Memory Map*

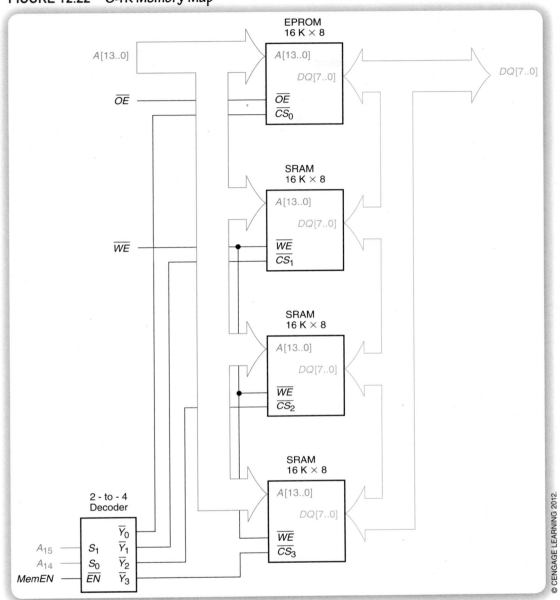

Your Turn

12.6 Calculate the number of 128K memory blocks that will fit into a 1M address space. Write the start addresses for the blocks.

FIGURE 12.23 **Memory Map for Figure 12.22**

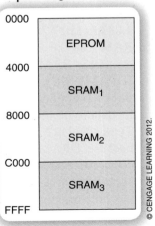

SUMMARY

1. A memory is a device that can accept data and store them for later recall.

2. Data are located in a memory by an address, a binary number at a set of address inputs that uniquely locates the block of data.

3. The operation that stores data in a memory is called the write function. The operation that recalls the stored data is the read function. These functions are controlled by functions such as write enable ($\overline{WE}$), chip select ($\overline{CS}$), and output enable ($\overline{OE}$).

4. RAM is random access memory. RAM can be written to and read from in any order of addresses. RAM is volatile. That is, it loses its data when power is removed from the device.

5. ROM is read-only memory. Original ROM devices could not be written to at all, except at the time of manufacture. Modern variations can also be written to, but not as easily as RAM. ROM is nonvolatile; it retains its data when power is removed from the device.

6. Memory capacity is given as $m \times n$ for m addressable locations and an n-bit data bus. For example, a 64K $\times$ 8 memory has 65,536 addressable locations, each with 8-bit data.

7. Large blocks of memory are designated with the binary prefixes K ($2^{10} = 1024$), M ($2^{20} = 1,048,576$), G ($2^{30} = 1,073,741,824$), and T ($2^{40} = 1,099,511,627,776$).

8. RAM can be divided into two major classes: static RAM (SRAM) and dynamic RAM (DRAM). SRAM retains its data as long as power is applied to the device. DRAM requires its data to be refreshed periodically.

9. Typically, DRAM capacity is larger than SRAM because DRAM cells are smaller than SRAM cells. An SRAM cell is essentially a flip-flop consisting of several transistors. A DRAM cell has only one transistor and a capacitor.

10. RAM cells are arranged in rectangular arrays for efficient packaging. Internal circuitry locates each cell at the intersection of a row and column within the array.

11. For packaging efficiency, DRAM addresses are often multiplexed so that the device receives half its address as a row address, latched in to the device by a $\overline{RAS}$ (row address strobe) signal and the second half as a column address, latched in by a $\overline{CAS}$ (column address strobe) signal.

12. Read-only memory (ROM) is used where it is important to retain data after power is removed.

13. Flash memory is a type of programmable ROM that is organized into sectors that are erased all at once.

14. Flash memory is often configured with one sector as a boot block, where primary firmware is stored. A bottom boot block architecture has the boot block at the lowest chip address. A top boot block architecture has the boot block at the highest chip address.

15. Sequential memory must have its data accessed in sequence. Two major classes are First In, First Out (FIFO) and Last In, First Out (LIFO). FIFO is also called a queue and LIFO is called a stack.

16. Dynamic RAM chips are often configured as memory modules, small circuit boards with multiple DRAMs. The modules usually have the same number of address locations as the individual chips on the module, but a wider data bus.

17. Memory systems can be configured to have the same data width as individual memory devices comprising the system, but with more addressable locations than any chip in the system. The additional addresses require additional system address lines, which are decoded to enable one chip at a time within the system.

BRING IT HOME

12.1 Basic Memory Concepts

12.1 How many address lines are necessary to make an 8×8 memory similar to the 4×8 memory in Figure 12.4? How many address lines are necessary to make a 16×8 memory?

12.2 Briefly explain the difference between RAM and ROM.

12.3 Calculate the number of address lines and data lines needed to access all stored data in each of the following sizes of memory:

   a. $64K \times 8$

   b. $128K \times 16$

   c. $128K \times 32$

   d. $256K \times 16$

   Calculate the total bit capacity of each memory.

12.4 Explain the difference between the chip enable $(\overline{E})$ and the output enable $(\overline{G})$ control functions in a RAM.

12.5 Refer to Figure 12.7. Briefly explain the operation of the $\overline{W}$, $\overline{E}$, and $\overline{G}$ functions of the RAM shown.

12.2 Random Access Read/Write Memory (RAM)

12.6 Briefly explain the difference between software and firmware.

12.7 Explain how a particular RAM cell is selected from a group of many cells.

12.3 Read-Only Memory (ROM)

12.8 A flash memory has a capacity of 8 Mb, organized as $512K \times 16$ bit. List the address range for the 16-KB boot block sector of the memory if the device has a bottom boot block architecture and if it has a top boot block architecture.

12.4 Sequential Memory: FIFO and LIFO

12.9 State one possible application for a FIFO and for a LIFO memory.

12.5 Dynamic RAM Modules

12.10 A SIMM has a capacity of $32M \times 64$. How many $32M \times 8$ DRAMs are required to make this SIMM? How many address lines does the SIMM require? How should the DRAMs be connected?

EXTRA MILE

12.2 Random Access Read/Write Memory (RAM)

12.11 How many address lines are required to access all elements in a $1M \times 1$ dynamic RAM with address multiplexing?

12.12 What is the capacity of an address-multiplexed DRAM with one more address line than the DRAM referred to in Problem 12.8? With two more address lines?

12.13 How many address lines are required to access all elements in a $256M \times 16$ DRAM with address multiplexing?

12.3 Read-Only Memory (ROM)

12.14 Briefly state why flash memory is unsuitable for use as system RAM.

12.6 Memory Systems

12.15 A microcontroller system with a 16-bit address bus is connected to a $4K \times 8$ RAM chip and an $8K \times 8$ RAM chip. The 8K address begins at 6000H. The 4K address block starts at 2000H. Calculate the end address for each block and show address blocks for both memory chips on a 64K memory map.

12.16 Draw the memory system of Problem 12.15.

12.17 The memory map of a microcontroller system with a 16-bit address bus is shown in *Figure 12.24*. Make a table of start and end addresses for each of the blocks shown. Indicate the size of each block.

12.18 Sketch the memory system described in Problem 12.17.

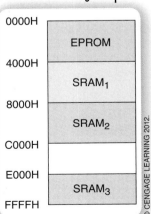

FIGURE 12.24 *Problem 12.17: Memory Map*

0000H	
	EPROM
4000H	
	SRAM$_1$
8000H	
	SRAM$_2$
C000H	
E000H	
	SRAM$_3$
FFFFH	

© CENGAGE LEARNING 2012.

GLOSSARY

1's complement A form of signed binary notation in which negative numbers are created by complementing all bits of a number, including the sign bit.

2's complement A form of signed binary notation in which negative numbers are created by adding 1 to the 1's complement form of the number.

8421 code A BCD code that represents each digit of a decimal number by its 4-bit true binary value.

A

Active HIGH An active-HIGH terminal is considered "ON" when it is in the logic HIGH state, indicated by the absence of a bubble at the terminal in distinctive-shape symbols.

Active level A logic level defined as the "ON" state for a particular circuit input or output. The active level can be either HIGH or LOW.

Active LOW An active-LOW terminal is considered "ON" when it is in the logic LOW state, indicated by a bubble at the terminal in distinctive-shape symbols.

Addend The number in an addition operation that is added to another.

Address A number, represented by the binary states of a group of inputs or outputs, uniquely defining the location of data stored in a memory device.

Address decoder A circuit enabling a particular memory device to be selected by the address bus of a larger memory system.

Address multiplexing A technique of addressing storage cells in a dynamic RAM that sequentially uses the same inputs for the row address and column address of the cell.

Address space A block of addresses in a memory system.

Adjacent cell Two cells in a Karnaugh map are adjacent if there is only one variable that is different between the coordinates of the two cells. For example, the cells for minterms ABC and $\overline{A}BC$ are adjacent.

Alloy A combination of metals which gives it a lower melting point than other metals.

Alphanumeric code A code used to represent letters of the alphabet and numerical characters.

Amplitude The instantaneous voltage of a waveform. Often used to mean maximum amplitude, or peak voltage, of a pulse.

Analog A representation of a physical, continuous quantity. An analog voltage or current can have any value within a defined range.

AND gate A logic circuit whose output is HIGH when all inputs (e.g., A AND B AND C) are HIGH.

Aperiodic waveform A time-varying sequence of logic HIGHs and LOWs that does not repeat.

ASCII American Standard Code for Information Interchange. A 7- or 8-bit code for representing alphanumeric and control characters.

Asynchronous Not synchronized to the system clock.

Asynchronous inputs The inputs of a flip-flop that change the flip-flop's Q outputs immediately, without waiting for a pulse at the CLK input. Examples include preset and clear inputs.

Atom Fundamental building block of molecules; smallest particle of an individual element.

Atomic number The number of protons in the nucleus of an atom.

Augend The number in an addition operation to which another number is added.

Axial leaded components A component package with a cylindrical body and leads from both ends.

B

b Bit.

B Byte.

Ball grid array (BGA) A square surface-mount IC package with rows and columns of spherical leads underneath the package.

Bidirectional counter A counter that can count up or down, depending on the state of a control input.

Bidirectional shift register A shift register that can serially shift bits left or right according to the state of a direction control input.

Binary coded decimal (BCD) A code in which each individual digit of a decimal number is represented by a 4-bit binary number. (e.g., 905 [decimal] = 1001 0000 0101 [BCD]).

Binary counter A counter that generates a binary count sequence.

Binary number system A number system used extensively in digital systems, based on the number 2. It uses two digits, 0 and 1, to write any number.

Bit *Binary digit.* A 0 or a 1.

Bit-organized A memory is bit-organized if one address accesses one bit of data.

Binary point A period (".") that marks the dividing line between positional multipliers that are positive and negative powers of 2 (e.g., first multiplier right of binary point = 2^{-1}; first multiplier left of binary point = 2^0).

Block diagram A high-level diagram showing a complex circuit as a combination of smaller subcircuits.

Boolean algebra A system of algebra that operates on Boolean variables. The binary (two-state) nature of Boolean algebra makes it useful for analysis, simplification, and design of combinational logic circuits.

Boolean expression, Boolean function, or logic function An algebraic expression made up of Boolean variables and operators, such as AND, OR, or NOT.

Boolean variable A variable having only two possible values, such as HIGH/LOW, 1/0, On/Off, or True/False.

Boot block A sector in a flash memory reserved for primary firmware.

Borrow A digit brought back from a more significant position when the subtrahend digit is larger than the minuend digit.

Bottom boot block A boot block sector in flash memory placed at the lowest address in the memory.

Breadboard A circuit board with hidden interconnections for wiring temporary circuits, usually used for prototypes or laboratory work.

Bubble A small circle indicating logical inversion on a circuit symbol.

Bubble A circle in a state diagram containing the state name and values of the state variables.

Bubble-to-bubble convention The practice of drawing gates in a logic diagram so that inverting outputs connect to inverting inputs and noninverting outputs connect to noninverting inputs.

Buffer An amplifier that acts as a logic circuit. Its output can be inverting (with the output inverted) or noninverting (where the output is the same as the input).

Bus A common wire or parallel group of wires connecting multiple circuits.

Bus contention The condition that results when two or more devices try to send data to a bus at the same time. Bus contention can damage the output buffers of the devices involved.

Bus form A way of drawing a logic diagram so that each true and complement input variable is available along a continuous conductor called a bus.

Byte A group of 8 bits.

C

Capacitor A circuit component which stores charge and is often used to smooth voltage levels.

Carry A digit that is "carried over" to the next most significant position when the sum of two single digits is too large to be expressed as a single digit.

Carry bit A bit that holds the value of a carry (0 or 1) resulting from the sum of two binary numbers.

CAS (*Column Address Strobe*) A signal used to latch the column address into the decoding circuitry of a dynamic RAM with multiplexed addressing.

Cascade To connect an output of one device to an input of another, often for the purpose of expanding the number of bits available for a particular function.

Case shift Changing letters from capitals (uppercase) to small letters (lowercase) or vice versa.

Cell The smallest unit of a Karnaugh map, corresponding to one line of a truth table. The input variables are the cell's coordinates, and the output variable is the cell's contents.

Cell content A 0 or 1 that represents the value of the output variable (e.g., *Y*) of a Boolean expression for a particular set of coordinates.

Cell coordinate A variable around the edge of a K-map that represents an input variable (e.g., *A*, *B*, *C*, or *D*) for the Boolean expression to be simplified.

Chip An integrated circuit. Specifically, a chip of silicon on which an integrated circuit is constructed.

Clear Reset (synchronous or asynchronous).

Clock A special case of a symmetrical, periodic waveform with a specified frequency.

CLOCK An enabling input to a sequential circuit that is sensitive to the positive- or negative-going edge of a waveform.

CMOS analog switch A CMOS device that will pass an analog or digital signal in either direction, when enabled. Also called a transmission gate or bilateral switch. There is no TTL equivalent.

Coincidence gate An Exclusive NOR gate so called because its output is HIGH when its two inputs are the same.

Combinational logic (Combinatorial logic) Digital circuitry in which an output is derived from the combination of inputs, independent of the order in which they are applied.

Command lines Signals that connect the control section of a synchronous circuit to its memory section and direct the circuit from its present to its next state.

Common anode display A seven-segment LED display where the anodes of all the LEDs are connected to the circuit supply voltage. Each segment is illuminated by a logic LOW at its cathode.

Common cathode display A seven-segment display in which the cathodes of all LEDs are connected together and grounded. A logic HIGH illuminates a segment when applied to its anode.

Complement form Inverted.

Complementary metal-oxide-semiconductor (CMOS) A family of digital logic devices whose basic element is the metal-oxide-semiconductor field effect transistor (MOSFET).

Conditional transition A transition between states of a state machine that occurs only under specific conditions of one or more control inputs.

Conductor A material that allows current to easily flow through it, with a very low resistance.

Continuous Smoothly connected. An unbroken series of consecutive values with no instantaneous changes.

Control input A state machine input that directs the machine from state to state.

Control lines Signals to a device input that functions to control the device rather than determining the outputs. Typical examples of control lines include *enable* signals.

Control section The combinational logic portion of a synchronous circuit that determines the next state of the circuit.

Count enable A control function that allows a counter to progress through its count sequence when active and disables the counter when inactive.

Count sequence The specific series of output states through which a counter progresses.

Counter A sequential digital circuit whose output progresses in a predictable repeating pattern, advancing by one state for each clock pulse.

Count-sequence table A list of counter states in the order of the count sequence.

Current The flow of electricity.

D

Data Binary digits (0's and 1's) that contain some kind of information. The digital contents of a memory device.

Data book A bound collection of datasheets. A digital logic data book usually contains datasheets for a specific logic family or families.

Data inputs The multiplexer inputs that feed a digital signal to the output when selected.

Datasheet A printed specification giving details of the pin configuration, electrical properties, and mechanical profile of an electronic device.

DDR (Double Data Rate) SDRAM that uses both edges of the clock to transfer data.

DDR2 (Double Data Rate) SDRAM that uses both edges of the clock and a different internal clock speed to transfer data twice as fast as DDR RAM.

Decimal number system Positional number system with ten digits, 0 to 9, and positional multipliers that are powers of ten.

Decoder A digital circuit designed to detect the presence of a particular digital state.

DeMorgan's equivalent forms Two gate symbols, one AND-shaped and one OR-shaped, that are equivalent according to DeMorgan's theorems.

DeMorgan's theorems Two theorems in Boolean algebra that allow us to transform any gate from an AND-shaped to an OR-shaped gate and vice versa.

Demultiplexer A circuit that uses a binary decoder to direct a digital signal from a single source to one of several destinations.

Difference The result of a subtraction operation.

Difference gate An Exclusive OR gate so called because its out is HIGH when its two inputs are different..

Digital A representation of a physical quantity by a series of binary numbers. A digital representation can have only specific discrete values.

Digital waveform A series of logic 1's and 0's plotted as a function of time.

Discrete Separated into distinct segments or pieces. A series of discontinous values.

Distinctive-shape symbols Graphic symbols for logic circuits that show the function of each type of gate by a special shape.

Don't care state: An output state that can be regarded as either HIGH or LOW, as is most convenient. A don't care state is the output state of a circuit for a combination of inputs that should never occur under stated design conditions.

Double-rail inputs Boolean input variables that are available to a circuit in both true and complement form.

Double-subscript notation A naming convention where two or more numerically related groups of signals are named using two subscript numerals. Generally, the first digit refers to a group of signals and the second to an element of a group (e.g., $X03$ represents element 3 of group 0 for a set of signal groups, X).

DOWN counter A counter with a descending sequence.

Download The process of sending a circuit schematic drawing into a programmable device.

Dual in-line memory module (DIMM) A memory module with DRAMs and connector pins on both sides of the board.

Dual in-line package (DIP) A type of IC with two parallel rows of pins for the various circuit inputs and outputs.

Duty cycle (DC) Fraction of the total period that a digital waveform is in the HIGH state. $DC = t_h/T$ (often expressed as a percentage: $\%DC = t_h/T \times 100\%$).

Dynamic RAM A random access memory that cannot retain data for more than a few milliseconds without being "refreshed."

E

Edge The part of the pulse that represents the transition from one logic level to the other.

Edge detector A circuit in an edge-triggered flip-flop that converts the active edge of a *CLOCK* input to an active-level pulse at the internal latch's *SET* and *RESET* inputs.

Edge-sensitive Edge-triggered.

Edge-triggered Enabled by the positive or negative edge of a digital waveform.

EEPROM (or E²PROM) Electrically erasable programmable read-only memory. A type of read-only memory that can be field-programmed and selectively erased while still in a circuit.

Electric current Flow of charge.

Electrocution Death from electrical shock.

Electron Negatively charged particle orbiting the nucleus of an atom.

Enable A logic gate is enabled if it allows a digital signal to pass from an input to the output in either true or complement form.

Encoder A circuit that generates a digital code at its outputs in response to one or more active input lines.

End-around carry An operation in 1's complement subtraction where the carry bit resulting from a sum of two 1's complement numbers is added to that sum.

EPROM Erasable programmable read-only memory. A type of ROM that can be programmed ("burned") by the user and erased later, if necessary, by exposing the chip to ultraviolet radiation.

Even parity An error-checking system that requires a binary number to have an even number of 1's.

Excitation table A table showing the required input conditions for every possible transition of a flip-flop output.

Exclusive NOR (XNOR) Gate A 2-input logic circuit whose output is the complement of an Exclusive OR gate.

Exclusive OR (XOR) Gate A 2-input logic circuit whose output is HIGH when one input (but not both) is HIGH.

F

Fall time (t_f) Elapsed time from the 90% point to the 10% point of the falling edge of a signal.

Falling edge The part of a signal where the logic level is in transition from a HIGH to a LOW.

Farad (F) Unit of capacitance.

FIFO (First In, First Out) A sequential memory in which the stored data can be read only in the order in which it was written.

Firmware Software instructions permanently stored in ROM.

Flash memory A nonvolatile type of memory that can be programmed and erased in sectors, rather than byte-at-a-time.

Flip-flop A sequential circuit based on a latch whose output changes when its *CLOCK* input receives an edge.

Flux A rosin-based material contained in or used with solder to clean the surfaces as the solder joint forms.

Frequency (f) Number of times per second that a periodic waveform repeats. $f = 1/T$ Unit: hertz (Hz).

Full adder A circuit that will add a carry bit from another full or half adder and two operand bits to produce a sum bit and a carry bit.

Full-sequence counter A counter whose modulus is the same as its maximum modulus ($m = 2^n$ for an n-bit counter).

G

G 1,073,741,824 ($= 2^{30}$). Analogous to the metric prefix "G" (giga-).

Gated SR latch An SR latch whose ability to change states is controlled by an extra input called the *ENABLE* input.

Glue logic Gates or simple circuitry whose purpose is to connect more complex blocks of a larger circuit.

H

Half adder A circuit that will add two bits and produce a sum bit and a carry bit.

Hardware The electronic circuit of a digital or computer system.

Hexadecimal number system (Hex) Base-16 number system. Hexadecimal numbers are written with sixteen digits, 0–9 and A–F, with power-of-16 positional multipliers.

Hierarchical design A design that is ordered in layers or levels. The highest level of the design has components that are complete designs in and of themselves. These lower-level blocks might contain yet-lower levels of design components.

High-impedance state The output state of a tristate buffer that is neither logic HIGH nor logic LOW, but is electrically equivalent to an open circuit; seemingly disconnected from the circuit. (Abbreviation: Hi-Z.)

I

In phase Two digital waveforms are in phase if they are always at the same logic level at the same time.

Inhibit (or disable) A logic gate is inhibited if it prevents a digital signal from passing from an input to the output.

Insulator A material that does not allow current to pass through, with a very high resistance.

Integrated circuit (IC) An electronic circuit having many gates or other components, such as transistors, diodes, resistors, and capacitors, in a single package.

Inverter Also called a NOT gate or an inverting buffer. A logic gate that changes its input logic level to the opposite state.

J

Johnson Counter A serial shift register with complemented feedback from the output of the last flip-flop to the input of the first. Also called a *twisted ring counter*

K

K 1024 (= 2^{10}). Analogous to the metric prefix "k" (kilo-).

Karnaugh map: A graphical tool for finding the maximum SOP or POS simplification of a Boolean expression. A Karnaugh map (or K-map) works by arranging the terms of an expression so that variables can be canceled by grouping minterms or maxterms.

Kirchhoff's Current Law (KCL) The sum of all currents entering a node must be zero; in other words, any current entering a node must also leave the node.

Kirchhoff's Voltage Law (KVL) The sum of all voltages around a path in a circuit must equal zero.

L

Large-scale integration (LSI) An integrated circuit having from 100 to 10,000 equivalent gates.

Latch A sequential circuit with two inputs called *SET* and *RESET*, which make the latch store a logic 1 (set) or logic 0 (reset) until actively changed.

Lead (pronounced "leed") The leg of a device: the connection piece of the component to be attached to the circuit board.

Leading edge The edge of a pulse that occurs earliest in time.

Least significant bit (LSB) The rightmost bit of a binary number. This bit has the number's smallest positional multiplier.

Left shift A movement of data from the right to the left in a shift register. This is usually defined as moving from the LSB toward the MSB.

Level-sensitive Enabled by a logic HIGH or LOW level.

Levels of gating The number of gates through which a signal must pass from input to output of a logic gate network.

LIFO (Last In, First Out) A sequential memory in which the last data written are the first data read.

Light-emitting diode (LED) An electronic device that conducts current in one direction only and illuminates when it is conducting.

Lightning Discharge of static electricity between the atmosphere and ground.

Logic diagram A diagram, similar to a schematic, showing the connection of logic gates.

Logic gate An electronic circuit that performs a Boolean algebraic function.

Logic gate network Two or more logic gates connected together.

Logic HIGH (or logic 1) The higher of two voltages in a digital system with two logic levels.

Logic level A voltage level that represents a defined digital state in an electronic circuit.

Logic LOW (or logic 0) The lower of two voltages in a digital system with two logic levels.

Logical product AND function.

Logical sum OR function.

M

M 1,048,576 (= 2^{20}). Analogous to the metric prefix "M" (mega-).

Magnitude bits The bits of a signed binary number that tell us how large the number is (i.e., its magnitude).

Magnitude comparator A circuit that compares two *n*-bit binary numbers, indicates whether or not the numbers are equal, and, if not, which one is larger.

Mask-programmed ROM A type of read only memory (ROM) where the stored data are permanently encoded into the memory device during the manufacturing process.

Master reset An asynchronous reset input used to set a sequential circuit to a known initial state.

Maximum modulus (mmax) The largest number of counter states that can be represented by n bits ($m_{max} = 2^n$).

Maximum SOP simplification The form of an SOP Boolean expression that cannot be further simplified by canceling variables in the product terms. It may be possible to get a POS form of the expression with fewer terms or variables.

Maxterm A sum term in a Boolean expression where all possible variables appear once, in true or complement form [e.g., $(\overline{A} + \overline{B} + C)$; $(A + \overline{B} + C)$].

Mealy machine A state machine whose output is determined by both the sequential logic and the combinational logic of the machine.

Medium-scale integration (MSI) An integrated circuit having the equivalent of 12 to 100 gates in one package.

Memory A device for storing digital data so that they can be recalled for later use in a digital system.

Memory map A diagram showing the total address space of a memory system and the placement of various memory devices within that space.

Memory modules A small circuit board containing several dynamic RAM chips.

Memory section A set of flip-flops in a synchronous circuit that holds its present state.

Metric system The system of units based on multiples of 10.

Minterm A product term in a Boolean expression where all possible variables appear once in true or complement form (e.g., $\overline{A}\ \overline{B}\ \overline{C}$; $A\ \overline{B}\ C$).

Minuend The number in a subtraction operation from which another number is subtracted.

Modulo arithmetic A closed system of counting and adding, whereby a sum greater than the largest number in a sequence "rolls over" and starts from the beginning. For example, on a clock face, four hours after 10 A.M. is 2 P.M., so in a mod-12 system, 10 + 4 = 2.

Modulo-n (or mod-n) counter A counter with a modulus of n.

Modulus The number of states through which a counter sequences before repeating.

Molecule Combination of atoms; the smallest particle of a substance other than a pure element.

Moore machine A state machine whose output is determined only by the sequential logic of the machine.

Most significant bit (MSB) The leftmost bit in a binary number. This bit has the number's largest positional multiplier.

Multiplexer A circuit that directs one of several digital signals to a single output, depending on the states of several select inputs.

Multiplicand The number in a multiplication problem that is multiplied by each digit in the multiplier.

Multiplier The number by which the multiplicand is multiplied to get the result of a multiplication problem.

N

NAND gate A logic circuit whose output is LOW when all inputs are HIGH. (A combination of NOT and AND.)

Negative logic A system in which logic LOW represents binary digit 1 and logic HIGH represents binary digit 0.

Neutron Particle in the nucleus of an atom having no charge (or a neutral charge).

Next state The desired future state of flip-flop outputs in a synchronous sequential circuit after the next clock pulse is applied.

Noble elements Elements with full orbital paths. It is very difficult for these elements to combine to form molecules.

Node A junction of electronic components.

NOR gate A logic circuit whose output is LOW when at least one input is HIGH. (A combination of NOT and OR.)

Nucleus Center of an atom, containing neutrons and protons. Electrons orbit around the nucleus.

O

Octet A group of eight adjacent cells in a Karnaugh map. An octet cancels three variables in a K-map simplification.

Odd parity An error-checking system that requires a binary number to have an odd number of 1's.

Ohm (Ω) Unit of resistance.

Ohm's Law $V = R \times I$: the voltage across a component equals its resistance multiplied by the current through it.

Operand A number or variable upon which an arithmetic or Boolean function operates (e.g., in the expression $x + y = z$, x and y are the operands).

OR gate A logic circuit whose output is HIGH when at least one input (e.g., A OR B OR C) is HIGH.

Orbital path Paths of electrons orbiting the atom's nucleus. Each orbital path holds a certain number of electrons.

Order of precedence The sequence in which Boolean functions are performed, unless otherwise specified by parentheses.

Out of phase Two digital waveforms are out of phase if they are always at opposite logic levels at any given time.

Output decoding A feature in which one or more outputs activate when a particular counter state is detected.

Overflow An erroneous carry into the sign bit of a signed binary number that results from a sum or difference larger than can be represented by the number of magnitude bits.

Oxidation The process of oxygen reacting with the heated tip of a soldering iron causing discoloration and affecting the ability of the tip to transfer heat.

P

Pad Copper on a circuit board meant to be a point where a component is attached. On a through-hole board, this copper will surround the hole where the lead passes through.

Pair A group of two adjacent cells in a Karnaugh map. A pair cancels one variable in a K-map simplification.

Parallel Resistors in parallel are connected to each other at both ends. They have the same voltage across them, but any current flowing into the parallel combination of these resistors splits among them and joins back together after passing through.

Parallel binary adder A circuit, consisting of n full adders, that will add two n-bit binary numbers. The output consists of n sum bits and a carry bit.

Parallel load A function that allows simultaneous loading of binary values into all flip-flops of a synchronous circuit. Parallel loading can be synchronous or asynchronous.

Parallel-load shift register A shift register that can be preset to any value by directly loading a binary number into its internal flip-flops.

Parallel transfer Movement of data into all flip-flops of a shift register at the same time.

Parity A system that checks for errors in a multi-bit binary number by counting the number of 1s.

Parity bit A bit appended to a binary number to make the number of 1s even or odd, depending on the type of parity.

Period (T) Time required for a periodic waveform to repeat. Unit: seconds (s).

Periodic waveform A time-varying sequence of logic HIGHs and LOWs that repeats over a specified period of time.

Polarized part A component which must be inserted in a circuit with correct polarity: a component with a positive and negative side.

Portable document format (PDF) A format for storing published documents in compressed form.

Positional notation A system of writing numbers where the value of a digit depends not only on the digit, but also on its placement within a number.

Positive logic A system in which logic LOW represents binary digit 0 and logic HIGH represents binary digit 1.

Potential energy Stored energy that has potential to do work.

Present state The current state of flip-flop outputs in a synchronous sequential circuit.

Preset An asynchronous set function.

Presettable counter A counter with a parallel load function.

Printed circuit board (PCB) A circuit board in which connections between components are made with lines of copper on the surfaces of the circuit board.

Priority encoder An encoder that generates a binary or BCD output corresponding to the subscript of the active input having the highest priority. This is usually defined as the active input with the largest subscript value.

Product term A term in a Boolean expression where one or more true or complement variables are ANDed (e.g., $\overline{A}\,\overline{C}$).

Product-of-sums (POS): A type of Boolean expression where several sum terms are multiplied (ANDed) together (e.g., $(\overline{A} + \overline{B} + C)(A + \overline{B} + \overline{C})(\overline{A} + \overline{B} + \overline{C})$).

Programmable device A device which can be reprogrammed or "rewired" internally, allowing the testing of digital circuits without wiring new components.

Programming or burning ROM Storing a program or code in a read-only memory (ROM) device.

Proton Positively charged particle found in the nucleus of the atom.

Prototype A circuit built functionally the same as a circuit we wish to test.

Pull-up resistor A resistor connected from a point in an electronic circuit to the power supply of that circuit.

Pulse A momentary variation of voltage from one logic level to the opposite level and back again.

Pulse width (t_w) (of an ideal pulse) The time from the rising to falling edge of a positive-going pulse, or from the falling to rising edge of a negative-going pulse.

Pulse width (t_w): (of a nonideal pulse) Elapsed time from the 50% point of the leading edge of a pulse to the 50% point of the trailing edge.

Q

Quad A group of four adjacent cells in a Karnaugh map. A quad cancels two variables in a K-map simplification.

Quad flat pack (QFP) A square surface-mount IC package with gull-wing leads.

Queue A FIFO memory.

R

Radix point The generalized form of a decimal point. In any positional number system, the radix point marks the dividing line between positional multipliers that are positive and negative powers of the system's number base.

RAM cell The smallest storage unit of a RAM, capable of storing 1 bit.

Random access memory (RAM) A type of memory device where data can be accessed in any order, that is, randomly. The term usually refers to random access read/write memory.

RAS (Row Address Strobe) A signal used to latch the row address into the decoding circuitry of a dynamic RAM with multiplexed addressing.

RBI Ripple blanking input.

RBO Ripple blanking output.

Read Retrieve data from a memory device.

Read-only memory (ROM) A type of memory where data are permanently stored and can only be read, not written.

Recycle To make a transition from the last state of the count sequence of a digital counter to the first state.

Redundant term An extra, unneeded product term in a simplified Boolean expression.

Refresh cycle The process that periodically recharges the storage capacitors in a dynamic RAM.

RESET (1) The stored LOW state of a latch circuit. (2) A latch input that makes the latch store a logic 0.

Resistance Characteristic of electrical components that resists current flow.

Resistor A circuit component which resists the flow of current.

Response waveforms A set of output waveforms generated by a simulator tool for a particular digital design in response to a set of stimulus waveforms.

Right shift A movement of data from the left to the right in a shift register. This is usually defined as data moving from the MSB towards the LSB.

Ring counter A serial shift register with feedback from the output of the last flip-flop to the input of the first.

Ripple blanking A technique used in a multiple-digit numerical display that suppresses or "blanks" leading or trailing zeros in the display, but allows internal zeros to be displayed.

Ripple carry out or ripple clock out (RCO) An output that produces one pulse with the same period as the clock upon terminal count.

Rise time (t_r) Elapsed time from the 10% point to the 90% point of the rising edge of a signal.

Rising edge The part of a pulse where the logic level is in transition from a LOW to a HIGH. In an ideal pulse, this is instantaneous.

Rosin A key component in flux; made from pine tree sap.

Rotation Serial shifting of data in a shift register with the output(s) of the last flip-flop connected to the synchronous input(s) of the first flip-flop. The result is continuous circulation of the same data.

S

Schematic A drawing of a circuit being designed that uses special symbols for circuit components.

SDRAM (Synchronous DRAM) Dynamic RAM whose data are synchronously transferred to and from the data bus.

Sector A segment of flash memory that forms the smallest amount that can be erased and reprogrammed at one time.

Select inputs The multiplexer inputs that select a digital input channel.

Sequential circuit A digital circuit whose output depends not only on the present combination of inputs but also on the history of the circuit.

Sequential logic Digital circuitry in which the output state of the circuit depends not only on the states of the inputs, but also on the sequence in which they reached their present states.

Sequential memory Memory in which the stored data cannot be read or written in random order, but must be addressed in a specific sequence.

Serial shifting Movement of data from one end of a shift register to the other at a rate of one bit per clock pulse.

Series Resistors in series require all current flowing through one resistor to also flow through the next resistor, and so on.

SET (1) The stored HIGH state of a latch circuit. (2) A latch input that makes the latch store a logic 1.

Seven-segment display An array of seven independently controlled light-emitting diode (LED) or liquid crystal display (LCD) elements, shaped like a figure-8, which can be used to display decimal digits and other characters by turning on the appropriate elements.

Shift register A synchronous sequential circuit that will store and move n-bit data, either serially or in parallel, in n flip-flops.

SI system The International System of units; the modern form of the metric system.

Sign bit A bit, usually the MSB, that indicates whether a signed binary number is positive or negative.

Sign extension The process of fitting a number into a fixed size of 2's complement number by padding the number with leading 0's for a positive number and leading 1's for a negative number.

Signed binary arithmetic Arithmetic operations performed using signed binary numbers.

Signed binary number A binary number of fixed length whose sign is represented by one bit, usually the most significant bit, and whose magnitude is represented by the remaining bits.

Simulation The verification, using timing diagrams, of the logic of a digital design before programming it into a CPLD or an FPGA.

Single in-line memory module (SIMM) A memory module with DRAMs and connector pins on one side of the board only.

Small-outline dual in-line memory module (SoDIMM) A smaller version of a DIMM often used in laptop or tablet computers.

Small outline IC (SOIC) An IC package similar to a DIP, but smaller, which is designed for automatic placement and soldering on the surface of a circuit board. Also called gull-wing, for the shape of the package leads.

Small-scale integration (SSI) An integrated circuit having 12 or fewer gates in one package.

Software Programming instructions required to make hardware perform specified tasks.

Solder A metallic alloy comprised of a mixture of tin and lead or other metals with a melting point lower than the metals to be joined.

Solder joint A soldered connection between an electronic component and a circuit board.

Soldering The act of attaching electronic components to a circuit board using solder.

Soldering iron A device with a heated tip used to manually solder components to circuit boards.

SRGn Abbreviation for an n-bit shift register (e.g., SRG4 indicates a 4-bit shift register).

State diagram A diagram showing the progression of states of a sequential circuit.

Stack pointer The address in memory of the top of the stack: the stack pointer itself is also stored in memory.

Stack A LIFO memory.

State machine A synchronous sequential circuit, consisting of a sequential logic section and a combinational logic section, whose outputs and internal flip-flops progress through a predictable sequence of states in response to a clock and other input signals.

State variables The variables held in the flip-flops of a state machine that determine its present state. The number of state variables in a machine is equivalent to the number of flip-flops.

Static electricity A stored charge in an item that dissipates quickly when the item comes into contact with something at a different electrical potential.

Static RAM A random access memory that can retain data indefinitely as long as electrical power is available to the chip.

Status lines Signals that communicate the present state of a synchronous circuit from its memory section to its control section.

Steering gates Logic gates, controlled by the *ENABLE* input of a gated latch, that steer a *SET* or *RESET* pulse to the correct input of an SR latch circuit.

Stimulus waveforms A set of user-defined input waveforms on a simulator file designed to imitate input conditions of a digital circuit.

Subcircuit A block of circuit components that can be used as a single component in a higher level of a hierarchical design.

Subtrahend The number in a subtraction operation that is subtracted from another number.

Sum The result of an addition operation.

Sum bit (single-bit addition) The least significant bit of the sum of two 1-bit binary numbers.

Sum-of-products (SOP): A type of Boolean expression where several product terms are summed (ORed) together (e.g., $\overline{A}\,B\,\overline{C} + \overline{A}\,\overline{B}\,C + A\,B\,C$).

Sum term A term in a Boolean expression where one or more true or complement variables are ORed (e.g., $\overline{A} + B + \overline{D}$).

Surface-mount technology (SMT) A system of mounting and soldering integrated circuits on the surface of a circuit board, as opposed to inserting their leads through holes on the board.

Synthesis The process of creating a logic circuit from a description such as a Boolean equation or truth table.

Synchronous counter A counter whose flip-flops are all clocked by the same source and thus change in synchronization with each other.

Synchronous inputs The inputs of a flip-flop that do not affect the flip-flop's Q outputs unless a clock pulse is applied. Examples include D, J, and K inputs.

Synchronous Synchronized to the system clock.

T

T 1,099,511,627,776 $(= 2^{40})$. Analogous to the metric prefix "T" (tera).

Terminal count The last state in a count sequence before the sequence repeats (for example, 1111 is the terminal count of a 4-bit binary UP counter; 0000 is the terminal count of a 4-bit binary DOWN counter).

Thin shrink small outline package (TSSOP) A thinner version of an SOIC package.

Through-hole A means of mounting DIP ICs on a circuit board by inserting the IC leads through holes in the board and soldering them in place

Time HIGH (t_h) Time during one period that a waveform is in the HIGH state. Unit: seconds (s).

Time LOW (t_l) Time during one period that a waveform is in the LOW state. Unit: seconds (s).

Timing diagram A digital waveform, typically with multiple signals on one plot.

Tinning the tip The process of applying a light coat of solder to the tip of a soldering iron for storage to prevent oxidation.

Tip The heated end of the soldering iron – the tip should make contact with the lead and pad when soldering.

Toggle Alternate between opposite binary states with each applied clock pulse.

Tolerance Allowable variability in the value of an electronic component.

Top boot block A boot block sector in a flash memory placed at the highest address in the memory.

Trace A copper line on a circuit board connecting one component to another. A path for current to flow on a circuit board.

Trailing edge The edge of a pulse that occurs latest in time.

Transistor-transistor logic (TTL) A family of digital logic devices whose basic element is the bipolar junction transistor.

Transparent latch A latch whose output follows its data input when its *ENABLE* input is active. (Gated D Latch)

Tristate buffer A gate having three possible output states: logic HIGH, logic LOW, and high-impedance.

True form Not inverted.

True-magnitude form A form of signed binary number whose magnitude is represented in true binary.

Truncated-sequence counter A counter whose modulus is less than its maximum modulus ($m < 2^n$ for an n-bit counter).

Truth table A list of output levels of a circuit corresponding to all different input combinations.

U

Unconditional transition A transition between states of a state machine that occurs regardless of the status of any control inputs.

Units A term representing a precise quantity or measure.

Universal shift register A shift register that can operate with any combination of serial and parallel inputs and outputs (i.e., serial in/serial out, serial in/parallel out, parallel in/serial out, parallel in/parallel out). A universal shift register is often bidirectional, as well.

Unsigned binary number A binary number whose sign is not specified by a sign bit. A positive sign is assumed unless explicitly stated otherwise.

UP counter A counter with an ascending sequence.

V

V_{CC} The power supply voltage in a transistor-based electronic circuit. The term often refers to the power supply of digital circuits.

Very-large-scale integration (VLSI) An integrated circuit having more than 10,000 equivalent gates.

Volatile A memory is volatile if its stored data are lost when electrical power is lost.

Voltage Electrical potential.

Volt Unit of electrical potential

W

Wire-wrap A circuit construction technique in which the connecting wires are wrapped around the posts of a special chip socket or PCB connector, usually used for prototyping or laboratory work.

Word Data accessed at one addressable location.

Word length Number of bits in a word.

Word-organized A memory is word-organized if one address accesses one word of data.

Write Store data in a memory device.

INDEX

A

Active HIGH, 69–70
Active level, 69–70
Active LOW, 69–70
Addend, 222
Addition
    calculator design for, 264–266, 264*f*–266*f*, 264*t*,
        268–269, 268*f*, 269*f*
    signed binary arithmetic, 226–227
    unsigned binary arithmetic, 222–223
Address, 406–408
    8K X 8 memory, 409, 409*f*
Address decoder, 422, 424, 424*t*
Address multiplexing, 414
Address space, 422
Adjacent cell, 126
Alloy, 11
Alphanumeric codes
    ASCII, 233–235, 235*t*
    BCD, 153, 170, 171*t*, 178–179, 179*f*, 179*t*, 232–233
    defined, 233
    8421, 233, 233*t*
    Amp. *See* Ampere
Ampere (Amp), 20*t*
Amplitude, 56–57, 57*f*, 318, 319
Analog
    CMOS analog switch, 187–188, 188*f*
    defined, 42
    digital *vs.*, 42–44
    keywords associated with, 42
    signal, 43*f*
AND function, 65–66, 65*f*, 65*t*, 66*f*, 66*t*, 68, 68*f*
AND gate, 65–66
    defined, 65
    enable/inhibit properties in, 82–84, 82*f*, 84*f*, 84*t*
    Multisim example with, 68, 68*f*
    switches used with, 68*f*
    symbol for, 65*f*, 66*f*
    truth table for, 65*t*, 66, 66*t*
Aperiodic waveforms, 56, 56*f*
Arithmetic circuits, 221–256
    binary adders and subtractors, 236–251
    full adder, 236–243, 237*f*–243*f*, 237*t*
    half adder, 236–240, 236*f*–238*f*, 236*t*
    overflow detection, 249–251, 250*f*, 250*t*, 251*f*, 251*t*
    parallel, 243–251, 244*f*, 245*f*, 247*f*, 248*f*, 250*f*, 250*t*, 251*f*, 251*t*
ASCII code, 233–235, 235*t*
Asynchronous, 314–315, 314*f*, 315*f*
Asynchronous inputs, 315–316, 316*f*, 316*t*

Asynchronous load, 353–354, 353*f*, 354*f*
Asynchronous load element
    asynchronous clear with, 354, 354*f*
    Multisim example with, 354, 354*f*
Asynchronous pulse error, state machine with, 389*f*
Asynchronous reset, synchronous counter with, 317–318, 317*f*, 318*f*
Atom, 21, 22*f*
Atomic number, 21–22
Atomic structure, 21–24
Augend, 222
Axial-leaded component, 8

B

B. *See* Byte
b. *See* Bit
Ball grid array (BGA), 88, 90
Batteries, 6
BCD. *See* Binary coded decimal
BCD encoders, 178–179, 179*f*, 179*t*
    truth table for, 179*t*
BGA. *See* Ball grid array
Bidirectional counter, 352, 353, 355
Bidirectional shift register, 359, 364–365, 365*f*
Binary adders and subtractors, 236–251
    full adder, 236–243, 237*f*–243*f*, 237*t*
    half adder, 236–237, 236*f*–238*f*, 236*t*, 238–240
    parallel, 243–251, 244*f*, 245*f*, 247*f*, 248*f*, 250*f*, 250*t*, 251*f*, 251*t*
Binary coded decimal (BCD), 153, 170, 232–233
    defined, 170, 232
    encoder, 178–179, 179*f*, 179*t*
    truth table for, 171*t*
Binary counter, 336
Binary inputs, 46–48
    digital circuit 3-imput combinations of, 46, 46*t*
    digital circuit 4-imput combinations of, 47, 47*t*
    repetitive pattern in, 48
Binary number system, 45–46, 45*f*
    conversion between hex and, 51*t*, 53–54
    counting in, 48
    decimal conversion to, 49–50, 49*f*
    defined, 45
    hex equivalents for, 51*t*
Bit (b), 45, 406, 408
Bit-organized, 414
Block diagram
    calculator design, 262, 262*f*
    defined, 260
    synchronous counters, 339*f*
Boolean algebra, 63, 102–151